Conceptions of the Watery World in Greco-Roman Antiquity

Also available from Bloomsbury

Mountain Dialogues from Antiquity to Modernity edited by Dawn Hollis
and Jason König
The Ancient Mediterranean Sea in Modern Visual and Performing Arts
edited by Rosario Rovira Guardiola
The Horse in the Ancient World: From Bucephalus to the Hippodrome
by Carolyn Willekes

Conceptions of the Watery World in Greco-Roman Antiquity

Georgia L. Irby

BLOOMSBURY ACADEMIC
LONDON • NEW YORK • OXFORD • NEW DELHI • SYDNEY

BLOOMSBURY ACADEMIC
Bloomsbury Publishing Plc
50 Bedford Square, London, WC1B 3DP, UK
1385 Broadway, New York, NY 10018, USA
29 Earlsfort Terrace, Dublin 2, Ireland

BLOOMSBURY, BLOOMSBURY ACADEMIC and the Diana logo are trademarks
of Bloomsbury Publishing Plc

First published in Great Britain 2021
This paperback edition published 2023

Copyright © Georgia L. Irby, 2021

Georgia L. Irby has asserted her right under the Copyright, Designs, and Patents Act, 1988,
to be identified as Author of this work.

For legal purposes the Acknowledgments on p. xiii–xiv constitute an extension
of this copyright page.

Cover design: Terry Woodley
Cover image: Mosaic in Kykkos Monastery in Cyprus © Brand X Pictures/Getty

All rights reserved. No part of this publication may be reproduced or transmitted in
any form or by any means, electronic or mechanical, including photocopying, recording,
or any information storage or retrieval system, without prior permission in writing
from the publishers.

Bloomsbury Publishing Plc does not have any control over, or responsibility for,
any third-party websites referred to or in this book. All internet addresses given in
this book were correct at the time of going to press. The author and publisher regret
any inconvenience caused if addresses have changed or sites have ceased to exist,
but can accept no responsibility for any such changes.

A catalog record for this book is available from the British Library.

A catalog record for this book is available from the Library of Congress.

Names: Irby, Georgia L. (Georgia Lynette), 1965- author.
Title: Conceptions of the watery world in Greco-Roman antiquity / Georgia L. Irby.
Description: London: Bloomsbury Academic, 2021. |
Includes bibliographical references and index.
Identifiers: LCCN 2021004296 (print) | LCCN 2021004297 (ebook) |
ISBN 9781784538293 (hardback) | ISBN 9781350136458 (ebook) |
ISBN 9781350136465 (epub)
Subjects: LCSH: Hydrology–History–To 1500. | Water–History. |
Water–Mythology. | Water–Religious aspects–Paganism. |
Water and civilization. | Civilization, Classical.
Classification: LCC GB659.6.I72 2021 (print) | LCC GB659.6 (ebook) |
DDC 306.4/50938–dc23
LC record available at https://lccn.loc.gov/2021004296
LC ebook record available at https://lccn.loc.gov/2021004297

ISBN: HB: 978-1-7845-3829-3
PB: 978-1-3502-3944-9
ePDF: 978-1-3501-3645-8
eBook: 978-1-3501-3646-5

Typeset by RefineCatch Limited, Bungay, Suffolk

To find out more about our authors and books visit www.bloomsbury.com
and sign up for our newsletters.

Contents

List of Illustrations	x
List of Abbreviations	xi
Acknowledgements	xiii
Introduction: Conceptions of the Watery World	1
Water's Fundamental Importance to Life as We Know It	1
Water as an Organizing Principle in the Greco-Roman Mediterranean Basin	1
Modern Studies	7
Conceptions of the Watery World	7
Methodology	8
Part One Interpreting the Watery Framework: Philosophy, Cosmogony, and Physics	11
1 Water and the Creation of the World	13
Introduction	13
Mesopotamian Origins	13
Hesiod	15
From Myth to Reason	16
The Presocratics	17
The Milesians	17
Thales	18
Anaximenes	19
The Rejection of Material Monism	20
Empedocles and the Four-element Theory	21
Plato and Aristotle	22
Plato	23
Aristotle	24
The Epicurean and Stoic Schools of Philosophy	26
Epicureanism	26
Stoicism	27
Hybrid Interpretations: Microvoids	29

	Plutarch, "Whether Fire or Water Is More Useful"	29
	Conclusion	32
2	Seas and Lakes	35
	Introduction	35
	Exploring the Atlantic	36
	Pytheas	36
	Roman Initiatives	37
	Greco-Roman Conceptions of "Hydrology"	38
	Ocean	39
	Presocratics	39
	Plato	41
	Aristotle	44
	Circumambient Ocean	45
	The Nature of the Oceans and Seas	46
	Salt	46
	Density	48
	Tides	49
	Sea Depth and Levels	54
	Sea Surfaces	56
	Other Characteristics of Salt Water	57
	Currents	59
	Whirlpools	60
	Reefs	61
	Conclusion	62
3	The Interplay Between Water and Land: Land, Rivers, and Springs	63
	Introduction	63
	Sea and Land	64
	Rising and Falling Sea Levels	64
	Effects of Siltation	66
	Sources of Terrestrial Waters	69
	Rivers	70
	Ister	72
	Phasis	73
	Indus	74
	Nile	74
	Underground Rivers	77
	Springs and Fountains	79

	Paradoxical Waters	79
	Floating Islands	81
	Conclusion	82
Part Two	**Explaining Watery Phenomena**	**83**
4	Watery Weather	85
	Introduction	85
	Climate and the Hydrological Cycle	85
	Initiatives and Limitations	86
	Aims of Weather Prognostication in Antiquity	87
	Sources and Ancient Methodologies	87
	Weather and the Stars	89
	Red Skies …	90
	"Things High Up"	91
	Clouds	91
	Rainbows	92
	Thunder and Lightning	93
	Precipitation	95
	Dew	96
	Rain	96
	Frost and Snow	97
	Hail	97
	Devastating Weather Events	100
	Earthquakes	100
	Tsunamis	101
	Storms and Superstorms	104
	Historical Storms at Sea	106
	Floods	107
	Conclusion	109
5	Water, Health, and Disease	111
	Introduction	111
	Water-related Diseases	111
	"Typhoid"	113
	Malaria	115
	Epidemics/"Plagues"	117
	Hydrophobia ("Rabies")	119
	Other Complaints Caused by Water	120

Illness and Travel on Water	121
Sailor Diseases	121
Healing Waters, Bathing, and Hot Springs	122
Conclusion	124
Part Three Imagining the Watery World	**125**
6 Biological Creatures of the Sea	127
Introduction	127
Metamorphosis	128
Anthropomorphism	129
Creatures of the Deep	130
Dolphins	131
Seals	135
Giant Sea Creatures	137
Whales	137
Sharks	139
Cephalopods	140
Conclusion	144
7 Mythical Creatures of the Sea: Sea Monsters and Sea Gods	145
Introduction	145
Imagined Sea Creatures	145
Capricorn	148
Hippocamp	148
Sea Monsters	149
Typhoeus	150
Scylla	150
Kete	151
Skiron's Turtle	152
Hybrid Gods of the Deep	152
Mermen	153
Old Men of the Sea	153
Conclusion	155
8 Water and the Divine: Unseen and Magical Forces of the Spiritual World	157
Introduction	157
Miasma and Ritual Purity	157

Acknowledgments

No work of scholarship is produced in isolation, and this project has its genesis in a course that I have the privilege of teaching to first year students at William & Mary, "Why Water Matters," where students learn about the fluidity of critical inquiry and the scholarly process as an interplay between primary sources, academic training, and the scholar's own cultural and political biases. Together we investigate the critical question about human engagement with the natural world, our sources of information, the advantages and disadvantages of those sources (often fragmentary, poorly preserved, inadequately contextualized, and insufficiently curated), our understanding, perception, and interpretation of those sources, how social, cultural, and political factors influence our perception of the environment, and how our own attitudes to the natural world shape our culture, art, literature, and politics. This book is very much the product of that conversation. In particular, I thank those students in my Fall 2018 class. Enduring an uncivilized 8:00 a.m. curtain call, this talented and enthusiastic group read drafts of the chapters in various stages of polish, bringing slips to my attention, and making countless salient suggestions for improving transitions, adding material, and clarifying arguments. They have earned their mention here: Greg Arrigo, Elizabeth Ashley, Theo Biddle, Jonathan Broady, Marcus Crowell, Zack Daniel, Daniel Gittings, Caileigh Gulotta, Christian Gulotta, Zack Johnson, Cole Kim, Rebecca Klinger, Abhishek Mullapudi, Freddie Nunnelley, Charlie Perry, Clay Shafer, Ben Sharrer, Percy Skalski, Ann Grace Towler, and Tammy Yin. I should also like to recognize Emma Grenfell, Abby Maher, Jake Morrin, Lindsey Smith, and Georgia Thoms, students from my Fall 2019 class who vetted the penultimate version of this manuscript, and brought to my attention a number of lingering, pesky infelicities.

My gratitude extends also to Alex Wright (now at Cambridge University Press), whose gentle prodding guided this project from scattered lecture notes to book manuscript, and to Alice Wright and Georgina Leighton, my editors at Bloomsbury, who took up the gauntlet midstream, and their assistant Lily Mac Mahon. The manuscript has benefited immensely from the helpful observations and suggestions of the anonymous reviewers. Thanks are also owed to senior production editor Rachel Walker at Bloomsbury, Merv Honeywood, the project manager at Refinecatch, copyeditor Susan Dobson, and the page setters whose combined efforts behind the scenes facilitated the publishing process and added sparkle to the text. I am grateful to the Inter-Library Loan Staff at Swem Library at William & Mary, who sometimes even seem to anticipate the needs of their patrons, as well as the cheerful, knowledgeable, and efficient staff in Information Technology who keep my digital resources in good working order. Also meriting recognition is my research assistant, Keegan Sudkamp-Tostevin, who painstakingly double checked many of the primary references, caught slips both great and small, and helped bring my vision into greater focus. His deep curiosity and enthusiasm were inspirational as my energy levels waned. Individual chapters have benefited from the scrutiny of Robert Nichols and Jessica Stephens, my colleagues at William & Mary. Andrew Ward's comments on Samothrace have been invaluable. The adroitly graceful comments of Molly Ayn Jones-Lewis, especially on the medical material, have resulted in many improvements, a more nuanced treatment, and a more coherent structure of the entire manuscript. Duane W. Roller read the entire manuscript in draft form. As always, his eagle-eye caught many infelicities, and his encyclopedic comments, leading to new avenues of inquiry, have improved the substance and spirit of the text. Finally, sincere thanks to the reader for thoughtful comments and astute suggestions that helped bring into focus the book's overarching vision. Any errors that remain are my own.

Many colleagues, students, and friends have cheered the project on from its inception. I would like to single out Tejas Aralere, Joyce Holmes with her endless supply of smiles and hugs, and John Oakley who always has time for a chat about Greek vases. I should also like to thank Jennifer Andrews-Weckerly, Cary Bagdassarian, Charlie Bauer, Michael Bryant, Mike Crookshank, John Donahue, Lu Ann Homza, Bill Hutton, Michele and Les Hoffmann, Martha Jones, Jasmane Ormand, Steve Otto, Jessica Paga, Huntley Polanshek, Linda Reilly, Jessamyn Rising, Rebecca and Marshall Scheetz, Wayne Shaia, Molly Swetnam-Burland, Joshua Timmons, Gene Tracy, Ben Zhang, the crew of the Godspeed at Jamestown Settlement, Va., and my mentors James C. Anderson and Christoph F. Konrad. I am, as ever, indebted also to John L. Robinson, my nautical mentor and best friend, and my mother Patricia A. Irby for her daily support and encouragement. My father, from whom I inherited my love of boats and capacious curiosity, is, as ever, woven deeply into the fabric of these pages.

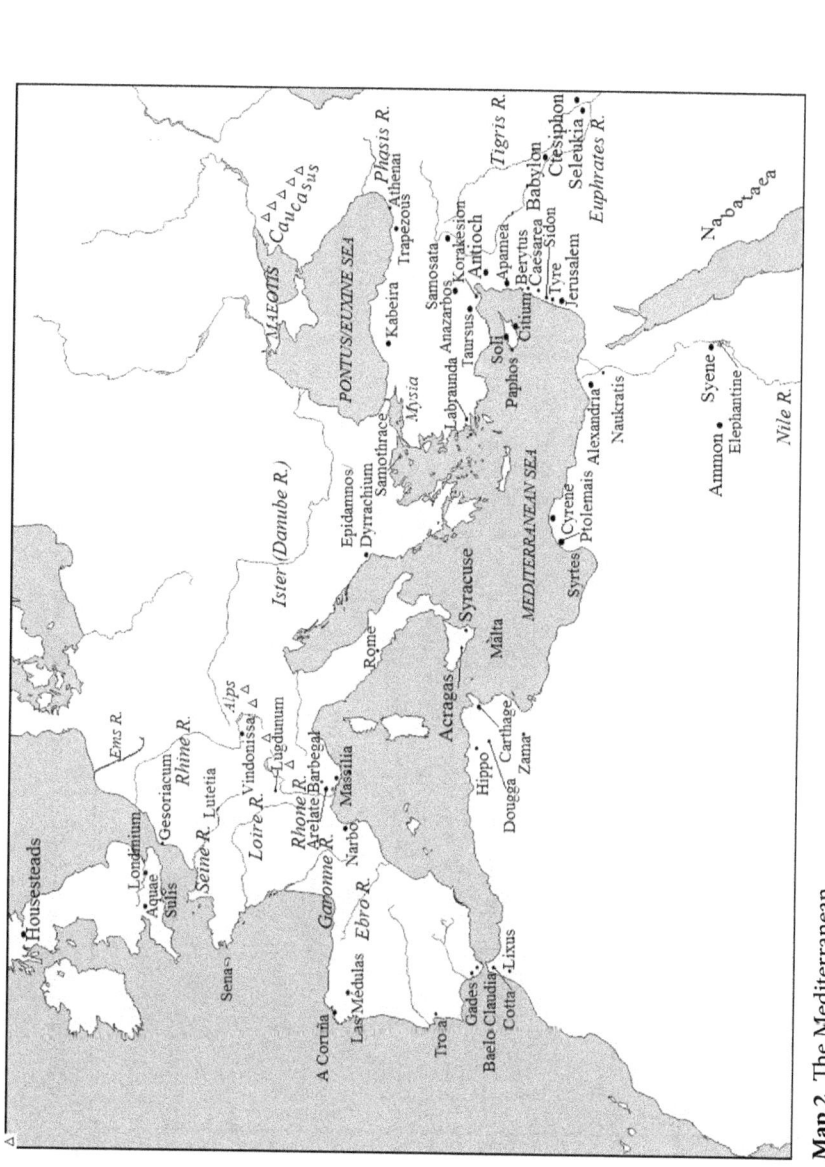

Map 2 The Mediterranean.

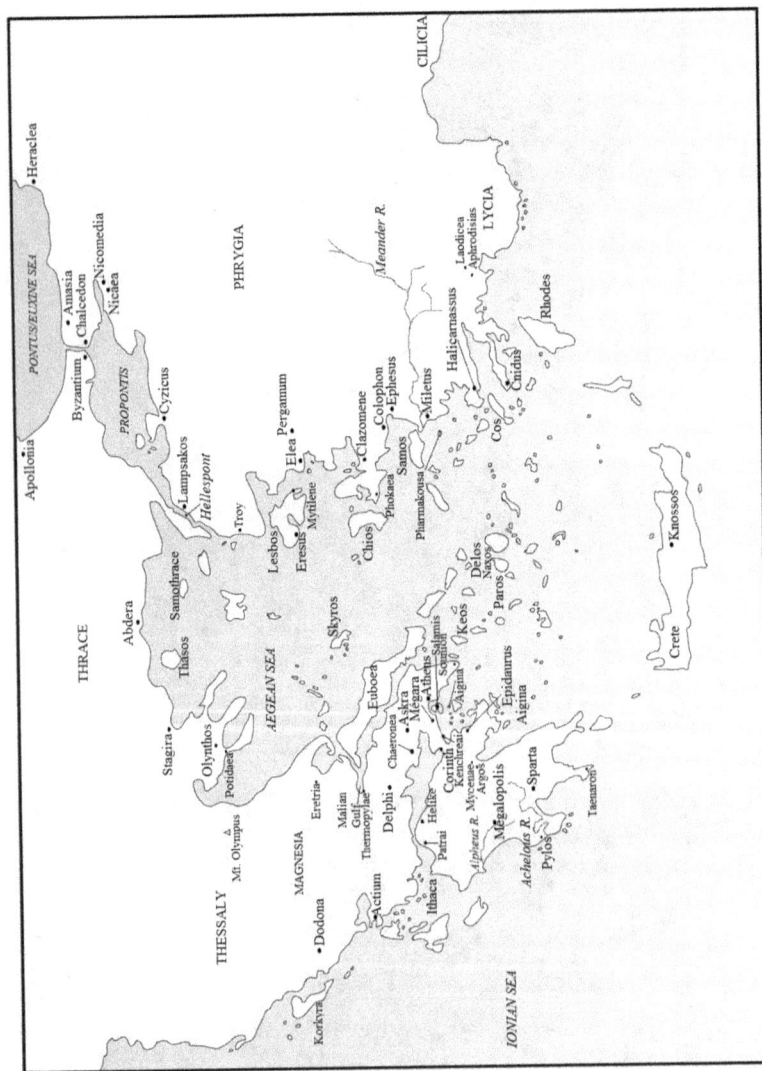

Map 1 Poleis and Regions of the (Greek) Eastern Mediterranean.

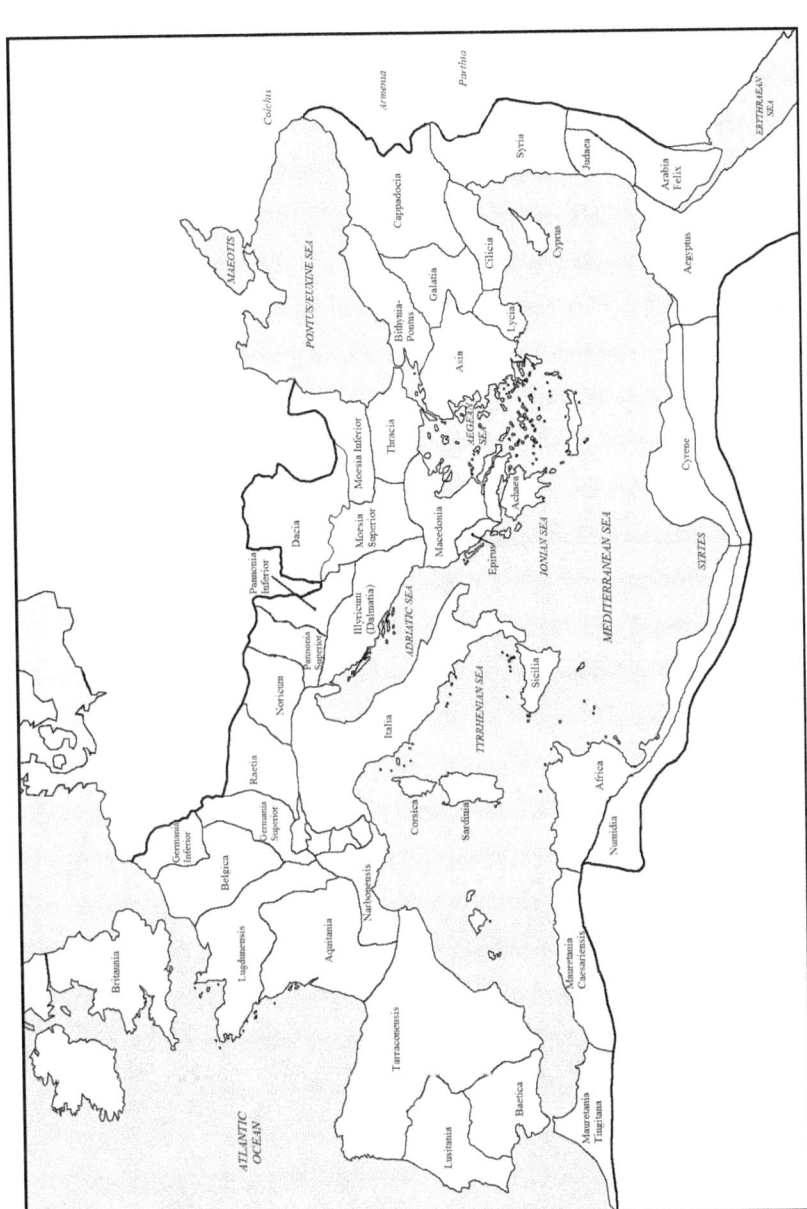

Map 3 Provinces of the Roman Empire.

ants or frogs around a marsh" (*Phaedo* 109b). Teeming with fish and traffic, delimiting and connecting places and people, rivers and seas served conveniently as lines on maps, marking boundaries and imposing order on early cartographic initiatives: Ocean frames the world, the Nile River flows between Libya (Africa) and Asia, the Tanais River and the Pontus (Black Sea) separates Asia from Europe, the Mediterranean Sea delimits Europe from Libya (Fig. 0.1).

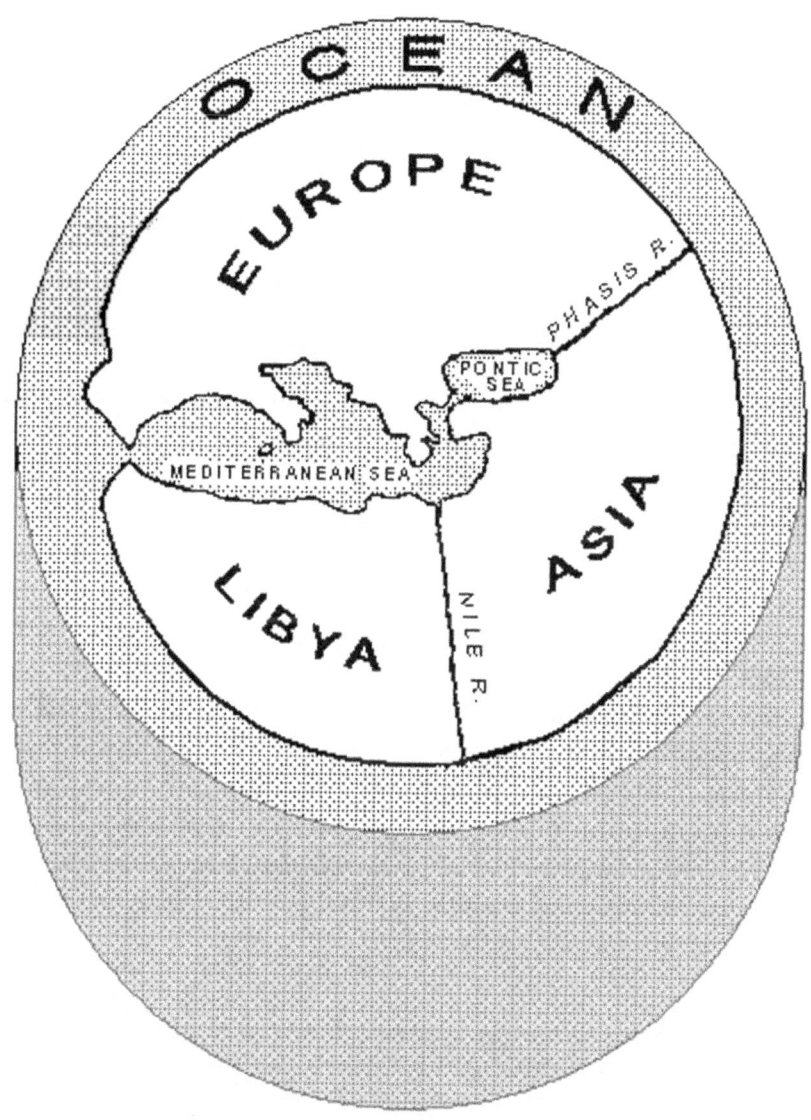

Fig. 0.1 Theoretical reconstruction of Anaximander's Map, early sixth century BCE.

Introduction: Conceptions of the Watery World

Water's Fundamental Importance to Life as We Know It

Water is an essential resource, permeating our world and enabling life as we know it. Seventy-one percent of the earth is covered in water (96 percent of which is the world's oceans; only 2.5 percent of the earth's water is fresh, most of which is prohibitively vaulted away in glaciers, ice caps, or remote, rocky subterranean aquifers). Water—the earth's only self-renewing vital resource, in a constant cycle of evaporation and precipitation (Solomon 2010: 10)—exists in the air as vapor and as moisture in the soil. The human body is comprised of more than 50 percent water, and a person cannot survive without water for more than a few days. Evidence suggesting the existence of liquid water near the surface of Mars has led scientists to conjecture whether life once existed on the red planet.[1] Moreover, rings comprised of significant ratios of aqueous ice revolve around the moons of Jupiter, Saturn, Neptune, and Uranus, planets whose atmospheres and interiors are believed to be replete with H_2O. NASA's Hubble Space Telescope, furthermore, has revealed a saltwater, sub-surface ocean on Jupiter's satellite, Ganymede, and evidence strongly suggests subsurface oceans on two of Jupiter's other icy moons (Europa and Callisto) and two of Saturn's (Enceladus and Titan).[2] Recent discoveries of exoplanets (beyond our solar system) that might have liquid water spark the imagination.[3] Finally, as the world's population grows, access to clean, safe water is becoming an increasingly politicized topic, and the next great war, so political theorists speculate, will be over water.

Water as an Organizing Principle in the Greco-Roman Mediterranean Basin

Water was no less important to the ancients. Plato described the peoples dwelling between the Phasis River and Pillars of Herakles as "living around the sea like

It is only natural, then, that the ancient peoples of the Mediterranean Basin would seek to understand the origins and effects of bodies of water, large and small, and that they would formulate complex and interesting explanations for hydrological phenomena. Aided by data collected from sailors, merchants, and explorers, they investigated sea and river depths, salinity, currents, tides, silting, estuaries, riverine anomalies (especially the annual summer-time Nile flood), the sources of rivers, such as the Nile and the Danube, and other aquatic phenomena.

Water fired the Greco-Roman imagination. An integral component of the Greek landscape, the sea was the backdrop for mythology and heroic quests: Herakles, Jason, and Odysseus traveled by sea. Herakles saved Hesione, princess of Troy, and Perseus rescued the Aithiopian beauty, Andromeda, from sea monsters. Metamorphosis was the hallmark of water gods, like their ever-flowing protean medium that naturally assumes the shape of whatever container might hold it. By changing shape, Thetis attempted to avoid Peleus' embrace (Ovid, *Metamorphoses* 11.221–228), and Proteus recoiled from prophesying to Menelaus.[4] Fanciful sea monsters were envisaged in both myth and zoology. In addition, Greek and Roman literature is abounding with tales of adventures by sea, from the pens of Homer, Apollonius of Rhodes, Vergil, Lucian, and many others. These authors had a deep knowledge of the sea in all its moods. Descriptions of storms at sea are vivid,[5] and the technical aspects of sailing are often realistically depicted. Odysseus, for example, understood celestial navigation (*Odyssey* 5.273–277), and Aeneas' helmsman Palinurus was a master of the art (Vergil, *Aeneid* 3.513–520). The sea and seafaring were the *lingua franca* of the ancient world, underpinning references, analogies, and nautical metaphors that sprinkled literature and popular culture, most notably Plato's ship of state. It was also a focus of cult: two Olympian gods held sway over the sea representing different aspects: Poseidon embodied the wild, unpredictable nature of the sea; Athena guided humanity's maritime conquest, subjugating Poseidon's rolling sea with her boats, as she had bridled her uncle's wild horse. Rituals accompanied sailing expeditions, and water was (and remains) a means of cultic purification and a meteorological agent of destruction in the hands of gods (or the forces of physics) who sought to destroy the miscreant human race through flooding.

The sea was explored by gods and natural philosophers, as well as kings and their agents. The Titan Atlas is said to have seen the "depths" of the entire sea (*Odyssey* 1.52–53), and Plato's *Phaedo* imagines the deepwater realm (109b–113d; cf., Chapter 1). *The Alexander Romance* (compiled in the third century CE, five centuries after Alexander's death) duly records an apocryphal anecdote of the great general's descent into the deep in a spherical glass submarine so that he could discover what was on the seabed. He purportedly reached a depth of about 154 meters (500 feet).[6] However fanciful Alexander's bathysphere might be, the legend does speak to intense curiosity regarding the dominant geographical feature of the Mediterranean world, the sea.

Alexander may not have descended into the sea, but his admiral, Nearchus, is an important source for the hydrology of India. Moreover, Arrian, who explored the Black Sea under the authority of Hadrian (ruled 117–138 CE), is invaluable for the hydrology of that area. At least two treatises entitled *On Ocean* are known, composed respectively by Pytheas of Massilia (Alexander's contemporary) and Posidonius of Apamea. Both, unfortunately, survive only in a few tantalizing fragments. Claiming to have reached "Ultima Thule," a distant, icy land in the northern Atlantic, Pytheas was an intellectual pariah whose works were redacted by hostile successors who (unfairly) delighted in showing up Pytheas' (perceived) mendacity. Pytheas' observations, furthermore, did not align with Aristotelian interpretations of the natural world. In contrast, Posidonius was much admired. Like so much Greco-Roman scholarship, Posidonius' hydrological treatise was broad in scope, covering not just hydrology (including tidal behavior), but also the earth's climactic zones, the size and extent of the inhabited world, celestial phenomena, terrestrial geography, ethnography, and history, as filtered through a Stoic lens.

Other authors wrote more broadly on water. For Thales (sixth century BCE), the first Greek natural philosopher, rational explanations of the world begin with water. For Empedocles, water is one of the four essential building blocks of the material world. Aristotle interrogated the nature of water (*Meteorology* 1.12–2.3), as did his student, Theophrastus. Theophrastus' *On Waters* survives partly in Arabic,[7] and his short treatise *On Fish* is extant in Greek.[8] But his *On the Sea* (Περὶ θαλάττης) is, sadly, lost. Finally, in the late first century CE, Plutarch debated the relative merits of water and fire in his short "Whether Fire or Water is more Useful" (*Moralia* 955d–958e; cf., Chapter 1).

Hydrological knowledge was seamlessly aligned with geography, astronomy, and meteorology, which were all informed by philosophical prejudice. Vacillating coastlines were long part of the scholarly dialogue, and geographers, from Homer onward, dutifully recorded the lengths of coastlines and the proximity of settlements to significant waterways. Ancient nomenclature, however, was equivocal. Is the Mediterranean a "sea" or a "lake" (see Chapter 2)? Additionally, Greek employs several words that can be understood broadly as an open expanse of water: ὠκεανός (*okeanos*: open sea: the primeval source of all other bodies of water) is typically reserved for the waters that were thought to frame the world (i.e., the Atlantic); πόντος (*pontus*: open sea), cognate with *pateo* (walk), πάτος (*patos*: something that is trodden or beaten, as in a path), and *pons* (bridge), suggesting water as a passageway;[9] πέλαγος (*pelagos*: high sea, open sea), referring sometimes to particular seas;[10] and θάλασσα/θάλαττα (*thalassa/thalatta*: sea, salt water), possibly cognate with ἅλς (*hals*: salt), and perhaps adopted from a pre-Greek word. *Thalassa* was usually reserved for the Mediterranean Sea in Greek sources. The Greeks referred to the Mediterranean as "the Sea," "this Sea," or "our Sea."[11] The Romans called that same body of water *Mare Nostrum* ("Our Sea"). "Mediterranean" did not come into common use until the sixth century CE, recorded as such by Isidore of Seville (*Etymologies or Origins* 9.1.8). This represents a paradigm shift as emphasis is transferred from sea (*mare*) to land (*terra*).

Although several important primary sources are lost, it is possible to tease out Greco-Roman knowledge, understanding, interpretation, and use of water in the world (including oceans and seas) from a long scholarly tradition that begins even before Thales. Curiosity about the natural world, especially the Mediterranean Sea and other large bodies of water, was deep. And here we shall explore that trajectory, the evolution of Greek and Roman conceptions of the sea, from Homer's circumambient river to unified philosophies of hydrology expressed in Aristotle, Seneca the Younger, and others. Other valuable sources include geographical authors (Eratosthenes, Pliny, Strabo), travelers' accounts (*Periploi*), and poets in Greek (Homer, Aeschylus, Pindar) and Latin (Vergil, Ovid, and others).

Knowledge of rivers, lakes, and the Sea was also gathered on a systematic, global scale. During the hegemony of Aristotle's famous student, Alexander of Macedon, as well as in the scientific "golden age" immediately following the general's death, theories and mathematical models came to be correlated with a growing body of facts about the world, including in hydrology. Scholars in numerous disciplines accompanied Alexander, who was eager to conquer and explore the entirety of the οἰκουμένη (*oikoumene*) east of the Aegean, in his endeavor to extend Greek culture as far east as the Punjab. His coterie included biologists, zoologists, physicians, historians, geographers, and surveyors who were instructed to collect data and to produce full records of their observations. Alexander's admiral, Nearchus (*FGrHist* 133), and his helmsman, Onesicritus (*FGrHist* 134), were valuable eye-witness sources for Eratosthenes,[12] Strabo, and Arrian on India. Nearchus overruled Onesicritus' order to sail directly to Cape

Maceta, admonishing the helmsman for not understanding Alexander's purpose in sending out the fleet:

> not because of any difficulty in getting his whole army safely through by the land route, but because he wanted to investigate the beaches along the line of the coast and the anchorages and islets.[13]

Onesicritus, nonetheless, claimed that his team "examined many things about nature" (f17). To what degree Onesicritus engaged with hydrological theory is difficult to ascertain from the brief fragments. Cited by Strabo more than thirty times, Nearchus commented on weather patterns (f18), riverine siltation (f17), the annual flooding of the Nile as compared with rivers in India (f20), and the large sea creatures in the Indian Ocean, including whales up to 23 *orguiae* (43 meters [140 feet]) who were easily chased off by loud noises:

> what was most troubling was the spouting that produced great streams and a large body of mist from the eruptions, so that they could not see the area in front of them.[14]

With reconnaissance and a shore party, Nearchus was also able to disprove his crew's conception that sailors seemed to disappear near a certain island (f1c), an island that Arrian reported as inhabited by a Nereid who turned sailors into fish.[15] Nearchus also listed the shoals along the coast of Sousis (f25), and he remarked on the lack of anchorages between Babylonia and India (f26). Another member of Alexander's coterie, the historian, Callisthenes, wrote a *periplus* (coasting guide) of the Black Sea (and probably beyond). Following Juba II's precis of Onesicritus' *Indika*,[16] the Roman encyclopedist, Pliny (6.96), criticized both Nearchus and Onesicritus for omitting toponyms and distances. But, advantaged by nearly four centuries of exploration and expansion under Rome's growing empire, Pliny simply had more data, and subsequent accounts were enhanced as knowledge accrued. Arrian, for example, referred to seasonal winds that prevented sailing (*Indika* 21.1) and the violent action of the open sea against the coast at the Indus' mouth (*Indika* 21.5). He gave particular attention to safe harbors and good anchorages (*Indika* 26.2, 29.1, 39.6). Although Arrian seems not to have visited India, he did in fact command an expedition in the Euxine. His account of the Black Sea is rich in hydrological detail, including the direction and effect of the winds (*Periplus* 3.2, 6.1), harbors that protected ships from the Thracian winds (*Periplus* 4.2, 18.3), and storm-churned waves washing over the sides of his ships (*Periplus* 3.3–4, 6.1).

Modern Studies

The waters of antiquity have not escaped the notice of modern scholars. Albin Lesky (1947) explored the evolution of the ancient Greek relationship with the sea, including mythology, as a process of adaptation from land-locked peoples to coastal settlers. Wachsmuth (1967) has marshaled evidence for nautical cults. More recently, Corvisier (2008) has broadly explored the ambivalent Greek attitude to the sea, both an inhospitable realm and the source of livelihood, commerce, and as the inspiration for mythology. Horden and Purcell (2000) cast a wide net in their investigation of the link between people and the micro-ecologies of the Mediterranean region (which cannot be disentangled from the sea). Broodbank treats geological, topographical, and ecological changes in the Mediterranean's basin at length (including the effects of the Ice Age of 21,000–18,000 BCE) in his magisterial 2013 volume. Beaulieu (2016) focuses on the symbolic resonance of aquatic mythology.

Conceptions of the Watery World

Our project falls into two volumes: *Conceptions of the Watery World* (Volume 1) and *Using and Conquering the Watery World* (*UCWW*) (Volume 2). In three chapters, we tackle the "science" of water as it provides a physical framework for the cosmos as understood by Greek and Roman thinkers: e.g., water and theories of the creation of the world (Water and the creation of the world), the nature of large bodies of water, ocean, and the seas—as understood by Greco-Roman natural philosophers—including salinity and tidal behavior (Seas and lakes), the disposition of terrestrial waters, such as rivers, springs, and natural fountains, and their interactions with land (The interplay between water and land). We then turn to consider how Greco-Roman thinkers explained watery phenomena, including their attempts to comprehend aqueous meteorological events (Watery weather),[17] and disease as a factor of visible and invisible natural forces (Water, health, and disease). In four chapters we survey the imagined watery world: the religious, the mythological, and the fantastical. Despite careful observation of and fascination with deep-sea creatures, they remained little understood. Ancient conceptions of marine fauna straddled the border between the biological and the fantastical, land and sea, and reflected the metamorphic qualities of their watery realm (Biological creatures of the sea). Like their blue-water habitat, marine animals remained mysterious, inspiring tales of even more terrifying mythical sea monsters who shared the abode of marine deities, who were in their turn metamorphic and transformative like water (Mythical creatures of the sea: sea monsters and sea gods). The divine was never completely divorced from ancient perceptions of the natural world, and water was a central factor in maintaining, understanding, and interrogating the delicate balance between human and divine: water could purify, water could be defiled, water could affirm purity, and water could reveal the will of the gods (Water and the divine: Unseen and magical forces of the spiritual world). Finally, the watery world was dangerous, and those who were compelled to travel on it sought safety from its hazards by supplicating gods who protect those who voyage on water with appropriate prayers, sacrifices, and festivals (Sailor cults and cults of sea gods).

Methodology

We rely heavily on written sources (literary, historical, and documentary) in order to reconstruct the ancient conception of water as a profound natural force. Although much of the written evidence comes from literary accounts of myth and legend, the stories were rooted in experience, rendering the tales of gods and heroes believable and feasible (e.g., the quarters of the winds). Artistic and archaeological evidence is brought to bear where it might be illuminating in our reconstruction of this slice of the intellectual life of the peoples who inhabited the Greek and Roman. How did these thinkers understand water as providing a framework of the cosmos? How were watery phenomena explained? How did water interact with the unseen and magical forces of the spiritual world?

Greek and Roman thinkers did not restrict themselves to the Mediterranean Basin. Nor do we. The familiar and the mundane did not hold the same interest for our sources as the unusual and the marvelous. Greco-Roman thinkers often defined themselves by contrast with distant and poorly understood peoples and locales, where peoples are lawless (Centaurs, Amazons) or especially beloved of the gods (Hyperboreans). By focusing on the edges of the Greco-Roman intellectual world (where natural phenomena do not behave according to the usual rules: e.g., remarkable rivers and paradoxical waters [Chapter 3]), we can begin to comprehend the center.

A vast amount of material has been produced over an expanse of time; by rhapsodes from the heroic age envisioned by Homer, and Greek and Roman thinkers active from the Archaic heyday of independent Greek *poleis* to the height of the Roman Empire. Natural philosophy evolves and is refined as questions become more sophisticated, answers grow more complex, and empirical data accrue from distant and exotic lands. We naturally follow the evidence (including Near-Eastern accounts where those traditions have formed the basis of Greek cosmogonic thought or where they provide useful comparanda). We aim to be chronological in our treatment. The sources are myriad, and an Appendix of major thinkers and writers provides a table of the approximate floruit dates of our most important primary sources as a handy reference for the reader (the reader is not expected to recall, for instance, the first mention of Posidonius of Apamea).

Although many are quick to point out the errors of ancient thinkers—who often did not understand what they were describing—the reader is reminded that our ancient authors were handicapped by a lack of technology, which modern scientists take so blithely for granted. Our own science and technology advances at a lightening pace, and some advanced or more enlightened society may consider us risible in the distant future. What is significant is not to what degree Greek and Roman writers were "right" or "wrong," but instead that they were asking interesting questions and proposing thoughtful responses to those queries.

Much of the evidence survives only in fragments, paraphrase, or redaction: e.g., the Presocratics, culled and distilled by Aristotle and others, for example, are neatly collected, translated, and annotated in the *TEGP*. Preservation and transmission of ancient thought began with the Greeks themselves. Their efforts resulted in a distorted and often uncontextualized selection—as thinkers aimed to refute and discredit their sources—but a selection nonetheless that was considered worthy to be copied numerous times over the span of 2,500 years (in many cases translated into Arabic and other languages).

The transliteration of Greek names is a subject that has become contentious: whether to retain the traditional Anglicized spellings or to Hellenize with the view to fidelity to the original language. In the interest of accessibility, we employ "traditional" Anglicized orthography of most Greek names. Although Sokrates (Socrates), and Empedokles (Empedocles), or Theophrastos (Theophrastus) should cause no problems, Aristoteles (Aristotle), Epikouros (Epicurus), and Dioskourides (Dioscorides) may raise eyebrows, and Herakleitos (Heraclitus) may seem like a mystifying string of unpronounceable letters. Despite existing conventions and every effort at consistency, T. E. Lawrence's thoughts on the transliteration of Arabic names is as revealing as it is entertaining, and remains applicable to any effort at transliteration:

Arabic names won't go into English, exactly, for their consonants are not the same as ours, and their vowels, like ours, vary from district to district. There are some 'scientific systems' of transliteration, helpful to people who know enough Arabic not to need helping, but a wash-out for the world.
Seven Pillars of Wisdom. New York: Doubleday reprint 1991, 21

All translations are my own, except for Seneca the Younger's *Natural Questions*, for which I rely on the superb translation of Harry M. Hine (2010), Strabo, expertly translated by Dwayne W. Roller (2014), Ancient Near Eastern material (Foster 1995; Heidel 1942), and where a previously published translation simply could not be bettered (as cited). Cross-references between chapters are indicated by chapter number (e.g., Chapter 1).

Part One

Interpreting the Watery Framework: Philosophy, Cosmogony, and Physics

1

Water and the Creation of the World

Introduction

The Greek mainland is mountainous and dry, with an average annual rainfall of 20–50 inches (51–130 cm), increasing with latitude. Rivers tend to be small, drying out in the summer and filling up in the wetter, colder months. Observed throughout the Mediterranean Basin, this cycle between the parched, hot summer and the cooler, damp winter provides context for the development of ancient theories about the creation of the world, firmly rooted within a watery framework. Cosmogonies penned by poets and philosophers relied on a paradigm where opposite properties were kept in balance, vacillating between arid and moist, warm and cool. The earliest recorded Greek hypotheses of cosmogony, expressed most fully by Hesiod in the seventh century BCE (*Theogony* 116–138), share striking resemblances with the earlier Mesopotamian tradition where the world was created by the separation of discrete elements from a primal quagmire, nor is it unreasonable to surmise that the Akkadian account traveled westward, influencing subsequent versions.[1]

Mesopotamian Origins

Let us begin "between the rivers" where the rhythms of the waterways were recognized as guaranteeing and regulating both human life and the land's fertility. Water, the "life of the land," is essential to life, ensuring fertility of the crops, which in turn sustains human and animal populations. But even more, water is the generative principle that initiates and shapes the Mesopotamian world as we see in the *Enuma Elish* (dating from as early as the nineteenth century BCE), where creation began with the intermingling of pre-existing fresh and salt water (1.1–8). A watery chaos existed before creation, held in check by the power of other deities.[2] From this primeval, thoroughly mixed watery state, the physical world emerged, envisioned as a human family whose progenitors were associated with two types of waters: salt (the wild and destructive, "chaotic" Tiamat) and fresh (Apsu/Abzu).[3] This blending of fresh and salt waters naturally occurs where bodies of water meet, as in the Persian Gulf where fresh waters from aquifers mix with the salt water of the Arabian sea.[4] Such hydrological effects are especially noticeable at Dilmum, the island nation where the Sumerians believed creation began and whose modern Arabic name, Bahrain, means "two seas."[5] The myth thus articulates an observable hydrological phenomenon.

The mingled waters were anthropomorphized. The Akkadian Abzu ("deep water") represented the fresh waters flowing in subterranean aquifers, which—so the Sumerians and Akkadians believed—fed the lakes, rivers, springs, wells, and other waterways that guaranteed life in their harsh, desiccated climate. Abzu, as a building block of Mesopotamian life and as a character in the cosmogonic account, was consequently envisioned as a primeval freshwater sea that lay below the realm of the human world (earth and underworld). Tiamat, likewise, embodied water both linguistically and physically: her name derives from the Akkadian *tiamtu/tâmtu*, "the normal word for sea."[6] Burkert additionally connects "Tiamat" to Tethys, the wife of the Greek Okeanos (below), strongly suggesting the Akkadian influence on early Greek mythology. Burkert further links Tiamat/Tethys to *thalassa*, a Greek word for sea whose origins are not Greek.[7]

From the brackish mingling of salt- and freshwater deities came the younger generation of gods, a genealogical structure repeated in the *Iliad*, where Okeanos (salt water) and Tethys (fresh water) were identified as the source of the gods in Homer[8] and Plato (*Timaeus* 40e). In turn, the younger Mesopotamian generation colluded to murder their father, Apsu, in order to usurp his authority (a motif of generational conflict repeated in Hesiod's *Theogony*, below). In revenge, Tiamat generated terrifying, hybrid creatures to wreak her revenge: sharp-toothed monster-serpents, "merciless of fang," whose bodies were filled with poison instead of blood, terrifyingly huge "fierce monster-vipers":

> And hurricanes and raging hounds, and scorpion-men,
> And mighty tempests, and fish-men and rams;
> They bear cruel weapons, without fear of the fight.[9]

Tiamat was a goddess of water, and her parthenogenic children—including hurricanes, tempests, and fish-men—were naturally aqueous. In revenge, the salt-water goddess created monsters to reclaim her authority, hybrids that anticipate the composite, serpentine Typhon, the son of Gaia and Tartarus, and other sea monsters (Hesiod, *Theogony* 823–835; cf., Chapter 7).

Slain by her son, Marduk (himself a god of waters and storms), Tiamat provided the generative materials for the physical world, graphically forged by Marduk:

> He (Marduk) split her up like a flat fish into two halves;
> One half of her he established as a covering for heaven.
> He fixed a bolt, he stationed a watchman,
> And bade them not to let her waters come forth.
>
> 4.137–140

The imagery here draws on the watery nature of both mother and son, as the child vanquishes and controls his parent. Marduk has dammed his mother's waters, a clear reference to the importance of hydraulic infrastructure in the Mesopotamian world, described with admiration by Herodotus (1.185–186), and foregrounded in the Akkadian cosmogony, the *Atrahasis*, where the gods dig watercourses (Tigris, Euphrates, springs) and canals (21–8). The creation of the world as we know it thus speaks to the importance of water—and of controlling it—in the dry Mediterranean littoral. Tiamat became the bounding principle of the physical world. Marduk employed the body of Tiamat, both the benign source of life and a corporealization of raw nature, to delimit the Mesopotamian lands. Tiamat's waters were restrained—dammed—by her very body, a physical barrier between creation and the contemporary world of human beings.

Hesiod

A millennium later, the motifs of Mesopotamian creation tales would be distilled by Greek poets. In his *Theogony*, Hesiod (from Askra, an inland town near Mt. Helicon) provided a vision of the creation of the Greek cosmos that replicates the separation motif expressed in the Babylonian cosmogony.[10] As in the Near Eastern tradition, mythology transitioned into physics in early Greek cosmogony, and the gods became the building blocks of the cosmos. In short, the physical infrastructure was constructed, literally, by pulling substance after substance from the primeval *chaos* (χάος: a yawning gape, as close to nothingness as possible): Gaia (earth), the solid foundation; then Tartarus, the primordial edge of creation beneath Gaia; and Eros, the cosmic principle of Love, necessary to bring together male and female. From Gaia were drawn the physical components of the orderly universe: Ouranos (Sky), Mountains, and Pontus (Sea). From the sexual union of Gaia and Ouranos were born the first generation of anthropomorphic deities, the Titans, who included Okeanos (*Theogony* 133) and Tethys (*Theogony* 136), deities of salt- and fresh water whose fertility was so enhanced that their offspring numbered 6,000—3,000 daughters and 3,000 sons—in order that each body of water might have its own discrete divine patron (*Theogony* 346–370). Like Tiamat, the de-anthropomorphized Okeanos framed the human world in Hesiod and Homer. Water is thus both a bounding principle and a generative one. Sharing in their father's fecundity, Okeanos' children, Nereus ("the old man of the sea") and Doris, together were the parents of a prodigious fifty nymphs.[11]

Whether the early poets, including Homer and Hesiod, viewed these "elemental" gods as the folkloric sources of agricultural prosperity, or as the theoretical, scientific framework of the cosmos, we cannot know. Herington argues for the former, but in Hesiod it is clear that elemental gods (Gaia, Ouranos, and Okeanos) serve a dual role.[12] Gaia is both the anthropomorphized mother of creation and the solid foundation of the human world. Ouranos is both the consort of Gaia and the physical vault of heaven, separated from the earthly realm by his son, Atlas (*Theogony* 517–518). Okeanos is the father of lesser waterways as well as the river that limits the earth.

From Myth to Reason

These charming tales were quickly rationalized, and a new field of human knowledge (Natural Philosophy) was created. The Greeks were hardly the first to make "scientific" advances. The Babylonian achievements in mathematics and astronomy were remarkable, including a place-value number system (like the system of Arabic numerals, in contrast with the discreet signs in the Egyptian, Greek, and Roman numeric systems), and advances in arithmetic, algebra, and even quadratic equations. From the second millennium BCE, Babylonian astronomers observed and recorded celestial data, tracking, for example, the appearance and disappearance of Venus. The Egyptian achievement in the higher (mathematical) sciences includes "the only intelligent calendar in human history,"[13] a 365-day year of twelve months of thirty days each, plus five additional days to synchronize the solar and lunar systems.

By the sixth century BCE, Greek thinkers went beyond the observational, delving into the theoretical, and began to ask new questions: what are the ingredients, composition, and operational paradigms of the natural world? Is *phusis* (nature) one thing or many? What is the process of change by which things come into being and pass away? What is the process by which matter is transformed from one thing to another? No longer was nature personified, and supernatural intervention was now considered specious. Greek thinkers at this time developed a fully articulated system of inquiry into the natural world, including methodology, a system which sweeps across the whole of what we consider "natural science." Here we survey those early thinkers for whom water was a significant component of natural philosophy.

The Presocratics

The Milesians

Because of a favorable combination of geography, wealth, and cultural infrastructure, Miletus was uniquely situated to foment intellectual inquiry, and the *polis* produced important early thinkers in natural philosophy, including Thales, Anaximander, and Anaximenes. Miletus was a vigorous center of commercial and colonial activity on the western coast of Turkey.[14] As such, it was a wealthy trade center through which the scientific and artistic achievements of the Near East (including Mesopotamia and Egypt) percolated into the Archaic Greek world. It would also have been a hub for sailors in port disseminating ideas and data from abroad.[15] A fully alphabetic writing system further enabled the promulgation of knowledge, which resulted in a burst of creativity throughout the Greek-speaking world (e.g., in Greek lyric poetry: Alcman, Alcaeus, Sappho). Citizen debate and open discussion resulted in legal codes, imposed not from above, but according to the political will of a majority of the citizen body, providing a political analog to philosophical critical inquiry.[16]

In this spirit of debate and persuasion, the Milesians offered different solutions to the same questions. Building one upon the other, they recognized the difficulties and new questions that arose from the solutions of their predecessors. Additionally, they understood the necessity not just of stating an idea but of defending it against critics. In their inquiry into nature, Greek thinkers applied the principles of rational criticism and debate, focusing not on specific instances (e.g., an individual earthquake) but rather on classes of natural phenomena. They did not seek to explain the causes of a single earthquake, but instead they aimed to understand the principles that were common to all earthquakes. They asked penetrating questions concerning the nature of physics and life, and they determined upon an orderly, predictable, mechanistic world in which things behave according to their nature.

Thales

Thales was the first to ask such questions. If he wrote anything at all, his words have been completely lost, and his thoughts were selectively redacted some two centuries later by Alexander of Macedon's famous tutor, Aristotle of Stagira, and by later, sometimes hostile, commentators. The earliest natural philosophers proposed systems of material monism, suggesting one element or idea as the *arche* (ἀρχή: origin or beginning) of the cosmos. Thales posited water:

> Thales, the originator of this kind of theory, says it (*arche*) is water (and that is why he asserted that the earth floats on water), perhaps getting this conception from observing that everything derives its nourishment from what is moist and that the hot itself arises from and lives off it (and the thing from which the hot comes to be is the source of everything else). He gets his conception both from this fact and from the fact that the seeds of all things have a moist nature, and also the fact that water is the source of growth for moist things.
>
> TEGP 15, cf., Vitruvius 8.1.1

> Thales of Miletus, one of the Seven Sages, is said to have been the first to pursue natural philosophy. He said the beginning and end of the world was water. For from this the world is composed when it is condensed and in turn dissolved, and the world is borne on it. From it <come> earthquakes, windstorms, and the motions of the stars ...
>
> TEGP 20

This is a sophisticated theory, and Aristotle, our source for the first passage, seems to interpret *arche* broadly as a primeval element from which the cosmos arose, as the element that constitutes the physical world, and as an explanation of natural phenomena.[17] Water would thus be both the source of stuff and the force by which matter could transform from one substance into another.

What Thales really proposed we shall never know, but, if we are to trust Aristotle, Thales' theory derived from his own critical observations of the natural world. This theory, in some way, began with water. Furthermore, although offering mechanistic explanations of natural phenomena (thunder or earthquakes) that were often ascribed to theistic causes, the Milesians maintained the divine nature of their first principles (consequently positing a "reformed" theology).[18] Thales, nonetheless, was admired for this first step, counted among the "Seven Sages," and he was acknowledged as the first (Hellenic) natural philosopher (*TEGP* 20). Although his theories were dismissed by his successors,[19] Thales' students and detractors would build on his theory, replacing "water" with other elements and building blocks.

Anaximenes

We turn now to Anaximenes, who studied under Thales' student, Anaximander. Anaximenes was credited with an even more sophisticated system that employed air as *arche*:

> Anaximenes, son of Eurystratus, of Miletus, was an associate of Anaximander, who says, like him, that the underlying nature is single and boundless, but not indeterminate, calling it air. It differs in essence in accordance with its rarity or density. When it is thinned, it becomes fire, while when it is condensed it becomes wind, then cloud, while still more condensed, water, then earth, then stones. Everything else comes from these.
>
> *TEGP* 3

On this theory, material properties (density or pressure)—as determined by environment (heat and cold)—explain how one substance (air) can generate different substances (from fire to stone). One material, air, thus can give the appearance of different states of matter (fire, wind, cloud, water, earth, stone). Just like water, air assumes different states of being—solid, liquid, or vapor—depending on temperature. Anaximenes even offered empirical proof:

> ... For he says what is contracted and condensed is cold what is thin and loose (using this very expression) is hot. Hence the saying that a man blows hot and cold from his mouth is not inappropriate. For the breath is cooled when it is compressed and condensed by the lips, but when the mouth is relaxed it becomes hot as it leaves the mouth because of being rarefied. Aristotle attributes this argument to the man's (Anaximenes') ignorance. For when the mouth is relaxed we exhale the heat from our own body, but when we blow through constricted lips, it is not the air from our mouths, but the air in front of our mouths, which is cold, that is pushed forward and falls on us.
>
> *TEGP* 7

Plutarch (our source for the passage) may be describing an early experiment (comparing the effects of blowing on one's hand through opened or puckered lips).[20] The exercise, nonetheless, shows an interest in empirical investigation, and the effect of temperature on matter would remain a guiding principle of Greek physics (see Chapter 4).

The Rejection of Material Monism

Subsequent thinkers would soon question the validity of material monism as an adequate paradigm of the physical world, raising new questions and perceiving new problems. Unlike other early philosophers who preferred prose, the widely-travelled Xenophanes of Colophon (near Ephesus) used poetry to interrogate the conventions and assumptions of human knowledge (and perhaps also to earn a living as an itinerant minstrel). For Xenophanes, who rejected the material monism posited by Thales and Anaximander, two elements together (water and earth) provide the generative material for all sensible matter: "all things which come to be and grow are earth and water" (*TEGP* 51).[21] Water is thus an essential component of Xenophanes' cosmogony and physics.

The Ephesian thinker, Heraclitus, who believed that all matter is in a constant state of flux and change, recognized the importance of rigor in intellectual inquiry, but he questioned the reliability of the typical human mind to arrive at a correct interpretation of the natural world:

> Many do not understand such things as they encounter, nor do they learn by their experience, but think they do.
>
> *TEGP* 10

> Learning many things does not teach understanding, else it would have taught Hesiod and Pythagoras, as well as Xenophanes and Hecataeus.
>
> *TEGP* 18

Heraclitus, nonetheless, gave preference to empirical data when tempered by rational thought: "The things of which there is sight, hearing, experience, I prefer." (*TEGP* 33)

In contrast, Parmenides of Elea in southern Italy, was perhaps the first to call into question the fundamental reliability of sensory perception, positing two coexisting versions of the cosmos: "the Way of Truth" (wherein change cannot occur) and "the Way of Persuasion" (the world of sensory perception in which humanity exists). Like Heraclitus, Parmenides recognized that scientific investigation is a process of interpretation, as did his contemporary, Protagoras of Abdera, who had argued that human sensory perception was the best and most credible guide to "truth," but that the sensory world appears differently to different people. There is consequently no baseline for determining what is "true."

Empedocles and the Four-element Theory

It was within this intellectual milieu that the "religious guru,"[22] Empedocles of Acragas (Sicily), was credited with the four-element theory first hinted at in Anaximenes and eventually embraced by Aristotle, thus enduring as nearly unimpeachable in the ancient world. Like Xenophanes, Empedocles wrote in verse. Like Xenophanes, Empedocles' interests were broad, and his theories were wide-ranging. For Empedocles, as his predecessors, sea (water) is an essential cosmogonic material that is significant in the development and framework of the earth and its *oikoumene* as the four elements vacillate between total mixture and total separation (*TEGP* 41). Despite critical gaps, sizable fragments of his poems, *On Nature* and *On Purification*, are extant.

According to Empedocles, four irreducible roots (ῥιζώματα: *rhizomata*)— fire, air, earth, and water—account for the material universe as they combine, variously yielding other substances, such as flesh and bone:

> Empedocles [posits] four [principles] adding earth to the three already mentioned [water, air, and fire]. For these always remain and do not come-to-be except by becoming more or less as they congregate or segregate to form or dissolve a unity.
>
> *TEGP* 28[23]

Empedocles thus recognized a distinction between "element" (root/ῥίζωμα) and compound. Pre-Empedoclean popular belief had posited a three-element paradigm of creation, promoting Gaia (earth), Ouranos (sky/air), and Okeanos (water) as the essential building blocks of the world, adding, occasionally, "underworld" (Tartarus) as a fourth.[24] These divine beings were among the deities who first came into existence, providing the armature of the cosmos.[25] Empedocles, furthermore, connected his own elemental roots to Olympian deities:

> Zeus is fire, life-giving Hera the earth which bears life-giving produce, Aidoneus (Hades) air, because although we see things through it, it is the only thing we do not see, and Nestis is water. For this alone (Nestis) serves as the vehicle of nourishment for all things that need it, while by itself it is not able to nourish them. For if it nourished things, Empedocles says, animals would never be seized by hunger, since there is always plenty of water in the world. For this reason he calls water Nestis, because though it is the cause of nourishment, it does not have the power to nourish things that need nourishment.
>
> *TEGP* 27

Empedocles has explicitly associated three of the elements with familiar Greek gods: fire (Zeus, the supreme god of the Olympian pantheon, the god of ordered law in the Greek cosmos), earth (Hera, Zeus' sister-wife whose power was almost equal to his own, especially with regard to the family), and air (Aidoneus/Hades, the god of the underworld who reigned over the breath-like souls of the deceased). Nestis ("the mortal spring"/water), however, is obscure. Elsewhere, Empedocles emphasized her elemental and cosmogonic significance:

> The four roots of all things hear first: shining Zeus, life-giving Hera, Aidoneus, and Nestis, who by her tears moistens the mortal springs.
>
> TEGP 26

In this fragment, as Kingsley observes, Nestis is treated in an entire hexameter line, whereas the other three elemental gods share a single line.[26] Nestis has long been identified with Hades' bride, Persephone (see Chapter 8). If the name is Greek, it would mean "fasting," and could thus refer to the abstinence observed by Demeter when she went in search of her missing daughter.[27] It is also likely that Empedocles has assimilated a Sicilian deity whose cult resembled Persephone's.

Empedocles' interpretation of the nature of water as an elemental cosmic root reflects the ambiguities and tensions that we have already seen in the archaic poets. Water both sustains and destroys life. Life cannot exist without it. On its own, however, water is barren. Empedocles intimated that not only are the ratios of the elements in a constant state of flux, but so are the very elements.[28] This theory was supported five centuries later by the Roman thinker, Seneca, who succinctly argued that there are "reciprocal exchanges between all the elements" (*NQ* 3.10.3) and that "everything is produced from everything: air from water, water from air, fire from air, air from fire" (*NQ* 3.10.1).

Plato and Aristotle

Directing their attention to the rules of reasoning, argumentation, and theory assessment, Presocratic thinkers succeeded in formulating questions (and answers) about the nature and creation of the material universe. These early theories were interpreted, adapted, and forged into comprehensive philosophical systems by Plato and Aristotle.

Plato

Plato (428/7–348/7 BCE) was the culmination of this development in many intellectual fields, including cosmology, ethics, epistemology, and the nature of *phusis* (the material world). Plato synthesized the theories of his predecessors and addressed the problems that those theories raised, offering his own explanations regarding plurality, the reliability of sensory perception, how change is possible, and how motion can occur (but only in a plenum [where corporeal air fills the gaps between more solid substances]). Plato's cosmos is guided by a benevolent, rational Demiurge (craftsman), who imposes order on the cosmos from the outside.[29] Like Heraclitus and Parmenides, who argued that sensory perception could be fallible and subject to misinterpretation, Plato distinguished between reality and "seeming" by postulating two planes of existence. The incorporeal, stable, eternal, unchanging Forms represent an underlying true reality, in contrast with sensible objects that belong to the changing world of becoming. Objects of a certain class (e.g., ship) have a share in the ideal of that object. Some characteristics are essential (the ability to float and move across water under sail or oar). Other features, however, are merely incidental (hull shape, deck length, number of masts or rowers, prow decoration). The fundamental characteristics link the physical object to its Ideal Form, however imperfectly. According to Plato, the carpenter who builds the ship is an analog of the Demiurge, the Craftsman who constructs the cosmos according to a divine plan. Consequently, the cosmos and everything in it are replicas (e.g., made from blueprints) and all things are imperfect because of the limitations inherent in available materials.[30] To answer Parmenides' utter denial that change could occur at all, Plato posited that change and stability are possible because each belongs to a separate realm of reality. The corporeal world is characterized by change, illusion, and imperfection, whereas the world of Forms is changeless, perfect, eternal, and stable. Consequently, both change and stability are genuine.

Plato also scrutinized the intrinsic nature of water, which was one of his three interchangeable elements. In the *Timaeus* (53c–57c), Plato explained a geometrical atomic theory that derives from Pythagorean, Democritean, and Empedoclean physics. On Plato's theory two-dimensional right-angled isosceles triangles and half-equilateral triangles combine to form the five three-dimensional building blocks of sensible objects. Four of these solids are connected to Empedocles' *rhizomata*:

tetrahedron = fire
cube = earth

octahedron = air
icosahedron = water
dodecahedron = quintessence

Each simple body exists in different forms according to the size of compositional units. Plato's system, however, allows for the transformation of one root into another, as constituent triangles recombine: an icosahedron of water is dissolvable into twenty constituent equilateral triangles, which can then recombine into two octahedra of air (eight equilateral triangles each) and one tetrahedron of fire (four equilateral triangles) through a process of cutting (elements with fewer faces are sharper, more angular, and therefore they "cut") and crushing (less acute solid angles are larger and so "crush" more acute objects). Water is the least mobile of the three interchangeable elements and the least sharp.

Aristotle

In pointed contrast, Aristotle (384–322 BCE) rejected abstract thought, but instead valued empirical observation and axiomatic deduction. He believed that objects could be understood only through the knowledge of their purpose or function, reducible to Four Causes (Formal, Material, Efficient, Final). The Final Cause overrules the other causes: a ship's purpose determines its form (boat-shaped), material (wood), and method of creation (carpentry). Aristotle accepted empirically observable change as genuine, the actuality of the potential *qua* such. Change can occur in quality (as a seed becomes a sapling), in substance (generation and corruption: trees grow and then are harvested to be used for ships' lumber), or form (i.e., the alteration of properties: colored sails might fade in sunlight). In the sub-lunar, terrestrial sphere—where birth, death, and change occur—Aristotle argued that traits that give character to individual objects belong to the object itself and do not exist independently as abstract ideals.[31] Attached to unqualified substrates, hardness, density, and other material properties are produced by the interaction of the senses with stuff (e.g., an object may be defined as hard or soft to the touch).

Recognizing the limitations of his predecessors, Aristotle viewed material monism as too limiting and the four-element theory as too absolute. He had soundly rejected atomism with its void, arguing that change, movement, and life are possible only in a plenum (continuum, the opposite of void: *Physics* 213b32–216b20). He thus adapted Empedocles' four-element theory. Each element, on

Aristotle's physics, is reducible to formal properties that are interchangeable through the exchange of qualities: (Fig. 1.1)

Air: wet and hot
Fire: hot and dry
Earth: dry and cold
Water: wet and cold

Combinations of these material properties—hot, cold, wet, dry (the idea underlying Hippocratic humoral theory)—allow for generation (coming into being), corruption (passing away), and change (growth) as one property transforms into its opposite: e.g., water is produced when warm air is cooled.[32] This is an economical explanation of elemental transmutation and change of state (solid, liquid, gas). By reducing the four elements to four contrary properties, the possibilities would be endless, as one element neatly changes into the next in a proto-chemical reaction. Aristotle could thus explain change in the world and the nature of the physical universe with a tidy theory that has so neatly cannibalized the achievements of his predecessors.[33]

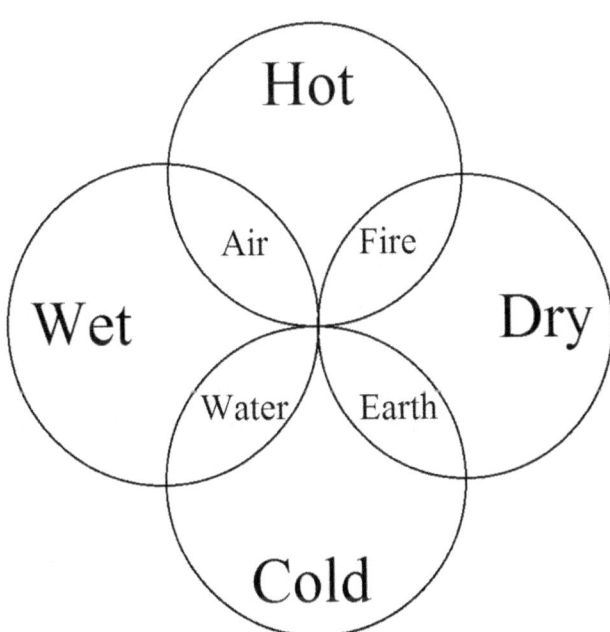

Fig. 1.1 The formal properties of Aristotelian elements.

The Epicurean and Stoic Schools of Philosophy

Like other avenues of inquiry, theories about water were shaped by philosophical prejudices, and we now briefly survey the two prevailing post-Aristotelian schools of philosophy that aimed to provide integrated views of the human and natural worlds: Stoicism and Epicureanism. Both schools of thought agreed on many guiding principles: the gods exist; irrational fears derive from ignorance of the causes of natural phenomena; the purpose of philosophy is to secure "happiness" (defined as *ataraxia*: "freedom from anxiety and fear"); knowledge of philosophy enables a person to live in harmony with natural law; the study of physics and ethics is essential for achieving the *summum bonum* ("the greatest good"). The schools, however, disagreed on the nature of the gods, the details of their physics, and what constitutes the *summum bonum*.

Epicureanism

Epicureanism was adapted from the atomic theory proposed by the pre-Socratic, Democritus, who argued that all objects are a combination of void and atoms ("un-cuttable," permanent and unchanging elemental particles that are infinite in number, size, shape, texture).[34] According to the school's eponymous founder, Epicurus (310–270 BCE), atoms have the properties of size, shape, and weight. In a void they move at constant speed in a downward motion, a "primeval cosmic rain," falling "quicker than the light of the sun."[35] A causeless, physical mutation, "swerve," compels atoms to collide at random, thus creating compounds or sensible objects. "Swerve" eliminates determinism and preserves "free will," and change is rendered possible by the re-arrangement of atoms. This causeless "swerve," however, was ridiculed in antiquity as specious.[36]

The fullest expression of the atomic theory is to be found two centuries later in the Roman poet, Lucretius (ca. 55 BCE), who argued that the interchangeability of the elements on Platonic and Aristotelian theory would render the elements perishable (and not eternal as the Greeks have argued: 1.782–802). For the Epicureans, water is soft by nature (1.281), and the atoms of all liquids and fluids (like water) are smooth and round, facilitating fluidity whereby water atoms do not dislodge other atoms:

> for a drink of poppy seeds is as easy as one of water, and the little globes are not viscous among themselves and likewise, when struck, a poppy seed is rolled downwards just as easily as water.
>
> 2.451–455

Lucretius used the example of clothes drying on a line to prove the existence of tiny particles (i.e., "atoms"). We cannot see the moisture soaking into the cloth nor can we observe heat drawing the water out: "therefore, water is sprinkled out into tiny parts which our eyes are unable to observe by reason" (1.305–310). Furthermore, Lucretius believed that the interaction of water with rocks proves the existence of void: "the liquid moistness of the waters in rocks and caves remains, and they all (rocks and caves) weep with plentiful drops" (1.348–349). Water atoms have "small, rollable shapes" (3.190: *volubilis*). Evoking Empedocles and others, Lucretius recognized water as the sustainer of life (*liquor almus*: 2.390), which can be deadly when amassed (1.282–289), an ambiguity that characterizes the earliest conceptions of water's dual nature.

The physics was strictly materialistic, and the Epicureans believed that, although the gods existed, they had no concern in human affairs (superstition, thus, is irrational).[37] The school also taught that superstition and unfulfilled desires cause pain, and that the *summum bonum* results from a state in which natural and necessary desires are satisfied. Because of its unqualified rejection of public service (which causes anxiety), Epicureanism was all the more distasteful to the duty-obsessed Romans.

Stoicism

Founded in Athens by Zeno of Citium, Cyprus, another school of philosophy, Stoicism, became pre-eminent in the Hellenistic and Roman worlds, leaving its mark on medical and technological theory. (Zeno met with his students at the Stoa Poikile in Athens, thus giving the school its name.) Synthesizing Aristotelian physics (plenum, mutable and transmutative elements, natural motion) with Heraclitus' ethical physics (fire, *Logos*), Zeno and his successors produced a coherent ethical and physical philosophy that was governed by a single first principle. That is to say, everything is in contact with the fiery, divine soul.[38]

On Stoic physics, the Empedoclean elements derive from fire, and they will return to fire. The elements are mutable and transmutative, as the ratios of heat, cold, wet, and dry cycle to transform fire (hot and dry) into earth (dry and cold), earth to water (cold and wet), water into air (wet and hot), and air back into fire. (see Fig. 1.1) This transformation is proved, for example, by the evaporation of liquid left in a pan, water bubbling up from the earth, and flames "dying" when deprived of air. Change and movement are possible by means of *pneuma* (breath), which binds the cosmos through kinetic tension (τονική κίνησις: *tonike kinesis*), a simultaneous motion in opposite directions that results in cohesion and equilibrium (compare Heraclitus' "Unity of Opposites"), analogous to the oscillating waves that occur when a stone is tossed into a confined pool (an uncanny, if superficial, resemblance to the theories of wave expansion and force field vibrations of modern physics). *Pneuma* thus provides a governing framework for physical phenomena that is fed by a divine, dynamic, universal, fiery soul, with which all matter is in contact.[39] For the Stoics, it was paramount to study both physics and ethics in order to achieve the *summum bonum*, i.e., eradication of irrational fears via understanding and correctly interpreting the causes of natural phenomena. The Stoic cosmos is a plenum, dynamic, and permanent, thoroughly mixed by the blending (κρᾶσις: *krasis*) of all elements in the continuum (thus eschewing the void that was so abhorrent to Aristotle). Outward movement bestows size, shape, and other qualities. Inward movement integrates materials causing them to be one, single substance.

The pantheistic, divine, organic cosmos is, on Stoic natural philosophy, the providential work of an active and rational God who gives shape to passive matter. This proves the existence of the gods.[40] The interconnectedness of all matter is the scientific basis of divination: i.e., every action affects the *pneuma*, through the concept of *sympatheia* wherein common affinity exists between the parts: whatever affects a single part, affects the whole. It is a system of absolute determinism.[41] This system earned the imprimatur of the Roman elite because of its tenet to conform private and political existence with an orderly cosmos (the important Stoic natural philosopher, Posidonius, had served in the public sphere), and its famous Roman adherents included Cicero who studied under Posidonius in the first century BCE, Nero's tutor, Seneca the Younger, a century later, and the emperor, Marcus Aurelius (ruled 161–180 CE).

The four-element theory espoused by Stoicism consequently pervaded Roman natural philosophy. In the first century BCE, the Roman architect and engineer, Vitruvius, who enjoyed the sponsorship of Augustus' sister, Octavia, advocated the four-element theory. Vitruvius explained the particular importance of water, without which living creatures dry up, becoming bloodless and parched (8 preface 3). He further asserted that, according to the opinion of unspecified learned men (physiologists, philosophers, and priests), "all things proceed from water" (8.1.1). Vitruvius, moreover, believed that all human learning is relevant to architecture, and that architecture (like medicine and other avenues of human inquiry) must be adapted to climate and topography in order to balance the elements and guarantee harmony with nature (6.1.2; cp., humoral theory).

Hybrid Interpretations: Microvoids

Although the four-element theory was widely promulgated, adapted, and embraced, some physical theories blurred the distinctions between the atomism and pluralism. In *On Fire*, Aristotle's pupil, Theophrastus, raised objections to classifying fire as an element because fire requires fuel and can, seemingly, be created.

Although the theory of nature as a plenum comprised of elements was widely accepted, the atomic theory and its void were not entirely abandoned. Despite his Peripatetic education, Strato of Lampsaskos—tutor of Ptolemy II Philadelphos ("Sister-loving:" ruled 285–246 BCE) in Alexandria, and the third scholarch of Aristotle's school in Athens—introduced an altered conception of void. Strato suggested that non-continuous "microvoids" permeated all matter, thus allowing, for example, solar rays to penetrate to the bottom of a full vessel without causing the liquid to overflow, provable by simple experiments: one can easily see that wine disperses evenly when poured into water, and solar rays penetrate to the bottom of a glass full of water without causing the liquid to overflow.[42]

Plutarch, "Whether Fire or Water Is More Useful"

The debate over the nature of water and its fundamental, philosophical utility continued over the centuries, receiving scrutiny in the late first century CE from Plutarch, a Middle-Platonist who drew also on Peripatetic and Stoic thought to interrogate moral and ethical questions. Plutarch espoused the principle of the dyad, a binary explanation of the sensible world, as apparent in "Whether Fire or Water is more Useful" (*Moralia* 955d–958e). In the treatise, included in a compendium of essays on various topics in ethics, customs, and philosophy, Plutarch scrutinized the properties, benefits, and dangers of each element.

The essay falls into two contradictory parts, dealing first with the advantages of water and the perils of fire, before exploring the benefits of fire and the shortcomings of water. Water's utility is borne out by the fact that "there is not a time in human history when a person does not need water" (day and night, through all the seasons), and that there was no time when water was unknown. All creatures (animals and plants), furthermore, require water. "Watery nature engenders all things as budding, growing, and bearing fruit." Humans are the only living creatures who know the uses of fire. From an ecological point of view, water as a discrete element is thus essential to all life. Life, however, can exist without Prometheus' gift, fire *per se* (humans can consume raw food; cooking is thus not a necessary technology). Conveniently ignoring the effects of flooding and storms on property and ecosystems, Plutarch posited that water is never detrimental. Water is also economical, cheaper than labor-intensive fire, which requires expensive fuel and equipment. Water, moreover, is self-fulfilling and self-sufficient (Plutarch conveniently also overlooked chronic water stresses, including drought, that plagued the eastern Mediterranean). Plutarch further observed that—far from universally helpful—fire can be a destructive "all-consuming wild beast" (θηρίον παμφάγον: *therion pamphagon*). Water is rendered even more expedient in combination with fire. When heated, for example, water can be medicinal. Plutarch then digressed from the elemental to the human sphere, citing the importance of Ocean/sea to human culture and progress. Accepting the four-element theory, Plutarch added a fifth element, not Aristotle's quintessence, but rather a sub-element of water to further emphasize its primacy. Water thus provides two essential elements: water *per se* (τὸ ὕδωρ: *hudor*) plus the sea (ἡ θάλασσα: *thalassa*), which links communities and facilitates trade, co-operation, and friendship: "if the sea did not exist, then human beings would be the most savage and needy living creature of all." The sea, Plutarch noted, is responsible for the transmission of essential human technologies including viticulture, grain agriculture, and literacy. There is, however, a disparity, since the water of the sea (*thalassa*) is neither a discrete building block of the physical world nor is it fundamentally or qualitatively different from water as an element.

Transitioning to his counter-argument, according to the Platonic view of earth where water is the foundation of the sub-lunar world, earth and water are the raw materials from which the physical world is generated, while fire and air serve as catalysts. This is key. Water, consequently, is inert and ungenerating without the agency of fire; life cannot exist without fire. At this point, fire signifies heat rather than the "all-consuming wild beast" over which water had just recently enjoyed pride of place. Plutarch then examined the intrinsic value of fire (i.e., heat): earth is barren without heat; without heat, water is stagnant and putrefies; motion and currents maintain heat in "living" waters, without which marine creatures would die. Corpses have moisture but no heat. Heat is thus the indicator of animal life. Moisture alone cannot generate growth, which can occur only when moisture has been warmed. Water without moisture is both less productive and even harmful. Although water is useless without fire, fire is nonetheless perfectly useful without water (in stark contradiction to his earlier arguments in favor of water's utility). Fire, thus, can be used without water, warm water is more advantageous than cold, and water always requires external fire (heat) to make it effective. Humans, Plutarch continued, cannot exist without fire, and even in environments where external fire may be lacking, humans are sustained by their own internal heat.

Plutarch's conclusion is anthropocentric:

> most suitable is the thing which we (people) alone employ especially since we (human beings) are able to comprehend what is better from our powers of reasoning.[43]

Art (skill: ἡ τέχνη: *techne*) is what is most profitable to human society. Technology is not possible without fire. Relevant here is Prometheus' soliloquy in the *Prometheus Bound*, where we learn the benefits bestowed to humankind by virtue of his gift of fire: carpentry and architecture, astronomy and the agricultural calendar, mathematics and literacy, animal husbandry and yoking, navigation, medicine, prophecy (divination), and metallurgy, a litany of skills and specialized knowledge that privileges the human race to overcome the capricious forces of the natural world,[44] a technology enabled by the knowledge and control of fire. Fire makes possible human control over nature (but, as Plutarch had observed, water enables the transmission of this technology).[45]

For Plutarch, technology was discovered and developed because of fire, and Hephaestus the god of fire (and Plutarch's analog to Aeschylus' Prometheus) continues to preserve human technological initiatives. For Plutarch, the greatest atrocity is time wasted in darkness, both actual and metaphorical:

> Someone might be awake through the night, but there would be no advantage to wakefulness unless fire offered us the benefits of daytime, and removed the distinction of day and night.
>
> §12

Fire removes this difference; fire enables study and work into the night; fire, consequently, is the "most useful of all things" (to human society) on the grounds of Platonic ethics. None of the senses, finally, can interact with unadulterated elemental water, but only with water that is combined/mitigated with other elements. On the other hand, since fire supplies vital energy, all the senses have a share of fire: "sight, which is the sharpest of the senses in the body, is an enflamed body of fire and that which allows faith in the gods." Plutarch eschewed comment on the inability of the senses to engage with pure, unmixed elemental fire. Plutarch concluded:

in addition, as Plato has said (*Timaeus* 471–b), we are able to conform our souls to the movements of those things in the sky on account of our faculty of sight.

§13

The argument is syllogistic: fire is an element; fire enables technology; technology enables vision; through fire human beings can improve their souls by observing the heavens. Plutarch has thus transitioned from an analysis of the characteristics of two of the elemental *rhizomata* into a morality lesson.

On a cosmogonic level, however, Plutarch seems to understand that both water and "fire" are essential and destructive to human life. Significantly, Plutarch recognized the symbiosis of the two elements and fire's catalytic properties. Moisture is what causes plants and other living things to grow, evocative of the Presocratic theories that placed the creation of life in a watery environment, as we saw above. In Parmenides, earth was water-rooted (*TEGP* 39), and Anaxagoras hypothesized that animals arose from the moist, hot, and earthy before the development of sexual generation (*TEGP* 37). Plutarch's axiomatic explanation of fire as a catalyst that enables material objects to appear from earth and water is hardly new; it is simply a recasting of ancient initiatives in natural philosophy.

Conclusion

The origin of life on earth remains a mystery, but water seems to have provided the raw material for life "as we know it" when cyanobacteria learned how to extract hydrogen from water in a process that resembles plant photosynthesis: an anaerobic process whereby unbound oxygen atoms are released into the environment "like toxic waste," thus "resulting in a holocaust far more profound than any human environmental activity."[46] Eventually, oxygen-breathing organisms evolved to consume the gas, thus averting the inevitable "oxygen crisis" (aerobic respiration is the opposite of photosynthesis). It is a great paradox that oxygen is a reactive and corrosive poison but also necessary for life (since there was little oxygen in the air of the early earth, the earliest organisms developed without it). Water is, likewise, essential for life, but its properties are destructive: water splits proteins (but the combination of proteins is necessary for life). Water, moreover, in the form of devastating floods and storms, can destroy homes, cities, landscapes, entire civilizations. This is a paradox that would have been appreciated by the Greeks who, for example, employed the same word for harmful poison and beneficial medicament (τὸ φάρμακον:

pharmakon), and who interrogated both the nurturing and destructive properties of water.

As the earliest natural philosophers recognized, water is a necessary ingredient for the germination of seeds and the sustenance of all life forms. Ancient cosmogonists thus envisioned a physical cosmos that arose from a watery chaos, giving shape to our world and serving as its boundary. Water is the generative principle that initiates and molds the human world in Mesopotamia and in Boeotia, a mythic paradigm that was interpreted rationally by Thales and others who viewed water as an essential building block of the material world, and perhaps even a mechanistic explanation for phenomena. It was nonetheless recognized that water is a powerful element of both growth and destruction—an ambivalence foregrounded in myth and natural philosophy—and it became one of the fundamental "roots" of the material world, whose properties could be changed (as wet gives way to dry or cold to heat), converting elemental water into earth or air. Water is generative, as illustrated by Aphrodite's aqueous origins (Hesiod, *Theogony* 191–192). But it is also a symbol of death, as indicated by its association with Hades' Sicilian bride (Nestis). Finally, despite attempts at rationalism, the divine is never fully divorced from the natural world. Although Plutarch conceded that water is fundamentally essential to life on earth, for him, the significance of water is elevated only when in service to human aims, that is to say, when it is purified and shaped by fire (in other words, by human technology) or as a conduit for technology. That is why Prometheus' gift was so dangerous: fire (technology) enables humans to manipulate, if not control, the natural environment. But it is a natural world that is represented by gods, one that is actualized by the gods. By aiming to manipulate the natural world, humans in turn controlled the divine world. The world that storytellers and rational thinkers created from water is thus most valuable when "improved" by human intervention.

2

Seas and Lakes

Introduction

Although Homer called the seas "barren" (e.g., *Iliad* 1.316), they were anything but. The Mediterranean (*Mare Nostrum*, "Our Sea" to the Romans) was central to nearly every facet of human life in classical antiquity (Introduction). Greek and Roman thinkers were naturally curious about the nature and behavior of bodies of water, differences between various types of watercourses (seas, rivers, springs), and how they interact with earth, sky, and each other. Our ancient sources investigated sea and river depths, currents, tides, silting, estuaries, riverine anomalies (most notably the Nile flood), density, and other factors.

The ancients were curious not just about the surface of the water and its capacity to facilitate travel, trade, and war, they also investigated the sub-marine realm. Knowledge of seabeds and sea life would have been collected and disseminated by divers. In Oppian we learn that men have explored the sea to a depth of 300 *orguiae* (about 550 meters [1,804 feet]), but the sea, where no fewer tribes or herds dwell than on the dry earth, is otherwise infinite and unmeasured, where "many things are hidden" (*Fishing* 1.82–89; cf., Aelian, *NA* 9.35). According to Oppian, fishermen have also mapped the seabed through autopsy (*Fishing* 1.9–12), despite the fact that ancient divers, whose vision naturally would have been occluded in the depths, lacked goggles. Nonetheless, keen eyesight was essential to the diver, a quality for which sponge divers in particular were renowned.[1] Divers even developed equipment that resembled modern gear in function and form: snorkels (Aristotle, *PA* 2.16.659a8–12); *lebes* ("cauldron") worn over the head, lip down, could extend the supply of oxygen, possibly doubling the length of a dive;[2] and lines might connect divers to crews topside to expedite their return to the surface (Oppian, *Fishing* 5.612–674). The ancients, nonetheless, lacked the means of exploring deep sea-floors and measuring salinity, temperature, or other hydrological data.

We note that ancient nomenclature is often ambiguous. What distinguishes a lake from a sea, for example? The landlocked Caspian, usually referred to as a "sea," fits Polyclitus of Larissa's criteria for lakes: it produces serpents and the water is sweet (as reported by Strabo, in whose day—the first century CE—it was believed that the Caspian did not debouche into the Ocean).[3] Herodotus also admired the Pontus (Euxine Sea), which is so large that lakes are embedded within it (4.85–86). Noting further the "lake-like" properties of the Pontus, the Aristotelian author of *Problems* observed that "lakes are whiter than the sea and whiter than rivers," in part "because of the many rivers that flow into it," that is to say, because of the foam generated by the inflow of water (23.6). Our evidence comes from glimpses in works on geography (Eratosthenes, Strabo, Pliny) and philosophy (Aristotle, Seneca the Younger), as well as travelers' accounts and coasting guides (*Periploi*). Here we explore how the ancients understood the characteristics of the bodies of water that provided the framework of their Mediterranean Basin, including salinity, tides, sea depth, whirlpools, and reefs.

Exploring the Atlantic

The ancient perception of the world was grounded within a philosophical context that was sometimes, but not always, tempered by empirical evidence. The Greeks and even the hydrophobic Romans knew the waters within the Mediterranean, but Ocean, beyond the Pillars of Herakles (Strait of Gibraltar), remained a source of mystery and terror, despite early expeditions beyond the Mediterranean and accurate, though marvelous, reports of natural phenomena. When Claudius prepared to invade Britain in 43 CE, his men responded with mutiny (as Gaius' troops likely had done, three years previously): the English Channel, opening into the vast Atlantic, beyond the calm waters of the Mediterranean, represented the edge of the known world, and crossing the rough waters of the Channel meant plunging into the unknown. Well before the Romans even imagined an empire, the Phoenicians under Necho II (ca. 600 BCE) ventured into the Atlantic and succeeded in circumnavigating Africa.[4] In the fifth century BCE, Hanno of Carthage led an expedition through the Pillars of Herakles.[5]

Pytheas

Contemporary with Alexander of Macedon, Pytheas of Massilia (Marseilles) was probably the first Greek to consider the Atlantic in any systematic way. A navigator and astronomer from a Greek colony, he explored the Ocean west of the European mainland and recorded his journey and observations in *On Ocean*.[6] The treatise is now lost but quoted and viciously criticized by Strabo, among others, who accused Pytheas of "shameless mendacity" for claiming to have explored "in person" the entire northern region of Europe "as far as the ends of the world" (2.4.1–2).[7] Other writers nonetheless used Pytheas' observations. Most modern scholars agree that the journey occurred, yet there is no consensus regarding its date or route. Pytheas may have reached the Shetlands or Orkneys, north of Scotland, or he may have landed on the shores of Norway, Jutland, or as far away as Iceland.

Pytheas may have set out from Massilia through the Pillars of Herakles or from Korbilon, perhaps on the Liger (Loire) River.[8] He stopped at the Cassiterides ("Tin Islands," whose location is contested) and then sailed across to Britain, up the east coast of Scotland, its Northern Isles, and the island of Thule (the Shetlands, Orkneys, or perhaps Iceland) before veering eastward towards the Baltic (see *UCWW*: Chapter 5 for Pytheas' discovery of latitude). Pytheas extended the knowledge of the west (as Alexander did of the east), reporting his observations of many phenomena, including hydrological. For example, Pytheas had described the mysterious northern land of Thule as a place where land, sea, and air lose their distinctive properties: "neither sea nor air—but something compounded from these, resembling a sea lung [probably comb-jellies] in which, he says, the earth, sea, and everything are suspended, as if it were a bonding for the whole, accessible neither by foot or ship" (Strabo 2.4.1, cf., 2.5.8, 43). The environment around Thule thus recalls the cosmogonic process where all matter is suspended in a soupy mixture, similar to Empedocles' cosmogonic state of total love (*TEGP* 41). The hypothesis of an animate ocean also found expression in Plato (*Phaedo* 111e–112e). What Pytheas observed may derive from many factors including exotic north Atlantic sea creatures, the behavior of coastal ice, or seemingly contradictory geothermal hot springs in an icy environment.[9] We shall return to Pytheas anon.

Roman Initiatives

The Romans eventually ventured into the Atlantic. During the subjugation of Carthage in 146 BCE, Scipio Aemilianus acquired the state library in which the vast Punic knowledge of the Atlantic was curated. His curiosity piqued, Scipio commissioned his own expedition, sending as its leader the polymath and historian, Polybius, who knew of Pytheas' work and was also hostile to it.[10] Polybius, who considered himself a modern and improved Odysseus,[11] was recognized in antiquity as a great explorer. Some three centuries later, Pausanias would cite an elegiac verse inscription, attached to a portrait of Polybius, commemorating that he, Polybius, "wandered over the earth and all the ocean" (Pausanias 8.30.8). Polybius explored the waters north and south of the Pillars, from ca. 150–146 BCE:

> ...We accepted the dangers and distresses thanks to our sailing in wandering beyond Libya and Hibernia (Spain) and also Galatia and the sea touching on the lands on the other side, in order that, correcting the ignorance of our predecessors in these matters, we might make these things, which are parts of the inhabited world, well-known to the Greeks.
>
> 3.59.7–8

With a fleet provided by Scipio, Polybius ventured into the African tropics, observing a forest replete with wild animals, "the sort produced by Africa," beyond a mountain (Mt. Atlas?) "toward the west" (Pliny 5.9–10).[12]

Polybius' curiosity about the lands south of the Mediterranean was apparently sparked by his African journey. He considered these areas further in his *On the Inhabited Parts of the Earth Under the Equator* (16.32), a work that was criticized by Strabo for its theoretical conception of climactic zones.[13] Polybius also wrote *On the Voyage of Odysseus, especially in the Neighborhood of Sicily*. Fragments suggest that our explorer inquired into coastal distances, navigation, tides, currents, rivers (Nile flood), and even marine life (34.2–4). Polybius explained that straits are difficult to navigate owing to reverse currents, and he reported on tuna who fell prey to larger predators in the Sicilian Strait, as well as the methods of fishing for swordfish near Scylla's lair.

Greco-Roman Conceptions of "Hydrology"

These fascinating expeditions, and more, are treated fully elsewhere.[14] But we now turn to Greco-Roman thought on the properties of waterways. We first consider "Ocean," identifiable, at least partly, with the Atlantic, beyond the Pillars of Herakles, and long known to the Mediterranean world, perhaps first discovered by the Phoenicians. Straddling the Atlantic and Mediterranean, Gadir/Gades [Cadiz] was founded by Phoenicians from Tyre as early as 1,100 BCE, as borne out by archaeological remains.[15]

Ocean

Our earliest view of Ocean in the Greek imagination comes from Homer who vividly depicted it on Achilles' shield. Encircling the inhabitable world (*oikoumene*) is the River Ocean (Okeanos: *Iliad* 18.607). Frequently "wine-dark," this "river," is sometimes "gently-flowing,"[16] sometimes violent "with a swell of current" (*Odyssey* 4.567), perhaps reflecting a distinction between the Atlantic (the "outer" Sea) and the much calmer Mediterranean ("inner" or "our" Sea). Ocean is, in contrast, sometimes merely a river with a flow and outlet. The poets described Okeanos as misty, unending, undraining, deep flowing, and unharvested. Originating "at the sun's rising," whence it flows around the earth, Okeanos was ultimately the source of all bodies of water as personified in Hesiod.[17]

Presocratics

In Thales, who suggested that the earth is a disc resting on water (*TEGP* 18–20), we have the first rational (de-anthropomorphized) view of ocean. Any theory that Thales might have posited on Ocean's origin or nature has not come down to us. A nascent hydrological theory does come from Thales' student, Anaximander, for whom Ocean is more than merely the physical frame of the world. Anaximander believed that matter, the material of which the physical world is comprised, derives from "the boundless" (*apeiron*), a mixture of all opposites from which the seeds of sensible objects separate (*TEGP* 16). Although rejecting water, Thales' *arche*, as the source of all matter (*TEGP* 10–11: he may have considered water too specific), Anaximander did suggest that the earth was entirely moist at first, slowly desiccated by "the winds and turnings of the sun and moon" (i.e., the heat of the sun and the winds that are generated by the revolution of celestial bodies around the earth) until solid earth was eventually revealed. Primeval moisture settled in the earth's hollows, consequently creating the seas. According to Anaximander, this primeval moisture continues to evaporate until "finally some day it will be completely dry" (*TEGP* 34, 35–36), a replication of the separation motif that we have noted in Hesiod and his near-eastern predecessors (Chapter 1). As for Thales, so for Anaximander, moisture was nourishing. As water and land grew warmer, animals—generated in moisture—were protected by bark or shells (or embryonic sacs within fishes, in the case of human beings) until they could make landfall (*TEGP* 37–39).

Xenophanes was perhaps the first Greek thinker to develop a coherent theory of Ocean. As the source of both water (rivers, rain) and weather (clouds, winds), the sea (*thalassa*) is the unifying lynchpin of his natural philosophy (*TEGP* 54–55). In the extant fragments, Xenophanes queried the nature of the sea, hypothesizing that "many mixtures" flowing into seawater account for its salinity (*TEGP* 59; see Chapter 3 for inland marine deposits). Finally, Xenophanes theorized connections between the sea and celestial bodies, including the sun and moon, which he defined as comprised of "incandescent clouds" (*TEGP* 60–61, 67), formed from evaporation like rainclouds, but manifested differently.[18]

Like his predecessors, Xenophanes considered empirical evidence when shaping his physics. With subsequent thinkers, the empirical had been expunged in preference for the abstract. In the thought of Heraclitus of Ephesus, "a humanist with scientific interests,"[19] epistemology forges the discourse on natural science. Although his fragmentary corpus does not indicate if he developed a unified theory of Ocean, Heraclitus, nonetheless, employed the essential nature of water in the physical world as a metaphor for his epistemology and the ephemerality of human reality. For Heraclitus, "existing things" are comparable "to the flow of a river" that is in a state of constant flux: "on those stepping into rivers staying the same other and other waters flow." This aphorism has been paraphrased famously as "into the same rivers you could not step twice...for other waters flow on" or more cryptically as "all things are in motion like streams" (*TEGP* 62–63, cp. 65–67). Although the appearance, size, and behavior of the river remains stable, its flow is endlessly replaced by new waters (*hetera kai hetera*), but the river itself remains constant.[20]

More concretely, on Heraclitus' natural philosophy, opposing binary forces—held together in a contentious equilibrium—govern the material world, which in turn is comprised of earth and sea (*thalassa*): "the turnings of fire: first sea (*thalassa*), and of sea, half is earth, half fire-burst" (*prester*: *TEGP* 51). According to Kahn,[21] this fragment may have been "intended to *suggest* some process of world formation or transformation." *Prester* is often translated as "tornado" or "waterspout," referring to a lightning storm or "fire from heaven" that occurs in conjunction with wind and storms.[22] Kahn concludes that Heraclitus' "half-earth, half- *prester*" must elicit binary equilibrial forces that are produced after the generation of sea. These binary forces desiccate the sea into earth and vapor, and the resulting vapors in turn nourish the celestial fire. *Prester*, then, seems to be the catalyst for the alternation of earth and sea, which together are in an eternal and diametric relationship: "<earth> is liquefied as sea and measured into the same proportion it had before it became earth."[23] For Heraclitus, the ratio between earth and water is absolute. Water (sea) is one of the foundational elements of the material world, and it is not converted into earth (as in material monism) but rather it maintains a "strong non-identity."[24] For Heraclitus, sea is both a physical component of the world and a metaphor for the paradoxical binary tension that governs it: "Sea (*thalassa*) is the purest and most polluted water: for fish (it is) drinkable and healthy, for men undrinkable and harmful."[25]

Two intriguing fragments suggest that Empedocles, like Xenophanes, was also asking questions about the nature of the sea (*TEGP* 99–100: e.g., why is seawater salty? See below.). Empedocles also surmised that the quantity of seawater is increasing, as evident from its encroachments on the land (*TEGP* 101). Surviving fragments intimate that Empedocles had developed a geological theory regarding the formation of terrestrial features (including cliffs and crags: *TEGP* 96), but we lack any further details of his theory of ocean. Did he interrogate the causes of the tides, sea depth, the nature of its surface, or seabed topography?

The Presocratic atomist and polymath, Democritus is ascribed with more than seventy titles on topics that treat the sciences (astronomy, geography, geology, etc.), music, the nature of the soul, ethics, and much more. The sources attest a well-developed theory of geology (*TEGP* 78–88), but Democritus' hydrology is poorly preserved. The fragments connect earthquakes, as in Thales, with the uneven distribution of terrestrial waters either because of winds or the incommensurate accrual of water in underground passages (*TEGP* 78–79). Democritus also hypothesized that the sea was receding and would eventually disappear, seemingly in accord with contemporary Empedoclean thought that viewed life on earth as a cosmic fluctuation of different states of matter. Aristotle found the theory risible.[26] It is unclear whether Democritus perceived a cycle of desiccation and hydration, like Anaximander, or if he believed in a teleological senescence of the sea.[27]

Plato

As with Presocratic thinkers, knowledge of the sea remained largely theoretical for Plato (428/7–348/7 BCE). In the *Phaedo* (109b–113d), Plato pitched his hydrology as a *katabasis* or descent into the underworld, as he explored the nature of the soul while recounting Socrates' last day. Plato's Socrates rhapsodizes on the nature of Our Sea—the waters between the Phasis River that debouche into the Black Sea at the eastern edge and the Pillars of Herakles in the west—and its relationship to the earth. Humankind inhabits only a small portion of this sea, which impinges on the littoral hollows. But our hollows are merely an impressionistic version of the true earth:

> It escapes our notice that we dwell in the hollows of the earth; we think that we are living upon it. Just like if someone inhabiting the depths of the sea thinks that he lives on its surface, seeing the sun and other stars through the water, he thinks that he sees the sky; on account of his sluggishness and weakness he would never arrive at the sea's surface, nor would he rise and lift his head from the sea to our world; nor would he see how much more spotless and beautiful it (the earth) happens to be than his (watery world).
>
> Phaedo 109c–d

In the *Phaedo* we have a watery analogue to the cave of Plato's *Republic* (7.514a–520a). *Phaedo's* Socrates invokes the wisdom of the epic poets with reference to the depth and expanse of the water-filled hollows that exist in Our Sea (but he did not speculate on their extent). Just as our atmosphere erodes the objects in it (stones, for example, are weathered and altered by exposure to air and weather), brine in like manner corrodes objects in the sea. Ignoring a thriving fishing industry[28] and plentiful (non-flowering, non-fruiting) marine vegetation, Socrates describes the sea as an imperfect (and unpleasant) place where nothing "worthy of mention" grows, the mud is measureless, and the swamps, at the confluence of sea and earth, are messy. In his perfect world, even mountains and stones are smooth, unpocked, and transparent, and the air- and water-filled hollows of the earth glitter with a seamless variety of color. Socrates imagines that other areas of human habitation exist beyond the Mediterranean, scattered around other hollows:

> some are deeper and broader than the one in which we live, others are deeper than those in our region but with a smaller gape, some are both shallower than ours and broader.
>
> Phaedo 111c

These hollows are linked by underground channels, and in them flow "a great volume of water" distributed in subterranean rivers of various qualities—hot, cold, fiery, muddy, clear, murky— evocative of the volcanic lava flows in Sicily. These subterranean rivers debouche into the deepest and largest chasm, Tartarus, as Socrates called it, the bottomless source of all waters. The waters exist in an unceasing fluctuation in this subterranean labyrinth, sloshing from one side of the earth to the other, with no fixed seat, but filling and depleting the rivers and lakes (including Our Sea) on the surface. Because of this breath-like ebb and flow (resembling tidal activity), bodies of water sometimes return to their sources. At other times they flow toward opposite channels. Four significant watercourses, each associated with death, populate this Socratic underworld: Ocean, the frame of Homer's world; Acheron (River of Woe), debouching into the Acherusian Lake, a rallying point for the souls of the dead; the fiery, lava-filled Pyriphlegethon (River of Fire); and the Cocytus (River of wailing) cascading into the Stygian Lake. Their currents and courses are geometrically symmetrical (Fig. 2.1). The Acheron flows in the opposite direction of Ocean. Appearing between Ocean and Acheron and mixing with no other rivers, the Pyriphlegethon spirals downward until it returns to Tartarus deep within the earth. Directly opposite the Pyriphlegethon between Ocean and Acheron, the Cocytus spirals in a contrary direction towards Tartarus, also mingling with no other waters.

Plato's physical world accords with his epistemology. With his dying words, Plato's Socrates explores the ambit between the imperfect physical world and the perfect world of pure knowledge, presented as an analogy of the border between the worlds of the living and the dead, where newly deceased souls receive their punishments or rewards at the appropriate rivers. Despite the apparent disjunction of the subterranean rivers, all the waters of the earth originate from and return to Tartarus on Plato's theory, and a unified sea washes over the entire earth, our *oikoumene* and others. We are in constant contact with those waters, a subtle reminder of our own mortality and Socrates' impending death.

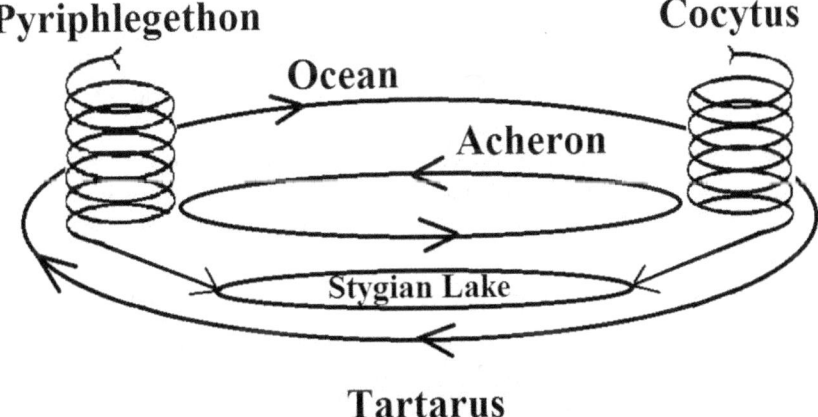

Fig. 2.1 The rivers of Plato's underworld, *Phaedo* 107a–115a.

Aristotle

Plato's hydrological *katabasis* is an integrated exploration of "ocean," its variable depth, expanse, and its relationship to other bodies of water. Aristotle's hydrology—bridging his discussions of lofty mountain springs and even loftier winds—is also cast as a sort of *katabasis*.[29] Aristotle (384–322 BCE) soundly repudiated Plato's vacillating, unified bodies of water as irrational and impossible (*Meteorology* 357a15–23). On Plato's theory, rivers would then be compelled to flow according to the surging of Tartarus. Would they flow up when Tartarus surges? That, of course, is impossible! And rivers cannot and do not return to their own sources. Aristotle explored many hydrological questions including the origins of the sea, its salinity, its distinction from "flowing" waters, and whether seas can be interconnected. *Meteorology* 2.1–3 is, furthermore, a rhetorical *tour de force* in which the author selectively analyzed the "silly" ideas of his predecessors in order to disprove them. For Aristotle, the Presocratics exaggerated the importance of Ocean as the source of nourishment for heavenly bodies (*Meteorology* 2.2.354b33–355a33). The Presocratics also asserted that the sea has "springs" (*pegai:* running water), which must belong to artificial (that is, not flowing) bodies of water.[30] Disputing this theory, Aristotle anachronistically filtered earlier conceptions through Plato's more deterministic view of Tartarus with its physical subterranean passages.[31] For Aristotle, Ocean/Sea is neither flowing nor artificial, nor is natural spring water ever found on such a large scale (*Meteorology* 352b29–353b30). Flowing waters *flow* from springs. Standing waters either have no source (lakes, swamps), or they are artificial (wells). The point is partly intended to buttress Aristotle's architectonic view of the world where the heavens assume the highest importance and the sea is relegated to "the bottom of the meteorological cosmos."[32]

Aristotle defined the sea as stagnant, unflowing, and entirely disconnected from rivers and the cycle of weather. Aristotle's predecessors believed

> that all the remaining heaven had been composed around this region (the earth) and for its sake, as being the most honorable and the beginning (*arche*) of it.
> *Meteorology* 2.1.353b3–5

The cosmological role of earth and sea had thus been elevated. Some seas are landlocked, such as the Hyrcanian and Caspian, which Aristotle correctly recognized as "separated" from the outer ocean. If these seas had sources, those dwelling on their coasts would have observed them (354a3–5). Nor can fresh rivers that debouche into salty seas serve as their sources because of the vast differences in the quality of river and seawater (fresh/salty: 354b20). Fresh river water, moreover, settles above salt water, according to the tenets of Aristotelian physics (and specific density), where the sea is restricted to a cup-like container. The Red Sea flows into the "ocean outside the straits" only through a narrow channel. These seas, consequently, have no sources. Aristotle resolved the debate by borrowing from Heraclitus' river. Bodies of water are one and the same. Uniform in shape and volume, the sea is also comprised of discrete parts that are in continual change. Some parts change more rapidly, others more slowly. Although the bulk remains constant, the parts do not.

Circumambient Ocean

Although rejected by Herodotus and Hipparchus of Nicea, owing to the lack of empirical evidence,[33] the circumambient Ocean model endured, endorsed by Aristotle, Eratosthenes, Posidonius, and Strabo.[34] Attempts to prove it, however, met with failure: Hanno's expedition reversed course before reaching the equator. Sent to circumnavigate Africa between 479–465 BCE, and detesting the isolation of his long voyage, the feckless Persian Sataspes doubled back in fear, asserting that his "ship was not able to advance forward but was kept back" (Herodotus 4.43). We shall never know if Eudoxus of Cyzicus' circumnavigation of Africa would have succeeded (late second century BCE). Shipwrecked off the coast of Spain, Eudoxus returned to Gades, resuming the voyage, but he may have been lost at sea: "what happened later, those from Gadeira and Iberia probably know" (Strabo 2.3.3–5). The expedition under Necho (609–594 BCE) may have been one of the few to achieve its goal, with reports of crossing the equator, faithfully reported by a disdainful Herodotus (4.42.2–3). Despite expeditionary failures, Eratosthenes was convinced that the external Ocean must be continuous because most of it had been sailed.[35] An outlier, Krates of Mallos (ca. 170–120 BCE) proposed two intersecting belts of Ocean that separate four symmetrical land masses, a stark divergence from previous (and subsequent) visions of the Ocean as a cartographic principle (Strabo 2.5.10). The circumambient ocean was eventually rejected by Ptolemy of Alexandria in the second century CE, who replaced it instead with an encircling *terra incognita* ("unknown land") of indefinite expanse, as inspired by new geographical data from the far east (*Geography* 7.5) (Fig. 2.2).

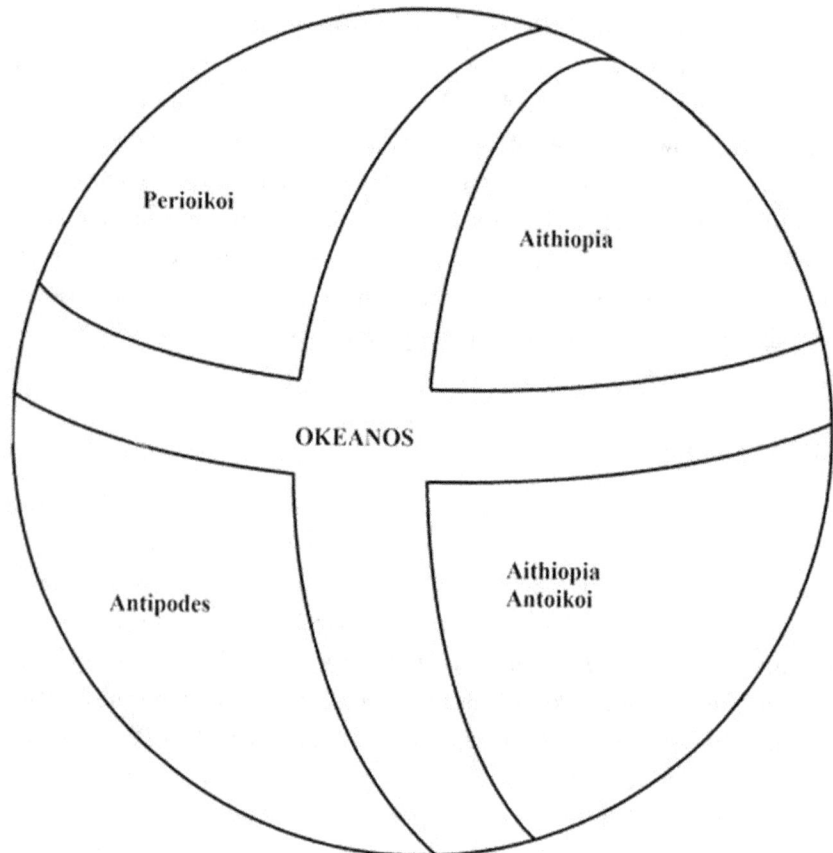

Fig. 2.2 Intersecting bands of Ocean according to Krates of Mallos.

The Nature of the Oceans and Seas

Salt

Early thinkers may have accepted the differences between salt and fresh (potable) water without question, suggesting simply that salt water is sweet (fresh) water seasoned by earth thus absorbing flavors "like things strained through ashes."[36] The first extant theory is credited to Xenophanes: over time, earth would be dissolved in water, since the two elements are constantly mixing. The sea's salinity, furthermore, is caused by the "many mixtures" that "flow into it" (*TEGP* 59). Empedocles cleverly suggested that Ocean water became salty as the sun heated it, as if the earth were sweating (*TEGP* 100), a hypothesis derided by Aristotle:

sweat implies a failure to digest, there is no need for the earth to digest anything, nor is there any indication that the earth is currently "sweating" (*Meteorology* 2.2.357b7–13). Natural place moreover dominates Aristotle's physics: only the lightest (fresh) water (with the tiniest quantities of salt) is evaporated, returning to the earth as rain. The heavier salt water remains in the ocean, settling beneath lighter fresh water. The sun's warmth, apparently, attracts the sweeter parts upwards. As liquid is desiccated by solar heat and sucked up into the atmosphere, salt is boiled back down into the sea.[37] Aristotle's experiments, tantalizingly left to the reader's imagination, prove, so he claimed, that vaporized salt water returns to its fresh state, not re-condensing into brine (*Meteorology* 2.2.358b12–17).

Why, then, is the sea salty? For Aristotle it is because of the amalgamation of moist (watery) and dry exhalations (earthy, containing residues that result from the natural process of generation).[38] As proof that earthy residue makes water salty, Aristotle noted a fabled Palestinian lake (Lake Asphaltitis, the Dead Sea, known to the Greeks from the time of Alexander[39]) where nothing sinks, waters are so bitter that no fish can survive, and clothes can be cleaned simply by dipping and shaking. Sent by Antigonus I Monophthalmos ("One-eyed," ruled 306–301 BCE) to investigate the Dead Sea, Hieronymus of Kardia may have written a treatise on this seemingly anomalous lake (Diodorus Siculus 19.100.1–3). In Strabo we have a vivid account of the Dead Sea (his "Lake Sirbonis"), whose waters are so heavy that people float and in which asphalt bubbles up, "as if the water were boiling."[40]

In Aristotelian *Problems* book 23, the peripatetic author investigated forty-one queries on the behavior of salt water and its differences with fresh water, sometimes giving different explanations for the same phenomenon. The author also sought to explain why salinity seems to differ from one body of water to the next (and even within layers of the same body). Perhaps, he thought, the force of the south wind driving waters under the earth causes southern waters to be saltier than those facing the north wind (23.25). Although salinity does vary, a sea's chemical composition is also affected by the amount of fresh water that debouches into it, either in the form of rain or river water. Waters further from land are likely to be saltier. As shown by satellite imagery from NASA's Aquarius Mission, the earth's saltiest waters (apart from the Dead Sea) are found in the North Atlantic between northern Africa and North America, where there is little rainfall but much evaporation.[41] This phenomenon—that open waters have greater salinity than coastal waters—was recognized in antiquity, and our author suggested that the greater motion of coastal waters renders the waters fresh (agitation was also considered a means of "freshening" stale water). Deeper water is, naturally, saltier (coastal waters tend to be shallower than open waters): "being heavy, the salty water is borne more into the depths." In this the author seems to contradict his own responses to queries 27 and 30: "why are the waters in the upper parts of the sea saltier and warmer than those below?" where he ascribed temperature as the cause of greater salinity at the surface, despite the fact that salt, as an earthy element, has a "tendency to sink." The sun, it would seem, "attracts" (evaporates) the lighter components (fresh water), leaving a high concentration of salt at the surface.[42] This is true, and salinity decreases until about 3,300 feet (1,000 meters), increasing moderately with the depth until reaching the halocline ("salt-slope") layer, where salinity rises sharply.

Density

Ancient thinkers observed that salt makes seawater denser, a fact that was demonstrated empirically. Aristotle noted that sea-going vessels nearly sink on a freshwater river with the same cargo, and eggs submerge in fresh water but float in brine (*Meteorology* 2.2.359a6–15). Regarding why ships seem more heavily loaded in harbor that at sea, Aristotle's students offered other explanations. Failing, however, to understand that the specific density of water can vary (greater for salt water than fresh or briny water), the peripatetic author of *Problems* surmised that, by virtue of its sheer volume, seawater could more effectively support ships from below (23.2):

> is it because the greater quantity of water stands up against the lesser quantity, and in the lesser amount (i.e., as in harbor) a vessel dips on account of its greater power? For it forces up the water below. In harbor, there is simply less resistance and ships move "with greater difficulty."

The greater density of seawater, nonetheless, accounts for several phenomena in our peripatetic source. Because of its thickness, the sea is better able to retain heat, and seawater is thus "less cold" than other waters (23.7).[43] Furthermore, dense seawater is earthy with "little moisture in it," which enables clothes washed in salt-water to dry more quickly (23.10). Once dried, however, the clothing will be stiff because of the salt, which can then be removed from the cloth only by a greater quantity of fresh water. (Salt molecules decelerate evaporation at the surface, fresh water consequently evaporates more quickly.) The density of seawater explains why it is easier to swim in seawater, which more easily supports the swimmer owing to the fact that it contains more "stuff" than fresh water (23.13: density is a factor of buoyancy, and on average there is 1.6 pounds more buoyancy in salt water than in fresh).

Centuries later in Rome, and distilling Posidonius' Stoic wisdom, Seneca did understand that the specific density of water could vary. Seneca cited a paradoxical lake in Syria where bricks could float, where timber drifts occasionally entirely above the water, half-submerged, or entirely beneath, and where the top would be level with the water, owing to the fact that "heavier water" holds up more of an object and the same objects sink to differing degrees in "lighter water" (*NQ* 3.25.5–6).

Tides

As long as ships have plied the open ocean, sailors have known about the tides, caused by the distortion in the shape of one body (the earth) by the gravitational pull of another, nearby object (the moon). The displacement of water in the earth's oceans is because of the moon's gravitational force, resulting in the small displacement of a large amount of water over a vast distance. The moon's gravitational force affects the center of the earth, not its surface, and the side of the earth that is facing the moon will experience a tidal bulge (water is attracted up with greater acceleration and is thus "piled up"), while the far side will also experience a high tide but with less amplitude. The cycle occurs about every 12.5 hours.

Praised by Strabo (1.1.7) for his knowledge of flood tides, Homer referred to the "ebb and flow of Ocean" when describing Scylla and Charybdis.[44] Tides are mild in the Mediterranean as compared with the Atlantic, and Homer described Ocean as "soft flowing" when the young Odysseus set out at dawn to hunt boar with his grandfather, Autolycus (*Odyssey* 19.434; c.f. Strabo 1.1.7), and when the Trojans, in the wake of a devastating battle, tended to their dead on the beach, again at dawn (*Iliad* 7.422). In the sixth century BCE, the Atlantic tides baffled Euthymenes of Massilia and Sataspes, an Achaimenid Persian.[45] Tides perplexed Alexander's troops, accustomed to the placid Mediterranean and unfamiliar with tides of the "Great Ocean" (Arrian, *Anabasis* 6.19.1). High tides in the English Channel at the full moon particularly frustrated Caesar's invasion of Britain in 55 BCE.[46] The tides in the Erythraean Sea were extreme:

> suddenly the depth is visible and some portion of the land, dry just now but sailed over just a short time ago; under the attack of a solid wheel of the entire

sea, very violently and forcibly compressed, the rivers are carried up against the nature of the flow for many stades.

Erythraean Sea 45

From the late fourth century BCE, Greek thinkers recognized that the moon influences tidal activity.[47] Careful observation of tidal behavior aided Scipio's attack of New Carthage in Spain (210 BCE),[48] and coastal peoples drew on their knowledge of local tides to design watercraft best suited to regional maritime conditions. The Venetians, a seafaring Celtic tribe who dwelled on the peninsula of Aremorica (Brittany, near modern Vannes, Roman Belgica), built broad-bottomed boats with high sterns and prows. Such boats could withstand dramatic Atlantic tides and swells,[49] and these craft likely inspired the design of the Viking long ships. Venetian towns, moreover, were constructed on headlands that jutted into the sea. The local population, consequently, took full advantage of the natural environment to protect themselves from enemy attack: invaders could not assault the towns at high tide on foot, and conventional boats would run aground in the shallows at low tide.

Tides were a cause of great curiosity, and tidal rivers were thought to border on the paradoxical. Without comment, Pliny reported rivers flowing back when Nero died in 68 CE (2.232.). Strabo described tidal rivers and estuaries at Tourdetania in southern Iberia (3.2.4), where in some areas the flood tide "is just like the flow of a river" and the sailing is "just like sailing down rivers" (since there is no resistance). Strabo incredulously but punctiliously documented that the Cimbrians were driven out of their territory by an excessive flood tide (110s BCE), "a phenomenon that is so regular and predictable" (7.2.1). Cicero offered tidal regularity as an example of nature's harmony:

> What more should I say about straits and maritime tides whose ebbs and flows are steered by the motion of the moon? Perhaps we could give six hundred examples of this same kind, that the relationship of distant objects appears natural.

Divination 2.34

Aristotle was accustomed only to the waters of the Mediterranean. But he was aware of Hanno's expedition and thus might have known about the tidal activity in the Atlantic (*Politics* 5.6.1307a5–6).[50] Whether he developed a theory of tides is unclear, but he did explain re-fluent currents in the Mediterranean straits:

> The sea appears to flow through the Straits, if anywhere on account of the surrounding land the sea is collected from a larger space into a small one, because of the frequent swaying here and there. This (motion) is unclear in the

great extent of the sea. But where topography presents a scant space through a narrow of land, the swaying which seemed slight in the sea necessarily seems great there (in the narrows).

Meteorology 354a7–10

The narrowness of the channels thus exaggerates the natural motion of the seas. As redacted by Strabo in his discussion of the western seaboard north of the Tagus River (Portugal), Posidonius seems to have misinterpreted Aristotle. Posidonius reprimanded Aristotle for positing a tidal theory based on the recoil of waves against sheer coastal cliffs: the high, rugged headlands off Spain and Morocco catch the ocean waves and hurl them back to sea. Posidonius had seen for himself the low, sandy beaches along that coast, where there is nothing against which the waves could bounce (f220Kidd). Posidonius' conjecture, however, is not substantiated in the *Meteorology*.[51] But Aristotle's authority was unimpugnable, and a popular belief that tides are caused by the winds may go back to him (Aëtius 3.17.1).

Pytheas was perhaps the first Greek thinker to consider tides in any systematic way. He was fascinated by the dramatic tidal activity around the British Isles, among the largest in the world. He allegedly reported tidal swells of up to eighty cubits (37 meters [about 120 feet]: Pliny 2.217), possibly caused by a storm surge.[52] Pytheas may also have been correlating the half-daily tides to phases of the moon ("fullness and faintness"), although it is more likely that he was observing a spring tide—when the sun and moon are both in the earth's equatorial plane and thus exert a greater gravitational pull in the same direction—coinciding with an equinoctial tide.[53] A spring tide occurring during an equinox would yield dramatic tidal amplitudes, as noted by Seneca (*NQ* 3.28.6). The author of the *Periplus of the Erythraean Sea* also remarked that *nocturnal* spring tides have the greatest force.[54]

Building on Pytheas' observations, Seleukos of Seleukia on the Tigris River was the first to write a monograph on tidal theory.[55] A rare proponent of heliocentrism, and working from a Stoic perspective,[56] Seleukos suggested that the moon's orbit and earth's rotation disturb the intervening *pneuma*, in turn causing fluctuations in sea levels. Tides, he argued, vary seasonally, are dependent upon the zodiac sign in the moon, and differ from sea to sea, an idea favored also by Hipparchus, but flatly rejected by Strabo, who staunchly maintained that the ocean in general and tides in particular behave uniformly.[57] Seleukos also noticed an inequality in the diurnal tidal activity of the Erythraean Sea: tides are "regular" (neap, with no great difference between high and low tide) when the moon is in equinoctial signs (Aries and Libra, the signs that rise after the equinoxes), but "irregular" (with extreme differences in amplitude) when in solsticial signs (Cancer and Capricorn). Seleukos surmised that this irregularity is proportional to the moon's distance from the earth's equinoctial (on the equator) and solsticial planes (greatest north/south declination). Seleukos thus concluded that, when the moon is on the equator (equinoxes), the Erythraean Sea has two daily cycles of equal highs and lows. When the moon is at 90° to the equator (solstices), the diurnal highs and lows are unequal. Darwin (1898: 86–7) confirms Seleukos' observation: the diurnal inequality in the Atlantic is practically nonexistent, whereas in the Erythraean Sea, trees that are "completely visible at low tide are completely covered at high tide" (Strabo 16.3.6). Seleukos likely saw the phenomenon firsthand.

Seleukos' bold statement must have piqued the curiosity of Posidonius of Apamea (in northern Syria on the Orontes River). "One of the most important and interesting figures of the first half of the first century B.C.,"[58] Posidonius aimed to explore, understand, and elucidate the entirety of the human intellectual achievement, and he wrote extensively on many topics of ethics and natural and mathematical sciences. Among the works ascribed to Posidonius is *On Ocean* (perhaps in imitation of Pytheas' title, above), extant only in scattered fragments in Strabo, Seneca the Younger, and other compilers. The treatise dealt with geography (climactic zones), the circumnavigation of Africa, voyages to India, and earthquakes (including Atlantis). Posidonius' investigations into phenomena are creative and grounded rigorously in observation. To this end he traveled broadly, scrutinizing natural phenomena and the peoples and customs of the world,[59] and he may have included Gades on his itinerary to test Seleukos' claims about dramatic tidal activity.[60] Posidonius spent thirty days on Spain's Atlantic coast where he studied tides whose swells exceeded those in the protected Mediterranean Basin. There he formulated a tidal theory that synthesized and quantified earlier hypotheses (f217bKidd) (Fig. 2.3). When the moon is at 30° (one zodiac constellation above the eastern horizon), the sea rises visibly until the moon reaches 90° (the meridian). Water levels then begin to recede until the moon is 30° above the western horizon, remaining steady until the moon is 30° below the horizon. Water levels again rise until the moon reaches the meridian below the earth (270°), when they start to ebb, holding level as the moon travels from 30° below to 30° above the eastern horizon. Posidonius also correlated varying tidal amplitudes to phases of the moon, noting that sea levels are greatest in conjunction with full and new moons, and smallest when coinciding with quarter moons. Posidonius' explanations of diurnal and monthly tides are substantially correct.

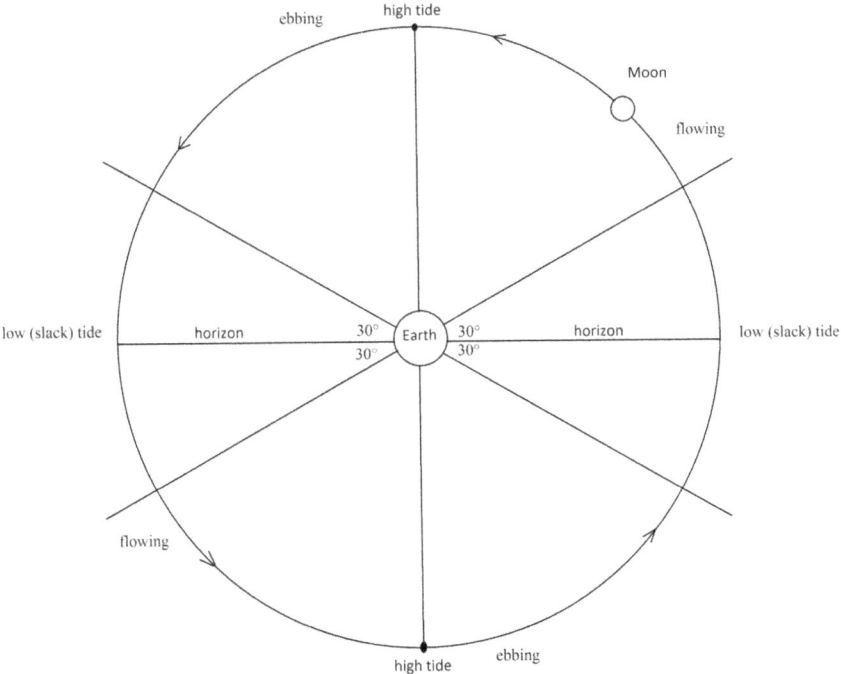

Fig. 2.3 The diurnal tidal cycle, according to Posidonius. Adapted from Kidd 1972: 283, Figure 13.

The good people of Gades, however, seem to have misled Posidonius. According to Strabo's distillation, Posidonius was told that the annual high tides occurred in alignment with the solstices (highest tides actually coincide with the equinoxes). Posidonius may have misunderstood the locals, or Strabo may have misinterpreted Posidonius. Posidonius, however, was deliberately looking for high tides at Gades during the solstice (Strabo 3.5.9), and his observation of another high sea coinciding with the summer solstice near Sicily may have confirmed in his mind the high solsticial tides (f227Kidd). In Sicily, the anomalous tides were because of a volcanic eruption.[61] Despite the slip, Posidonius succeeded in marshaling the empirical evidence into a unified theory of tides, a "remarkable contribution."[62]

This error, nonetheless, was not perpetuated. A century later, Seneca (*NQ* 3.28.6) and Pliny (2.215), who both presumably followed Posidonius, gave the maximum high tides at the equinoxes. This knowledge was put to practical use by Germanicus, the charismatic but headstrong nephew of emperor Tiberius, who tried to mitigate the effects of equinoctial tides in 15 CE. After campaigning against Arminius, chief of the Cherusci in Roman Germany, two legions returned to Gaul via a coast-hugging land-route between the Ems River and the shallow Ijsselmeer (whose waters do recede dramatically at low tide[63]) "so that, the fleet might float lighter in the shallow sea, or settle in the ebb" (Tacitus, *Annals* 1.70.2). The marching column, however, suffered from the combined forces of the north wind and equinoctial tide "when the Ocean swells exceedingly."

Pliny (2.212–221), furthermore, was the first to note the nearly constant lunitidal intervals that are specific to places (according to the risings of the "stars"), as well as the retardation of daily tidal intervals (or "lag" time). It is, however, the moon's declination, changing with time and place, that affects tidal amplitude and time, as Seleukos understood. Pliny also elaborated on Seneca's phase inequality (*NQ* 3.28: slight spring tides occur after new and full moons), and he reported the occurrence of lower tides when the moon is in the northern hemisphere (higher tides in the northern hemisphere are observed when the moon is in the southern hemisphere). Although true, the phenomenon is not tidal, but is caused instead by the earth's rotation (which most Greek and Roman thinkers rejected) due to acceleration because of the circular motion (which also causes an equatorial "bulge"). Pliny was, moreover, the first to record the eight-year tidal cycle of recurring lunar phases.[64]

Sea Depth and Levels

There were also attempts to measure the depths of the sea, which the Titan Atlas is credited with "discovering" (*Odyssey* 1.52–54). Herodotus (2.5.2) gave the earliest textual evidence of taking soundings, but archaeological evidence predates 2,000 BCE: Egyptian boat models from the tomb of Meketra, chancellor to Nebhepetra Mentuhotep II, feature crewmen holding sounding lines in the ships' bows.[65] With a lead-line attached to a rope (thousands of *orguiae* long [1,000 *orguiae* = 2 km]), Psammetichus, king of Egypt in the seventh century BCE, had reputedly hoped to prove that the sources of the Nile (between Elephantine and Syene in the Thebaid where the Nile enters Egypt) are bottomless (Herodotus 2.28.4). Either the lake was very deep, or its eddies and currents prevented the sounding line from reaching the bottom. At a distance of about 55 km from land (about 35 miles), the "Deeps of Pontus," were unfathomable according to Aristotle: "at any rate no one has yet been able to discover the limit (of the depth) by taking a sounding." Nero was also unable to take soundings in the Alkyonian Lake near Lerna, despite attaching his sounding weight to several lengths of rope (Pausanias 2.37.5).[66] Such readings so far from land may have formed part of a larger scientific project.[67]

Aristotle maintained that seawater flows to the deepest place, finding its natural place. The deepest parts of the earth are consequently filled by sea (*Meteorology* 2.2.355a33–b5, 355b17–19). Because of the intricate exchange between water and earth, rivers and ocean, there is a progression of the earth's seas from shallow to deeper: the Maeotis, Pontus, Aegean, Sicilian, and the Sardinian and Tyrrhenic Seas, the deepest of all (*Meteorology* 354a14–21). Water flows from the higher regions in the north, but intense alluvial activity makes the northern seas shallower and the outer seas deeper. Posidonius reported the Sardinian Sea as about 1,000 *orguiae* deep (about 1.4 miles, in reality closer to 3 km [about 2 miles]), the deepest of those that have been measured (f221Kidd). We do not know how the reading was taken (cf., Kidd 1972: 2b.794–5). In contrast, a shallow sea in the Arabian Gulf was estimated to be about 2 *orguiae* deep (3.75 meters [12 feet]), and it "appears grassy with seaweed and other weeds showing through" (Strabo 16.4.7), perhaps the Sargasso described by Avienus, through which vessels go slowly and sluggishly.[68]

The issue of sea depth and surfaces formed part of an atheistic, mechanistic, holistic theory of physics proposed by Strato (third century BCE). Explaining all phenomena as caused by the forces of weight and motion, Strato believed that the sea is wholly confluent but that levels could change and that the seabed is uneven, as demonstrated by the experience of sailors in the Greater and Lesser Syrtes, who could quickly run aground even in the "depths."[69] Strato had argued that the Internal (Mediterranean) and External Seas (Atlantic Ocean) had different seabeds, proved by an undersea ridge that extended from Sicily to Libya (Strabo 1.3.4).[70] These variable sea bottoms helped to explain a number of phenomena, including the creation and behavior of straits and the widespread impressions of marine animals on dry land and inland salt deposits.[71] Strato was among the sources used by Eratosthenes of Cyrene (in modern Libya) who argued that the Pillars of Herakles were broken open when rivers flooding into the sea uncovered seabeds that had once been obscured by shallow water (f15).

Eratosthenes also hypothesized that bodies of water had discrete surfaces and distinct levels, and he rejected the postulate proposed by his friend, Archimedes, that "the surface of all fluids that remain immovable will have the surface of a sphere, having the same center as the earth."[72] In 302–301 BCE, when Demetrios I of Macedon "Poliorcetes" ("Besieger of Cities") had hoped to construct a canal for his troops, his engineers at Kenchreai noted that the sea level was higher on the Corinthian Gulf side. Assessing the environmental risks, they advised against construction since cutting the canal would cause the isthmus to flood, rendering the proposed passage futile.[73] Some twenty years later, Ptolemy II Philadelphos' engineers calculated the surface of the Erythraean Sea as 3 cubits (about 1.5 meters [ca. 5 feet]) higher than the Nile's, and the Nile's surface as higher than the Mediterranean's (Pliny 6.166).

Sea Surfaces

The ancients also interrogated the nature of the surface of water. What is its geometry? Do different bodies of water cohere? Or are they discrete? The answers to these questions had ramifications for their greater understanding of the earth's watercourses.

In Aristotle we find the genesis of Archimedes' postulate:

> But because this surface of the water appears this way (spherical), they take as a principle the fact that water is always naturally disposed to stream together. The hollower spot is closer to the center. Therefore, let the lines AB and AΓ be drawn from the center, and let the line BΓ be joined. Therefore, the line at the base, from which is drawn AD, is shorter than those from the center. Now that place is hollower, so that water will flow around until there is a balance. But now, AE is equal to those lines from the center, such that it is necessary that there be water near those lines from the center. For then it (the water) will be at rest. For the line binding the lines from the center is circular. Therefore, the appearance (surface) of water from which BEΓ (is drawn) is spherical.
>
> *Cosmos* 2.4.287b4–14 (Fig. 2.4)

Aristotle thus showed geometrically that, at rest, water assumes the shape of a sphere. The theory was refined by Archimedes and invoked by Pliny to explain several empirical phenomena: hanging drops of liquid are always globular; liquid dribbled onto something soft (like dust) always appears spherical; liquid in a full cup is always highest at the cup's center; adding a few drops of liquid to a full cup does not cause it to overflow; land visible from a ship's masthead is not always visible from the deck; and distant ships seem to descend as they move further away. The natural sphericity of liquids explains how oceans can cohere at all:

> Finally, could ocean, which we confess is very far away, cohere in any other shape and not fall apart without any enclosing rim? This very issue brings us back to the marvelous: how, even if it is balled up, the edge of the sea does not fall off.
>
> Pliny 2.164–165

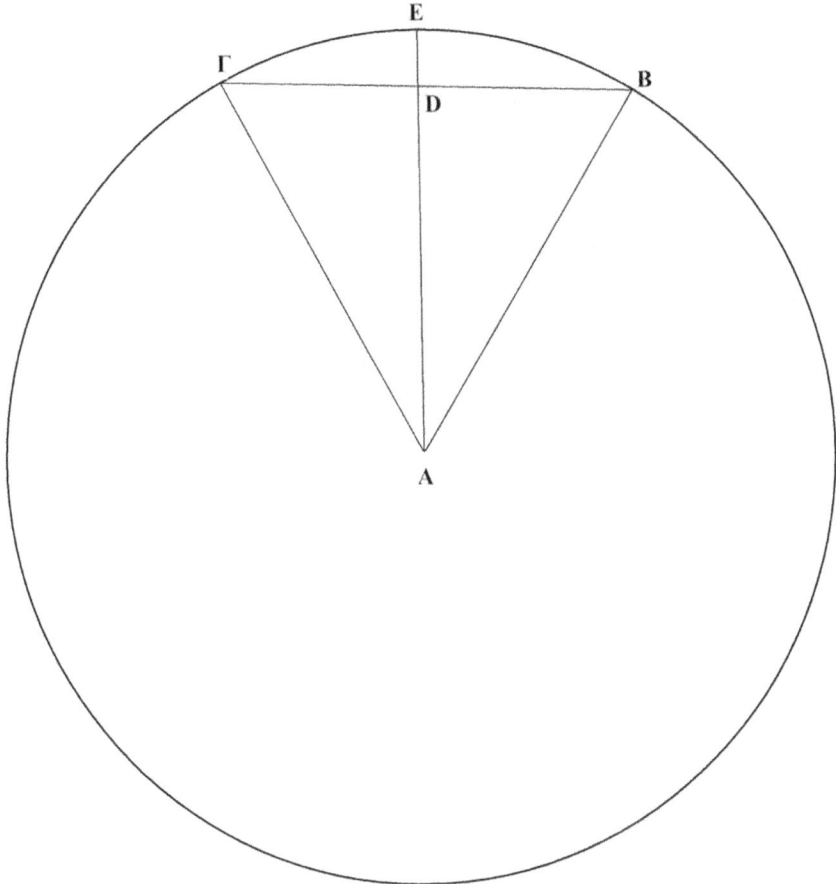

Fig. 2.4 Aristotle's proof of the sphericity of water, *Cosmos* 287b4–14.

Without citing Aristotle or Archimedes, Pliny cited "Greek inquirers" who have proved that the seas cannot be flat, but rather they must be spherical, because "waters from every part converge towards the center and for this reason they do not fall off, because they press on inwards." His attempt to replicate Aristotle's geometrical proof is, at best, sloppy.[74]

Other Characteristics of Salt Water

Pliny's "proofs" of the sphericity of liquid evoke the "Tell Me Why" tone of Aristotelian *Problems* book 23. According to our peripatetic author, the sheer volume of seawater explains why less rippling is observed in the open sea than in shallow, confined bodies of water: small quantities of water are more "divided" by air than larger bodies and, consequently, are beaten around and broken up more (23.1). The larger surface area of the open sea disperses breezes widely, thus weakening them (resulting in warmer breezes), while breezes from rivers are tunneled through narrow spaces and are therefore colder (23.16). Waves in large bodies of water "calm down" more slowly than in confined pools because of the greater amount of motion occurring within larger bodies of water where there is "more ebb and flow" (23.17), perhaps in reference to the greater swells in open water that are not broken by coastlines.

Air and water, furthermore, have mutual effects, in part determined by the nature of the water's shape and deployment, open sea, or confined pool. A second explanation for the temperature of sea breezes is offered. These are warmer than river breezes because rivers are naturally cold, whereas the sea is neither hot nor cold; the breezes, resulting from the evaporation of liquid, take on the temperature of the original liquid (23.16). Moreover, waves seem to originate before wind strikes the water. Why? Since the sea is a continuous body (and air is not) and since it is heavier and less easily disturbed, it is constantly subject to what we know as Newton's third law of physics "for every action, there is an equal and opposite reaction," expressed by our peripatetic thinker thusly:

> Is it because the sea having been forced in advance near the origin of the wind, always brings about the same effect? Consequently, since the sea is compressed (i.e., continuous), as if by a single blow the movement occurs on all (parts). This occurs at one time, so that the first and the last parts happen to be moved at the same time.[75]

Air also affects the appearance of seawater, according to our Aristotelian scholar. In contrast with the Aegean Sea where the air is clear "to a great distance" and reflected darkly off the water, the waters of the Euxine reflect the "thick and white" quality of the local air (23.6). When north winds blow over the sea, waters become more transparent (than when southerly breezes blow) because waters are warmed by the south wind, which in turn causes the "fatty juice" of salt water to melt and float on the surface "owing to its lightness" (23.9). This oily substance is opaquer than water. Water warmed by a southerly breeze, consequently, appears more colorful, while water cooled by a northern wind is more "thoroughly mixed" and "clear." This oiliness, a product of its salinity and "earthiness," also renders seawater combustible and less effective at extinguishing fires, so claims the author (23.15, 32). The salinity of seawater thus explains its greater transparency relative to fresh water, either creating "great voids" (23.8), or because of its oily composition (23.38). It is also possible, according to our

Peripatetic thinker, that the sea alone appears transparent, since fresh water is mixed with sediment.

Currents

Also of intellectual and practical interest is the behavior of the currents in the straits, most famously at Cape Pelorus (Strait of Messina), the treacherously narrow passage between the eastern tip of Sicily and the "toe" of Italy.[76] Here dwelt a sailor-snatching sea monster (Scylla), and here sailors found a dangerous natural whirlpool (Charybdis). Homer described the strait in harrowing detail, a pass between two sheer cliffs through which no ship could sail (except the *Argo*, and only with divine aid: *Odyssey* 12.59–110). The pointed peak of one cliff (against which is "the heavy swell of dark-eyed Amphitrite's crashing") extends to the sky, eternally crowned by a dark cloud. Midway up is Scylla's cave. At the lower cliff is the whirlpool, "deadly Charybdis," which promises certain death for Odysseus' entire crew.

Aristotle attempted to account for re-fluent currents in the Mediterranean Straits as resulting from the contraction of water within the narrows:

> The sea seems to flow in the narrowest places where through the framing coastline the open sea contracts from a large to a small space, because it (the water) often sways back and forth. In a large quantity of sea, this is unobserved. But in the narrows of the land, a scant passage restrains (the water), there the shores constrict the swaying which seemed small in the open sea but now seem great.
> *Meteorology* 2.1.354a5–10

In the Chalcidean Strait (Euripus) that separates Euboea from the Greek mainland, strong and swift tidal currents reverse direction about four times a day, a particularly troubling navigational hazard. Ancient biographers, moreover, suggested that this apparently inexplicable behavior (multiple daily reversals of direction), and the fact that a satisfactory solution was not forthcoming, induced Aristotle to commit suicide.[77] According to Eratosthenes, the currents here shift direction seven times a day (f16Roller). In Mela (2.108), Eratosthenes' seven shiftings are doubled: seven alternations occur both daily and nightly. In Strabo (9.2.8), the underlying reason for this ostensibly unusual behavior "must be investigated elsewhere."

Eratosthenes was possibly the only ancient thinker to explain the vacillation of currents in narrow straits, which he understood as owing to the discrete surfaces of each sea (f16Roller). Re-fluent currents in narrow straits resemble flow and ebb tides, which Eratosthenes thought were caused by the water sloshing between higher and lower levels. Strabo, our source here, dismissed the postulate as impossible. In accordance with Archimedes' proposition of the sphericity of liquids, the sea must be spherical. It cannot have an inclined surface. How then can the oscillation of the currents in the straits be explained? Eschewing scientific explanations ("a more scientific discussion than appropriate in this treatise") and dismissing Eratosthenes, Strabo preferred an *ad hoc* approach: "it is sufficient to say that there is no single explanation for the currents in straits that corresponds to their form" (1.3.12). The phenomenon, however, was well-known, occurring conspicuously at Messina, the Euripus at Chalkis on Euboea in eastern Greece, and the Hellespont.

Currents are complex hydrological phenomena that are generated by many factors including the variable densities of contiguous bodies of water that differ in salinity or temperature (e.g., the warm Gulf Stream). Surface currents are the only ones that ancient authors would have observed, and these are largely created by the winds together with coastal geography and the Coriolis force (inertial force) that results from the earth's rotation.[78] The interplay between the winds and coastal geography was described by Homer: e.g. "there the Southwest Wind thrusts the great wave towards the leftward headland" (*Odyssey* 3.295). The author of the *Periplus of the Erythraean Sea* (25), furthermore, acknowledged that winds blowing from mountains on Diodorus Island in the Erythraean Sea contributed to the strong currents in the strait there.

Whirlpools

Whirlpools occur naturally where contrary currents meet or where tides affect the fast currents of waters in narrow straits. The "notoriously treacherous" Charybdis was described as a deep eddy "into which the reflux of the strait cleverly pulls down boats, which are swept away with a twisting around and great whirling."[79] Charybdis was a common feature in myth,[80] spewing water and flames whenever the deadly sea monster, Typhon, whom the gods buried under Mt. Aetna, moved. In Pindar (*Pythian* 1.15–28), water surges up through interconnected subterranean cavities because the entire Strait is inflamed. Charybdis is duly cited and described in subsequent authors,[81] but there are no additional attempts to explain the phenomenon.

Homer and other poets recounted three daily cycles of Charybdis' ebbing and spouting. Polybius, Eratosthenes, and Strabo corrected this to two cycles.[82]

Strabo defended Homer, whom Vergil followed, by ascribing the erroneous triple gushing to either a copyist error or rhetorical hyperbole intended to induce greater fear (Strabo 1.2.16; 1.2.36). Wreckage would be conveyed from Charybdis to the Tauromenian shore, giving the place its name: *Kopria* (Refuse).[83] The whirlpool, together with a violent storm, was responsible for the destruction of Octavian's fleet in the Strait in 36 BCE.[84] Other whirlpools were described: at Genethlium in Argolis, where bridled horses were once drowned to propitiate Poseidon (Pausanias 8.7.2); at Cape Caldone on the Persian Gulf (Pliny 6.147); and in Lycia, in Apollo's sacred grove where worshipers sought oracular guidance (Athenaeus 8.333c–f). The locals believed that a fresh-water spring produced the whirlpool. Additionally, the *Erythraean Sea* author warned his readers of violent whirlpools in a bay near the Sinthos (Indus) River (40).

The Aristotelian author of *Problems* correctly ascertained that whirlpools are products of the currents, which he explained as factors of the winds:

> Now a current occurs when, after the earlier wind has been stopped, wind blows in the opposite direction over a sea flowing because of the earlier wind, and especially when the south wind blows in the opposite direction.
>
> 23.5

Reefs

When Aelius Gallus arrived in Arabia Felix where "Sebastos" (Augustus) Caesar had sent him to learn about the Aithiopians, Trogodytes, and the Arabian Gulf, he (Gallus) had hoped to "have dealings with wealthy friends or to subdue wealthy enemies." His reconnaissance was, however, sabotaged by the seemingly friendly Nabataeans, whose administrator, Syllaios, had suggested a water route full of typical maritime dangers, including shoals and reefs:

> He (Syllaios) did not indicate either a safe coastal or land route, but inflicted on him circuitous routes without roads and through places lacking in everything, with rocky shores, or full of undersea rocks and covered with shoals. In such places the flood-tides, as well as the ebb-tides, particularly caused distress.
>
> Strabo 16.4.23

Reefs are a notorious danger for ships, and some are named: the "Ass of Antron" below Antron in the Euboean Strait (Strabo 9.5.14); the "Midland" on the way to Lesbos, where colonists sacrificed live bulls to Poseidon and virgins to Amphitrite (Plutarch, *Dinner of the Seven Wise Men* 163a–b); and the "Ant" between Sciathus and the Magnesia archipelago, where Persian ships went aground in 480 BCE (a stone beacon was later erected to mark the reef: Herodotus 7.183). On his return from Troy, Menelaus lost half of his fleet to a reef (*Odyssey* 3.291–300), Sergestus ran afoul of a reef thus losing a ship race (Vergil, *Aeneid* 5.202–206), and Ovid lamented his historical shipwreck on a reef in the Pontus (*Pontus* 4.15–24). Fishing boats were destroyed when fishermen carelessly or ignorantly floated over reefs (Alciphron, *Fishermen* 1.7). Fishermen, nonetheless, took advantage of rich reef habitats to harvest anchovies (Oppian, *Fishing* 4.468–87) and the murex-purple snail (Pliny 9.131). Aegina was notorious for its treacherous approach to the harbors owing to underwater rocks and reefs, which Pausanias attributed to the workmanship of the *polis*' founder, Aeacus, as a safeguard against pirates and enemies who might launch an attack (2.29.5–6). The reefs were indeed purpose-built, as indicated by British admiralty charts and underwater surveys. Extending 1,700 meters (ca. 5,600 feet), and with an average height of 2.7 meters (ca. 9 feet), the reefs were constructed of irregularly shaped stones and were probably installed at a time when their tops would have just skimmed below the surface, creating an effective underwater hazard. These manmade reefs seem to be unique in antiquity, fitted perhaps between 491 and 459 BCE when Aegina was under threat from Athens.[85]

Conclusion

Although the episode of Alexander's bathysphere is likely apocryphal (Introduction), it does speak to intense curiosity about the underwater realm, the dominant geographical feature of the Mediterranean Basin. Greeks and Romans gathered data from the seas, which they marshalled into hydrological theories as well as useful compendia for sailors, merchant-mariners, and navies. Alexander's bathysphere aside, the ancients simply lacked the technology to gain a true understanding of the deep ocean. They could not measure salinity or temperature, and thus they were unable to develop an accurate understanding of currents and other hydrological phenomena. Furthermore, despite the fact that most marine fauna in "Our Sea" is non-poisonous, the perils of the submarine environment were exaggerated (Chapter 6). Many dangers, however, were real— reefs, storms, treacherous currents within straits—and the sea remained a place of danger and mystery: the abode of gods, and the stage for the heroes of epic and saga.

3

The Interplay Between Water and Land: Land, Rivers, and Springs

Introduction

Visitors to the Grand Canyon in Arizona have observed firsthand the spectacular power of water to shape the natural landscape. Together with the natural forces of weathering and erosion, in a mere 2 billion years, settling to its base level (sea level for a throughgoing waterway like the Colorado), the Colorado River has hollowed out more than a mile's worth of diverse rocky sediment to forge the canyon (6,093 feet [ca. 2 km]), whose width at the rim ranges from 600 feet (183 meters) to 18 miles (29 km). Long a focus of awe and inspiration, the indigenous peoples speculated on the Canyon's creation and purpose. According to the Navajo, excessive rains led to flooding and rising water levels until an outlet enabled the rushing waters to drain away, carving out the spectacular Canyon.[1]

Just as water has excavated the landscape to reveal 2 billion years of geological history in northern Arizona, so too does water, less dramatically, mold coastlines and create islands and isthmuses. Greco-Roman thinkers understood this process:

> The sea most of all describes the earth and determines its form, by producing gulfs, the open sea, straits, as well as isthmuses, peninsulas, and capes, with both rivers and mountains providing assistance.
>
> Strabo 2.5.17

Ancient Mediterranean thinkers sought to understand how bodies of water interacted with each other as well as with the lands around them. Ephorus of Cyme (fl. ca. 360–330 BCE) had employed coastlines as an organizing principle of his geography of the Greek world, "deciding on the sea as a type of guide to his topography" (Strabo 8.1.3). In his own account of the Greek islands, Strabo followed suit, declaring "thus it is proper for me to follow the nature of the places and to make the sea the advisor" (8.1.3). We now turn to the synergies of water and earth to investigate the qualities of terrestrial waters and the aqueous interplay by which the lands received their framework.

Sea and Land

From the earliest days, Greek thinkers employed their knowledge of waterways as guides for geographical and cartographical discourse. Writers such as Homer, Eratosthenes, Strabo, and Pliny inquired about the interconnections of land and water. They recorded the topographical relationship of settlements to neighboring waterways, the shape and nature of coastlines, and the proximity of communities to the sea. We know where places are by virtue of their disposition to well-known waterways, as we see meticulously registered in the Homeric catalog of ships (*Iliad* 2.494–759). The inland peoples of Phocian Hyampolis and Anamoreia dwelt near the "immortal river," Cephisus, and the Lilaians were established at its well-springs. Those who followed the Lesser Ajax hailed from "beside the waters of Boagrios," the "Ox-Hide" River in Lokris that debouches into the Aegean near Thronium.[2] Cerinthus was a sea-borne island. The Argive towns of Hermione and Asine were situated "down the deep (Argolic) gulf." Agamemnon's kingdom encompassed Corinth, with harbors at each side of the isthmus, together with the coastal settlements of Hyperesia, Gonoëssa, Pellene, and Aigion, "all about the shore and the wide headland of Helike."[3] Helos was a "sea-ward" city in southern Lacedaemonia. Soldiers from Doulichion and the Echinades islands "live across the water from Elis" near Olympia in western Greece. Chalkis and Antron were by the shore. Ormenios was distinguished by the spring Hypereia (cf., Strabo 9.5.18). Near wintry Dodona in northwestern Greece (the site of Zeus' famous oracle), the River Titaressus "casts his bright current into the Peneius, yet he is not mixed with the silver whirls of Peneius, but like oil is floated along the surface above him." Homer also recorded the number of ships provided by each region. From landlocked Sparta, Menelaus alone had no navy and was compelled to borrow "sixty ships marshalled apart from the others" (*Iliad* 2.587). Cartographers designated major waterways as boundaries between continents (the Nile and Phasis Rivers separate Libya and Asia) and countries (the Euphrates delimits Cilicia and Armenia)[4] (see Fig. 0.1).

Rising and Falling Sea Levels

The interplay between land and water was (and is) perennial, as noticed by Xenophanes who hypothesized an unending fluctuation between land and water.

Xenophanes cited inland marine depositions in support of his theory of a cycle of terrestrial desiccation and moistening:

> a mixture of earth and sea occurs and in time earth is dissolved by the moist, claiming to provide as evidence the fact that sea shells are found in the midst of earth and in mountains, and in the quarries of Syracuse impressions of fish and seaweed have been found, and in Paros the impression of coral [or: bay] in the depth of a rock, and in Malta fossils of all sea creatures. He [Xenophanes] says these things happened when all things were covered with mud long ago and the impressions in the mud dried out. The human race becomes extinct when earth is carried down into the sea and becomes mud, and then the process begins again, and this change occurs in all the world order.
>
> TEGP 59

It was a geological process whereby earth mixes with sea, a complex interplay between land and water that shapes and redraws the pliable coastlines. For Xenophanes, the terrestrial world cycles between moistness and total desiccation, his proof of which—inland marine depositions—is hardly isolated: evidence has come to light at Syracuse in Sicily, Paros in the central Aegean, and Malta, south of Sicily. Xenophanes' travels took him westward from Anatolia, and he spent much of his adult life in Sicily and southern Italy (among his works is a poem on the founding of Elea in southwestern Italy). Xenophanes likely saw these fossilized remains for himself. Furthermore, his curiosity about fossils "show[s] his kinship with modern scientific inquiry."[5]

Xenophanes was not unique in his observations. Among Strabo's sources, Xanthos of Lydia (480–440 BCE) concluded that Armenia and Phrygia had once been under water, also on the strength of inland marine deposits.[6] During the reign of Artaxerxes (465–425 BCE), a severe drought had left rivers, lakes, and cisterns dry, revealing mollusk shells, the impressions of scallop shells, a salt lagoon, and possibly baleen ("sherds like combs"), which he (Xanthos?) "had often seen" (Strabo 1.3.4). Strabo interpreted the account as evidence that at one time the Euxine's mouth was above shore level. Rivers issuing into the Euxine, he suggested, had broadened the sea's opening and forced water in from the Propontis through the Hellespont. A similar explanation was offered for the creation of the strait between the Pillars of Herakles, that rivers breaking through filled the seas and drained the coastal shallows.[7] Eratosthenes, likewise, posited that water once covered the oasis at Ammon, where "one can see in many places oyster, mussel, and scallop shells."[8] Why else would one find such inland depositions, including wreckage from sea-going vessels, 2,000 to 3,000 stades from the sea? Long ago in

Homer's time, furthermore, Pharos island, the site of Alexandria's famous lighthouse, was in the open water (Strabo 1.3.17; cf., Homer, *Odyssey* 4.354–355). In Homer's defense (that Pharos was once an island), Strabo reported that he had witnessed firsthand a flood in the region of Pelousion (on the eastern extent of the Nile Delta) that had turned Mt. Kaison into an island, making roads navigable by boat.[9] Because of silting and alluvial accretion, it was widely recognized that the sea levels around Egypt were slowly receding while land levels were increasing.

Herodotus had also observed the phenomenon of inland marine depositions, finding sea shells in the mountains of Egypt. For Herodotus, this evolving reciprocity of land and sea raised questions about erosion and deposition, sea depth and changing levels, and even the age of the earth: a single mountain in Egypt, above Memphis, with sand and salt in its delta, for example, caused the erosion that had damaged the pyramids (2.12). Herodotus noted that the alluvial Delta was built up through a gradual accretion of silt along the Nile's banks (2.15). Egypt was thus famously called a "gift of the Nile" (2.5).

Just as alluvial deposition along the Nile Delta augments the fluxing riverbank, the enduring interchange between land and water, each rising or settling in their turns, was recognized by natural philosophers. This phenomenon was featured in the myth of Atlantis, where an island continent, "larger than Libya and Asia combined," beyond the Pillars of Herakles, sank into the sea because of seismic activity—violent earthquakes and floods—as reported to Solon by Egyptian priests (Plato, *Timaeus* 24e–25d). Plato surmised that the Pillars of Herakles were impassable, "impeded by the mud which the sinking island created."[10] By Plato's day, nonetheless, the Phoenicians and Egyptians had already ventured beyond the Pillars of Herakles (cf., Chapter 2; *UCWW*: Chapter 5). Plato may have borrowed the name Atlantis from Herodotus (4.184.4), whose Atlantians otherwise bore no resemblance to their Platonic counterparts. Although interrogating variant sea depths, Aristotle refused to speculate on the veracity of the Atlantis story (or even to mention the place), but he conceded that the body of water beyond the Pillars of Herakles must be shallow owing to the mud, but tranquil ("without breath") since it lies in a hollow.[11] Aristotle eschewed speculation on the cause of the hollow or the source of the mud. Posidonius would later presume that the Atlantis legend may have been based in fact.[12]

Effects of Siltation

The dramatic sinking of Atlantis, together with Plato's emergent shoal-mud, illustrates the delicate reciprocity of earth and water.[13] The exchange is almost imperceptible, but it endures as a core principle of natural philosophy. Although Aristotle dismissed as absurd Democritus' hypothesis that the sea was receding and drying up (*Meteorology* 1.13.353b6–11), he nonetheless subscribed to the cyclical interplay of water and earth (*Meteorology* 1.14.351a19–352b16). The earth grows and decays with heat and cold, and siltation occurs gradually as land encroaches on the sea, or progressive erosion ensues as the sea infringes on the land. Aristotle conjectured that Egypt was once a wet country accreted from its river, but was steadily drying out. Newly silted land would be marshy because of the retention of water. Over time, however, the water would evaporate as the alluvium aged. Aristotle was careful to disavow that such topographical changes could be construed as cosmogonic, since he firmly believed that the world was currently fully formed. Such changes are mere trifles "compared to the bulk of the size of the earth," and can be understood as analogous with rivers that change their courses (e.g., the Achelous and the eponymous, Meander Rivers).[14]

Siltation could change the nature of bodies of water, altering their degrees of salinity and affecting navigability, as with Lake Maeotis. By the second century BCE, Lake Maeotis (the Sea of Azov, a northeasterly extension of the Black Sea) had accreted to such a degree that it was unnavigable without a pilot who could steer larger boats through the deeper channels (4.40.8). According to Polybius, the Maeotis was once a salt sea, confluent with the Pontus, but in his day the Maeotis was sweeter (fresher) than the Pontus (and the Pontus is sweeter than the Mediterranean) owing to the fact that the salt had been forced out by the inflow of large rivers (4.40.4–10; 4.42.3). The Pontus, furthermore, continually receives alluvial deposits and would eventually become sweet. Considering rates of accretion and the ratio between the basins and alluvia, Polybius conjectured that the evolution of the Maeotis from lake to land (the basin would eventually become completely filled in) would occur as "is clear to anyone who is paying attention" (4.40.10). For this same reason—alluvial deposition—the Pontus was recognized as the sea with the freshest water: "since rivers flowing from the north and east are numerous and large and fill it [the sea] with sediment, while other rivers remain deep" (Eratosthenes f15Roller).

Alluvial deposition is a common feature of rivers. The Nile is merely a special example. Just as Egypt is the gift of the Nile, plains are the "offspring" of rivers, as Nearchus had designated the Indic plains (15.1.16). Riverine silt could fill up lakes and create deltas, such as the Maeotic and Pontic Seas, the Nile, and the rivers of India. This alluvial deposition could reach well out into the sea, as forecasted by Strabo for the Pyramus River in Cilicia. The Pyramus' growing delta might one day reach southward some 120 nautical miles across the

Mediterranean to Cyprus, so claims an oracle preserved in Strabo (12.2.4): "those in the future will experience the time when silver-eddying Pyramus will pour over its sacred beach and come to Cyprus."

Such extreme siltation, however, is untraceable in the open water because of the natural oscillation of the surf. Whatever sediment that might flow from rivers into the ocean would return to land because land-ward waves have more force: "casting all kinds of strange things on the land and 'pouring out much seaweed'" (Strabo 1.3.8, citing *Iliad* 9.7). This oscillation was interpreted as an analog to breathing (Strabo 1.1.8, 3.5.7), an image particularly resonant in the Stoic sources for whom the cosmos is governed by omnipresent *pneuma* (Chapter 1).

Although siltation is a natural phenomenon, it was (and still is) exacerbated by human activity, namely deforestation, whose effects are cascading and far-reaching. Forests are cleared to make way for agriculture and urbanization, and harvested wood is utilized for tools, building materials, and fuel. This human activity deeply influences the climate, topography, and fertility of the soil. Ancient authors, moreover, had a sense of the change resulting from large-scale deforestation. Plato lamented that mountains in Attica had only recently been heavily forested, but in his day those same mountains were "rocky skeletons" because of erosion that resulted from deforestation (*Critias* 111b). Eratosthenes, moreover, observed that Cyprus had once been overrun with vegetation but was no longer verdant in the third century BCE, owing to the demands of mining (f130).[15] In Livy, the Roman general, Fabius, in pursuit of the Etruscans in 310 BCE, found the Ciminian Forest in central Italy impassable (9.36.1), not the case in Livy's time, and Egypt once also boasted groves of giant trees that had disappeared by the mid-first century CE (Pliny 13.19). As trees are cut down in large numbers, water is no longer retained in the soil (as there are no plants to retain the water), and the climate becomes windier and more arid (e.g., Northern Africa: see Drake et al. 2011). In turn, the watershed is diminished and eventually desiccated. In the Roman province of Libya it was artificial irrigation that "made this arid region bloom."[16] As sediments were deposited downhill, siltation increased, clogging harbors (as at Miletus, Paestum, Ravenna).[17] Deforestation and siltation led to flooding (especially in Rome: Pliny 31.53, Chapter 4) and diminished water quality. Siltation, furthermore, profoundly affects the ratio between land and water: deposition from the Meander River, for example, had turned "the sea into dry land" between Priene and Miletus "in a very brief time," in contrast with the uninhabited and "untilled" regions around the Achelous River that "does not similarly bring mud on to the Echinades" (Pausanias 8.24.11; cp., Pharos, above).

Sources of Terrestrial Waters

Greek thinkers also inquired into the sources of water, whether all waterways derive from the same or different origins, whether all water comes from the sea, or if terrestrial waters create the sea. Noting the total volume of water on earth, Seneca observed that the inflow of water into the seas is as imperceptible as is the outflow from rivers (*NQ* 3.4, 8). Where does all that water come from, and how is it continuously replenished?

In Hesiod, Ocean is the physical, anthropomorphized origin of all waters (*Theogony* 337–370, cf., Chapter 1). Rationalizing Hesiod's conjecture, Hippon of Croton in southern Italy believed that the sea was deeper than bodies of fresh water. There must be more water in the sea, which, consequently, must be the source of other bodies of water. In the *Phaedo*, Tartarus—Homer's "deepest chasm"—is ascribed as the beginning of all waters.[18] Although it is clear that rivers flow into larger bodies of water, the origin of river water is another question altogether, and the earliest conception was twofold. In Anaxagoras, water is squeezed up from hollow cavities in the earth, producing the sea, and rivers are replenished by rain (*TEGP* 38). The conjecture endured, finding expression centuries later in Pliny who envisioned ocean water as being pressed up through underground channels to mountain tops (2.166). Pliny's hypothesis derives from the Stoic belief that, in order to cohere, the earth requires a constant flow of water, filtering throughout the environment (cf., Aristotle, *GC* 2.8.335a1, as anyone who has experienced both the humid tropics/subtropics and the dry, dusty desert could attest).

Arguing against Anaxagoras' hypothesis some six centuries later, Seneca maintained that rainfall could never be sufficiently copious to produce something as large as a river. Instead, rain could only dampen the ground's surface, and the ground could not absorb moisture to a depth of more than 10 feet (*NQ* 3.6-7: consequently, he thus dismissed the role of groundwater).[19]

Seneca also observed that waterways are neither eternal nor constant, but could become ill or healthy in turn, like fluids in the human body:

> So in the earth liquids often go bad: either a blow, or some upheaval, or the old age of the location, or cold, or heat corrupts their nature, a festering process forms a liquid, which may be either long-lasting or short-lived.
>
> *NQ* 3.15.4

Subterranean air might produce moisture, collecting like dew, and then trickling "from many directions into one place" yielding either a feeble flow "scarcely sufficient for a stream" or amassing in large reservoirs whence rivers pour out:

sometimes issuing gently, if the water just flows downhill under its own weight, sometimes violently and noisily, if breath is mixed in with it and forces it out.

NQ 3.15

Seismic activity could open or close conduits in the earth (like veins), causing rivers and springs to appear or disappear or to flow back and then be diverted along a new course through a new outlet. For example, on Mt. Corycus, overshadowing a harbor town in Lycia (Turkey), new springs emerged after an earthquake.[20] Theophrastus ascribed the appearance or disappearance of bodies of water also to human factors. Water simply appeared when the Macedonian king, Cassander (ruled 316–297 BCE), felled a forest on Mt. Haemus. Theophrastus' simplistic explanation, that the trees apparently no longer needed the moisture, was debunked by Seneca:

> generally the places with the most shade have the most water, and that would not be the case if trees dried up the water supply. They get their nourishment from near the surface, but rivers flow from deep within and are generated beyond the depth to which roots can extend. Then trees that have been cut down need more moisture: for they soak up enough not just to stay alive, but to grow.
>
> *NQ* 3.11.3

Theophrastus also reported odd streams in Crete that stopped flowing when the fields were abandoned, but, when the farmers returned, so too did the springs. Architectonically speaking, lack of cultivation caused the earth to harden, and rainwater could not, consequently, penetrate the ground. But, Seneca countered, how then could springs exist in unoccupied places?

Rivers

In Homer, two personified rivers distinguish the plain at Troy (*Iliad* 21.300–327):[21] sons of Ocean and Tethys in Hesiod (*Theogony* 345), Simois and his brother, Scamander, as the latter was called by mortals in Homer: the gods called this deep-eddying river, Xanthus (*Iliad* 20.74: "tawny," suggesting the agitation of mud after a storm: cp. Campbell 2012: 4). After the death in battle of his friend, Patroclus, Achilles slew many Trojans whose panoplies then glutted the Simois' riverbed. The enraged river emerged from his course to engage Achilles in battle, attempting to drown him (*Iliad* 21.299–327). However comically, the image, at any rate, speaks to the destructive power of fast-flowing, overflowing rivers.

Given the light rainfall in the Mediterranean Basin, largely limited to the cold, wet season, it is natural that sources of fresh water, especially rivers, would figure prominently in Greek and Roman lore. The Eridanos and Ilissos Rivers in Attica (now subterranean owing to the devastating flood of 1897), for example, could be "torrential" during the rainy season, but often dried up completely at the height of the summer heat (Plato, *Critias* 111a; Strabo 9.1.24). They are nonetheless small in accord with the dry climate.

Aristotle contended that rivers cannot originate from reservoirs (it would be impossible for a reservoir of sufficient size even to exist, he believed), nor do all rivers derive from a single source (*Meteorology* 1.13.349a11–53a26). Refining Anaxagoras' hypothesis, Aristotle argued for mountain-springs as the sources of rivers, and he compared high ground to saturated sponges, in contrast with the much drier plains. We have already seen that Aristotle regarded the notion of upward-flowing rivers preposterous (Chapter 2). Rivers consequently must flow from higher to lower ground: "the largest and most capacious rivers flow from greatest mountains" (*Meteorology* 1.13.350a3-4). Rivers could also originate at the feet of mountains or in marshes, or they might accrete slowly from water naturally contained in the earth, "transmitted by a gradual percolation, drop by drop." Nor do all rivers behave in the same manner, but each responds to the characteristics of its own particular environment (e.g., the size, density, and coldness of their mountains). Small, porous, stony, clayey mountains produce rivers that run dry more quickly. Finally, Aristotle also considered the effect of rivers on the oceans and seas into which they discharge. They seem to leave no trace, and the sea is in no way increased with the "innumerable and vast rivers flowing into it every day" because the waters are spread out over a vast area, "quickly evaporating" like a small cup of water poured over the top of a large table.

The explorer, Euthymenes of Massilia (ca. 500 BCE), claimed to have witnessed rivers that seem to flow inland from the Ocean.[22] He was likely observing the powerful ebb of a large tidal river, a phenomenon that fascinated Posidonius: "tidal flow is so powerful that it even turns great rivers in the opposite direction" (f219Kidd), particularly the Rhine and rivers in Spain and Britain.[23] According to Posidonius' redactor, the River Thames in Britain is filled up by the sea at flood tide and seems to flow away (rather than toward) the sea "for four days." The phenomenon of tidal rivers was ascribed to the influence of the moon: with its fire, the sun would quickly destroy any moisture that it removes from the earth, but the moon, with less vigorous heat, could only raise water and create waves but would not be able to affect the total volume of water.

Campbell (2012: 291) observes, "we are poorly informed about seemingly important rivers such as the Arnus in Etruria and the Liris and Volturnus in Campania." Writers focused on the unusual and the marvelous, not the familiar, and thinkers expressed their fascination with the large, exotic rivers that behaved differently from familiar waterways. They scrutinized navigable rivers that were deep and perennial, alluvial rivers, and rivers that, unusually, followed a natural south to north incline. Here we shall focus briefly on four distant rivers that sparked the Greco-Roman imagination: the Ister, Phasis, Indus, and Nile.

Ister

The Ister (Danube) was known to Hesiod (*Theogony* 349) and perhaps also to Homer who cited the Mysians (*Iliad* 2.858), the Thracian allies of Troy who dwelt along the banks of that river according to Strabo (1.1.10).[24] In Hecataeus, the Rhipaean (gusty) mountains lay north of the Ister, and beyond them abided the mythical Hyperboreans whom the gods visited regularly.[25] Herodotus regarded the Ister (which he construed as analogous with the Nile River) as the largest of the European rivers with five additional Scythian rivers flowing into it (4.48). According to legend it was through the Ister that Jason and his Argonauts made their escape from Medea's xenophobic father, vainly hoping that the river would afford passage to the Adriatic.[26] Apollonius' description, however, does suit the Ister's course from the Hungarian plains into the Balkans.[27] The ancient sources are contradictory. Aristotle assumed that the river's source could be in Mt. Pyrene in the Celtic territory, whence it flows through "all Europe into the Euxine (Black Sea)" (*Meteorology* 1.13.350b1). This origin was accepted by Eratosthenes and "corrected" by Strabo who gave the river's beginning as the mountains above the Adriatic, flowing only into the Euxine and not, as Eratosthenes had claimed, into both seas.[28] Strabo also referred to "many other torrential rivers that join the stream of the Istros" (4.6.9).

The Romans—for whom the Rhine and Danube Rivers essentially delimited their northern imperial frontier—would eventually collect improved data on northern Europe and its rivers (ten legions were in garrison along the Danube from the second to third centuries CE: Procopius, *Buildings* 4.5.2). Correctly placing the Danube's source in Germany in the Abnoua mountain range, at the confluence of the Brigach and Breg, where it streams east of Donaueschingen in southern Germany, Pliny (4.79) reported that the Hister (as it was called in Illyria in the Balkans) receives sixty tributary rivers, nearly half of which are navigable, before debouching into the Euxine in six channels (on a false analogy with the Nile).[29]

Wildlife, likewise, could be prodigiously large or impressive. Pliny observed that fish of the Ister were "very similar to a sea-pig," and they were so large that they must be harpooned with hoes and dragged from the river by a brace of oxen (Pliny 9.45). Following Aristotle (*HA* 8.13.598b15–16), Pliny also surmised that tuna reached the Danube from the Adriatic via an underground river (9.53). The Ister unfortunately receives nothing more than a passing mention in Arrian's *Periplus of the Black Sea* (20.3).

Phasis[30]

Known as early as Hesiod (*Theogony* 340), the Phasis River (modern Rioni), renowned for the unusual properties of its waters, was "broad-flowing and the utmost end of the sea" in Apollonius (*Argonautica* 2.1261). It was the easternmost navigable river or the "farthest journey for ships" in the popular imagination, a piece of folklore rejected by Strabo (11.2.16) who recorded many navigable tributaries attached to the river (Strabo 7f47 [48]; Campbell 2012: 299). Because of its exoticism, it remained mythically resonant. The most important river in Colchis, dividing Europe from Asia in Anaximander and Hecataeus (a continental border risibly dismissed by Herodotus: 4.45.2; Fig. 0.1), the Phasis was the eastern edge of Socrates' inhabited world (Plato, *Phaedo* 109a) and one of the limits of the Roman frontier (together with the Euphrates, Aithiopia, and Britain).[31] For Augustus and later Romans, command of the Phasis River ensured strategic control of the Black Sea. In the first century, it was crossed by 120 bridges, and was navigable for ships "of any size" for 38.5 Roman miles (about 57 km), even further for smaller vessels.[32] Originating in the southern slopes of the Caucasus mountain range, and filled by springs, the Phasis was routed through the narrows at Sarpana where it flows "roughly and violently" to Colchis (Strabo 11.3.4).

The Phasis River was renowned for its curiously peculiar properties. Arrian (*Periplus* 8) reported that the waters are light, floating above the sea and not mixing with it—just as the waters of the Titaressus float atop the Peneius River "like oil." The lightness of the water in the Phasis is because of the great number of rivers flowing into it, compounded by the narrow egress through the Bosporus, according to Arrian.[33] Modern hydrologists note that variant salinity (and striated specific densities) of the Phasis River, a phenomenon that was caused when the salty Mediterranean erupted into the river, as the denser salt water sank to the bottom.[34] Ancient authors commented on the river's unusual hue, as if colored by lead or tin (thus very dark), but they also claimed that the waters would clarify when left to stand, becoming even fresher over the years, so it was believed (Pliny 6.12). In order to preserve the presumed health-giving qualities of the river water, it was illegal to import water from abroad.

Indus

Herodotus is our earliest extant Greek source on India, and he credited Darius (ruled 522–486 BCE) with the discovery of the Indus River, which, like other Indian rivers, was alluvial (4.44; cf. Arrian, *Anabasis* 5.6.4).[35] Owing to the similarities of the Indus (and other great Indian rivers) with the Nile (alluvial rivers that flow south to north [Eratosthenes f74Roller], and strange, parallel wildlife[36]), India and Egypt were analogous in the popular imagination. Both rivers, the Nile and Indus, were considered prolific and nourishing "because of the moderating heat of the sun that preserves the nurturing elements while vaporizing what is superfluous." Because of this warmth, water (from the Nile) would boil at half the temperature required to boil water from other rivers, so it was thought. But the rivers in India were deemed more fertile because they "spread into plains that are large and broad, remaining for a long time in the same latitudes, and thus they are in such a way more nourishing." Indic rivers retained more heat as rain was "already boiling when it pours from the clouds" in India. It is the warmth of the waters that accounts for the size and quantity of the strange animals in India. The peculiar waters could also affect the quality of the fauna. Alexander's helmsman, Onesicritus, for example, had noted that water determines the color of the cattle who drink from the Indian rivers: "the color of foreign cattle who drink it changes to that of the native ones."[37]

Nile

Known to the Presocratics (Thales was the first to try to explain the river's "unusual" summer flooding), the Nile governed the rhythm of life in Egypt, sustaining agriculture and facilitating commerce. Ancient thinkers were intrigued both by its gradual summer flooding (rivers in Europe flood, usually more violently, during the wet winter season) and its mysterious origin. The Nile was thus more than just another river. Following Ovid (*Metamorphoses* 1.285–288), in Lucan, who addresses the river in the second person (10.285, 296, 317, 328), the elusive Nile is also anthropomorphized. Pliny the Elder, who, like Seneca the Younger and Lucan, also subscribed to a providential deity within the context of a teleological natural world, the Nile is anthropomorphized, assuming human qualities that are in alignment with human interests, including "playing the farmer" by means of its annual flooding (18.167).³⁸ In Xenophon of Ephesus' *Ephesian Tale* (second/third century CE), the Nile is invoked as a savior that rescues the protagonist (*Ephesian Tale* 4).

The source of the Nile was a topic of intense scholarly scrutiny. Juba II hypothesized that the Nile originated in the Mauretanian mountains of his country, flowing underground, eventually reappearing as a large lake in the Masaesulian territory, sinking once again beneath the desert "for a distance of twenty days' journey, before arriving at its known course south of Egypt."³⁹ Herodotus, who describes the river's course, correctly reports that the Nile's first cataract is in Elephantine, the island opposite Syene (Aswan), according to common opinion (2.17.2). The Greek toponym Κατάδουποι (*katadoupoi*: "thunderers") evokes the deep din of the rushing water. In Cicero, following Herodotus, the rushing water at *Catadupa* is so loud that the inhabitants are deaf (*Republic* 6.11).⁴⁰ Explorers into the nineteenth century consequently continued to seek the river's source in northwestern Africa.

The Greeks believed that the river's origin could not be located owing to its course through an uninhabitable desert, from a climate contrary (opposite) to our own (Diodorus Siculus 1.40.1). For Aristotle, passage through a hot territory is confirmed by the freshness (not brackishness) of Nile water (*Meteorology* 2.3.358b12–17). In Aeschylus, the Nile flows through the Bybline (Papyrus) mountains, unattested elsewhere (*Prometheus Bound* 810–812). Herodotus—whose authorities, with a single exception, professed ignorance—offered a third suggestion, which he had learned from an Egyptian official (2.28–29). Herodotus' knowledgeable priestly scribe, however, was certain that the river originated from a bottomless lake between the Syene and Elephantine mountains (Chapter 2). In the time of Amenhotep III (1411–1375 BCE), the Nile's inundation was recognized as somehow linked with the First Cataract, and the priest may have simply been repeating this ancient account.⁴¹ Herodotus' curiosity and research took him to Elephantine, but no further, because of the rugged terrain and powerfully treacherous current. He was consequently unable to confirm the report by autopsy. Alexander even thought that he had found the Nile's source when he saw crocodiles in the Hydaspes River in the Punjab (modern Jhelub River in Pakistan) and Egyptian beans in the Akesines River (modern Jihlam, originating in the Himalayas).⁴² Diodorus Siculus' Egyptian sources reported that the Nile emerges from the Ocean (1.37.7), a view also held by Hecataeus but rejected by Herodotus who staunchly opposed the theory of a circumambient Ocean (Chapter 2).⁴³

Eratosthenes explicitly compared the summer-time flooding of rivers in the Indus Valley with the Nile's annual flooding (f74Roller), the cause of which was one of another great intellectual debates of antiquity, whether owing to Etesian winds (Thales: *TEGP* 21), melting snow (Anaxagoras: *TEGP* 54–56), or summer rains (Agatharchides, probably on analogy with Indian rivers).[44] Various speculations about the Nile's flooding harmonized with contemporary systems of natural philosophy. The geometer and astronomer, Oinopides of Chios (fifth century BCE), suggested simply that water expands in the summer heat (Diodorus Siculus 1.41.1), while an anonymous commentator posited that the sun draws the earth's moisture into the river (Scholiast on Apollonius 4.269). Examining four theories, Herodotus was the first to consider the question systematically (2.19–24). He dismissed Thales' hypothesis that Etesian winds would prevent the Nile from flowing into the Mediterranean. These winds do not always blow, Herodotus countered, nor do they have the same effect on other rivers (as they should, were Thales correct). Although plausible, Anaxagoras' melting snow hypothesis is actually the most erroneous—so the historian thought—since the Nile flows through the warmest regions of the earth where snowfall would be impossible. Despite Herodotus' skepticism, "all of antiquity shared in this opinion" (snowmelt as the cause of the Nile flood), including Aeschylus, Sophocles, and Euripides.[45] Herodotus proposed yet a fourth explanation: that, directly overhead in Libya in the winter, the sun would dry out the Nile, which would then return to its regular level in the summer. The sun evaporates moisture that is then drawn up by winds and is eventually returned to the earth during the windy, southerly and southwesterly monsoons, a conjecture validated by the heavy annual rains over the sources of the Blue and White Niles, so claimed Herodotus.[46]

Seneca devoted an entire book to the question (*NQ* 4a), but "his preferred theory—if he had one—remains elusive."[47] After a lengthy moralizing preface on the pitfalls of flattery, Seneca disputed the long-held correlation of the Nile with the Danube (whose source was acknowledged to rise annually because of snowmelt). If we knew the Nile's source, Seneca surmised, we could then explain the annual summer flooding. Seneca's description of the river and its course, including the cataracts, is animated and deeply personified. The river meanders through deserts, sprawls onto marshes, and is dotted with large islands. At Philae,[48] the two Niles merge into a broad (but non-violent) river. Just beyond Aithiopia, the river meets the cataracts:

> There the Nile surges up among high crags that are sheer at many points, and it increases its violence. For it crashes over the rocks it encounters, struggles through narrows, and, wherever it is winning or losing, it seethes. There its waters are whipped up for the first time, when previously it had brought them without

disturbance down a gently sloping riverbed. Violent, raging, it rushes forward through resentful channels. It no longer resembles itself, for up to now its flow has been muddy and cloudy. But once it has struck the rocks and sharp crags, it foams, and its color derives not from its own nature but from the mistreatment it receives there, Finally, after battling through the obstacles, it is left hanging and suddenly plunges down a huge drop with an enormous din that fills the neighboring regions. A tribe settled there by the Persians could not endure it, since their ears were deafened by the continuous roar, so they moved to quieter parts.

NQ 4a.5

Local sportsmen would even ride the white-water rapids, plummeting head-first into the cataracts ("to the great terror of the spectators"), until they were catapulted towards calmer water. The manuscripts break off after Seneca's survey of the theories posited by his predecessors.[49]

Underground Rivers

In Greece, the nature of the landscape makes underground debouchments common, and there was speculation about the interconnections between those waterways, as we have already seen in the putative channel between the Adriatic Sea and Phasis River. The fact that aqueducts could collect water in pipes and trenches "as if the earth were sweating in higher places" was considered proof of the existence of subterranean lakes.[50] According to Aristotle, such underground channels are essential to prevent flooding. Because of the pressure from the waters above, "some rivers are swallowed up." Those lacking visible outlets find underground outlets instead, as noted on a small scale in Greece, but elsewhere on a larger scale (*Meteorology* 1.13.351a1–6). Seneca's account is the only extant one (*NQ* 3.7–10). He imagined underground rivers in comparison with large, terrestrial rivers. He offered three possible causes:

1) moisture is constantly expelled within the earth;
2) air is converted to water by the forces of darkness and cold (since, according to Aristotelian and Stoic physics, all elements have reciprocal properties);
3) or earth is converted to water.

Such rivers were observed, as, for example, by a crew sent into a derelict mine by a Philip (probably Philip II of Macedon), who had discovered enormous underground rivers and lakes in large subterranean caverns (Seneca, *NQ* 5.15.2).

Several examples of riverine games of peek-a-boo are recorded, as rivers are "swallowed up by some hollow in the earth," descending into caves, or gradually absorbed into the earth until they "return and recover their name and their course" (Seneca, *NQ* 3.19.4, 25.3–4). The Anias River around Pheneus in the north-central Peloponnese—which is alternately a plain or a lake (as in Eratosthenes' day)—carries water through underground channels (*zerethra*) for ca. 10 km (about 6 miles) to the southwest, to a depth of 400 meters (ca. 1,300 feet), before streaming into the Ladon and Alpheus Rivers.[51] At one time the lake receded, but there was flooding in the territory around the sanctuary at Olympia, near which the Alpheus flows, thus lending credibility to the river's underground propensity. The Alpheus itself was widely thought to disappear into the ground, only to reappear later on the western side of the Aegean Sea in the guise of the Arethusa Fountain on Ortygia, near Syracuse. Objects thrown into the Alpheus' mouth would reputedly resurface in the Arethusa.[52] The skeptical Strabo (6.2.4) argued that waters discharging into the Aegean would be salty, but the Arethusa was fresh. The Alpheus, therefore, could not flow into the Arethusa. The unusual behavior was nonetheless explained by a popular myth: the River Alpheus fell in love with the nymph Arethusa who escaped to Syracuse but was pursued by her aggressive lover.[53]

Likewise, the Tigris River flows through Lake Thopitis but then disappears, plunging under the earth with "a great noise and upward blasts," reappearing eventually near Gordaia.[54] The Thopitis, however, is not in the Tigris Basin, but, rather, near the Tigris' source. The terrain there was little understood, and even local residents might have thus believed that the river ran underground.[55] The historian, Theopompus of Chios (fourth century BCE), hypothesized an underground channel linking the Adriatic and Aegean Seas (Strabo 7.5.9). Finally, although classified as a "sea" because of its size, the Caspian was known to be landlocked, not mingling "with other seas" (Herodotus 1.203.4). With no visible outflow, the Caspian seemed unfathomable (Chapter 2), and Aristotle, who also recognized the Caspian as landlocked, deduced that there must exist an underground debouchment into the Black Sea (*Meteorology* 1.13.351a9–10). Other thinkers, however, presumed that the Euxine flowed into the Caspian, which in turn flowed into Ocean (Manilius 4.585–710). In the 280s BCE, Patrocles, who traveled there before venturing into India, postulated (erroneously) that the Caspian is linked with Ocean, and his (lost) account probably informed Eratosthenes' view of the Caspian as an inlet of the external Ocean.[56]

Occasionally, earthquakes could cause rivers to be swallowed. For example, the Melas at Orchomenos in Boeotia disappeared completely, either because it was diffused by the chasm into invisible channels, or because it was previously siphoned out by the marshes and lakes around Haliartos (Strabo 9.2.18).

Springs and Fountains

Springs and fountains provide a rich source of poetic inspiration, especially the famous Hippocrene, created by Pegasus and sacred to the Muses on Mt. Helicon in Boeotia.[57] Some thinkers suggested that the sea could contain springs, a supposition flatly denied by Aristotle on the grounds that springs must belong to artificial (not flowing) bodies of water (Ocean is neither flowing nor artificial, nor is natural spring water ever found on such a large scale: *Meteorology* 1.14.352b29–353b30). Five centuries later, Pliny noted the ubiquity of springs, from mountain ridges to the sea where one could draw fresh water from the Gulf of Gades, for example, and at Brundisium. Pliny synthesized his sources on the universal properties of springs before digressing into vague observations about hydrostatics and specific density. Springs, he suggested, are colder in summer than in winter, and chunks of bronze and lead sinks in springs, but sheets of lead float (Pliny 2.233).

Several remarkable springs were described, but authors did not speculate on the causes of their unusual properties.[58] In Seneca, we read about the cold fountain of Jupiter at Dodona in northwestern Greece, which could light torches brought close to its waters; its waters, incidentally, always stop flowing at noon. In Illyria, another paradoxical spring (perhaps containing naphtha) has the capacity of igniting clothing. Cold, sulfur springs give the appearance of bubbling because the collision of water with deep subterranean fires results in violent currents of air that force water upwards. The Fountain of the Sun is cold during the day but boils hotly at midnight. Some springs are self-healing, ejecting mud, leaves, ostraka, and putrid detritus (Seneca, *NQ* 3.26.6–7). Among these, so it was believed, the Arethusa disgorges impurities regularly, always during the quadrennial Olympic Games.[59]

Paradoxical Waters

Such springs are found in distant lands, far from our intellectual centers at Athens, Alexandria, or Rome. By the late first century BCE (when the Mediterranean Basin had fallen under Roman hegemony), authors began to relay similarly bizarre accounts of strange, paradoxical waters. As in the years following the campaigns of Alexander the Great, knowledge about the world accrued, sparking even greater curiosity about distant places as writers brought the world to their Roman readers. In so doing they emphasized the exotic and the unusual, often without explanation. The cultural milieu of the first century CE was imbued with *paradoxa* and fantasy,[60] and authors in many genres duly cataloged the bizarre properties of waters in distant, exotic lands at the edges of empire, or at least of Roman ethnic identity.

In Vitruvius' rapid survey of strange and wonderful waters (8.3.14–25), many unusual effects are singled out: some waters change the color of cattle (at Troy); the waters of the River Styx (in Arcadia) burst metals; acid springs dissolve kidney stones (as at Lyncestis in Macedonia); waters at Paphlagonia and the territory of the Medulli in the Alps cause inebriation and goiters; waters at Chios induce muteness; those at Susa (the capital of Persia) produce toothlessness; and those at Tarsus and Magnesia on the Meander, both in Turkey, precipitate beautiful singing voices. This "Cook's" tour of unusual waters is replicated in Ovid, who listed the paradoxical spring at Ammon (cold at noon, hot in the evening), insanity-inducing waters in Aithiopia, the petrifying waters of the Ciconian territory, the sobering waters of Clytor, and the Arcadian waters that are safe during the day but deadly at night.[61] In Strabo (5.4.5), we read about the toxic waters of Aornus Gulf (Lago d'Averno, a volcanic crater lake in southern Italy), whose (sulfurous) vapors kill birds that fly over it (sulfur is toxic to insects and rodents, but not to all species of birds). To cross the gulf, so it was assumed, travelers had to "sacrifice in advance and propitiate the infernal deities."

Seneca was also interested in paradoxical waters (NQ 3.25.1–4). He cited the deadly waters in Arcadia and Thessaly. Some springs also affect the coats of the livestock who come to drink from them: at Falerii (southern Etruria), for example, oxen grow white, while Boeotian springs turn sheep black. A spring on Andros, sacred to Liber (Dionysus), tastes remarkably like wine on the god's festival day in January. But other springs are poisonous, likely owing to toxic vegetation growing near the water. At Terracina, the locals had blocked up the mortiferous waters. Other waters can instill fine singing voices, or beauty, or abstemiousness. Vessels are burst apart when they are filled with the icy waters from the Styx at Nonacris in Arcadia. Springs at the mouth of the Timavus (northern Italy) grow and recede with the tides. Some rivers have similar effects: some waters in Thessaly are so noxious that all animals avoid them; some rivers can change the coat-color of sheep as if they had been dyed; but a river in Cappadocia alters only the markings of horses.

In Pliny (31.18–30), paradoxical waters are often diametrically balanced. At Hestiaeotis (Thessaly), the Cerona stream produces black fleeces in sheep; sheep who drink from the Neleus are white.[62] Those who drink from the waters at Sybaris (a Greek colony of southern Italy) are swarthy and curly-haired, but waters from the Crathis at Thurii (also in southern Italy) render imbibers fair- and strait-haired. In Boeotia, one spring causes forgetfulness, another improves the memory.⁶³ The *Lacus Insanus* (Lake "Frenzied") in Trogodytika (on the Egyptian coast of the Erythraean Sea), mentioned by Juba, is full of white serpents, and its waters alternately become bitter and fresh three times every day. There is a lake in India upon which nothing floats and another in Africa upon which everything floats. The spring of Marsyas in Phrygia casts out rocks. Waters at Crannon (Thessaly) keep drinks warm for three days. Mortiferous waters can be found in Armenia, Arcadia, and Thrace. Petrifying waters are attested at Eurymenae (Thessaly), Colossae (Phrygia), and Skyros (in the Sporades islands in the Aegean). Finally, two antithetical streams are confluent in Macedonia: the waters of one are wholesome, in the other flows a deadly poison.

Waters at the edges of the Roman Empire (Asia Minor, the Alps, Libya), thus, behave differently from those at its center. None of these unusual bodies of water is in Latium (Rome), and Vitruvius cited only two that are found at (unspecified) locations within Italy. But southern Italy was culturally Greek (and Etruscan), and thus "exotic" (non-Roman). Mela's *ad hoc* treatment of paradoxical waters also focused on the fringes: Ammon in Egypt (1.39); Scythia, where one river produces delicious water, but waters from a spring near the sea are very bitter (2.6); and Moesia, where a fresh-water spring exists within the brine of the Ister where it debouches into the Euxine (2.63). Pliny also largely restricted his catalog to the amorphous frontier (Arcadia, Mauretania, Phrygia, Thessaly, Thrace).

Floating Islands

A common phenomenon in marshes and other wetlands, floating islands are comprised of aquatic plants and mud, and they range in size from a few inches to several feet. Remaining a source of wonder, the best-known floating island was Delos, the birthplace of Apollo (Vergil, *Aeneid* 3.73–77). Seneca claimed to have seen a floating island at Cytiliae (central Italy), driven back and forth by even a "light puff," never keeping the same position, either because the water—full of medicinal minerals—is especially heavy, or by virtue of the island's texture, although solid enough to support trees, but composed of moveable, pliable, tenuous material:

> perhaps the dense liquid has assembled and bound together light tree trunks and leafy branches that were scattered across the lake.
>
> *NQ* 3.25.8–9

Pliny the Younger (*Epistle* 8.20) expressed great wonder at the floating islands in the sacred waters of Lake Vadimo, near Rome. These numerous tiny islands, sometimes joining together, sometimes floating apart, could even support cattle, which often mistook them for the lake's edge. Pliny's uncle, the eponymous encyclopedist, mentioned Vadimo and its floating islands, although without his nephew's wide-eyed wonder, as well as others, including the Dancing Islands, which seem to move according to the rhythms of chanted choral songs (Pliny 2.209; Seneca, NQ 3.25.8–9). In Mela (2.82), similar floating islands are recounted in the marshes of the Gallic coast (omitted by Pliny). Floating on an unnamed Egyptian lake and ruffled by the wind, Chemnis supported sacred groves and a large temple to Apollo (Mela 1.55).

Conclusion

The symbiotic, reciprocal effects of water and land are observed at both the cosmogonic level (with respect to the earth's life cycle) and the localized level (fluctuating coastlines and accreting deltas), and provide a framework for the landscape. Rivers change their courses, and river-ways become silted, as land and water replace each other over time. Water defines the boundaries of the land, and land affects the depth, salinity, turbidity, taste, salubriousness, and other properties of water. On ancient views, rivers and springs are the binding glue that allows earth to cohere, enabling its fertility. Rivers flow over the land and beneath it, carving out territory, defining culture, revealing geological prehistory where fossilized sea shells are now embedded far from shore, and anticipating future days when the earth will once again be covered with water. The Presocratics and their intellectual heirs sought to develop a comprehensive theory of earth and water, interrogating the synergy of the various pieces. How does Ocean contribute to inland waterways (if at all)? How does rain figure into the scheme of terrestrial waters? How do the fundamental elements of earth and water affect each other within the context of the sensible world inhabited by Greco-Roman thinkers? The solutions, though not always "correct," are thoughtful and imaginative, shaped by the principles of the dominant philosophical "schools" and the stories that define their cultures.

Part Two

Explaining Watery Phenomena

4

Watery Weather

Introduction

Climate and the Hydrological Cycle

Let us begin with a few words about the climate and hydrological cycle of the northern Mediterranean Basin, governed largely by the seasonal shifts of the "Westerly" wind belt, gaining strength in the winter as pressure over the pole decreases, and losing strength in the summer as pressure over the pole increases. The summer months (April to October) in the Greek peninsula average temperatures of 75°F (24°C), moderated by proximity to the Mediterranean Sea, the northwesterly Etesian breezes (conveying cooler air from the north), and the northerly shift of the "Westerlies" as they waft dry air from an overland course. Hot, dry summers are characterized by plentiful sunshine with occasional thunderstorms, precipitated by moisture and rapidly rising warm air. Athens, for example, enjoys abundant sunshine with an average annual rainfall of 14 to 35 inches (35–90 cm), increasing further north in Greece.[1] Mediterranean winters (November to March) are milder, windier, and rainier (40°F/4.4°C), owing to the southerly shift of the "Westerlies," bringing moisture from the Atlantic Ocean eastward over the Mediterranean Basin. Rivers were thus more likely to flood during the winter, with increased moisture transpiring into plants and soil or evaporating into the atmosphere, being returned as rain. Observed generally in the higher elevations during the colder, wetter months, snow was also recorded in Athens, which lay between the hotter southern zone and more moderate north, and Hesiod described frosty conditions, presumably in Boeotia (*Works*, 504–546).

A nuanced discussion of environmental change in the Mediterranean Basin in Classical Antiquity is beyond our scope, and the scholarship is growing. Hardly static, the climate was affected by fluctuating temperatures (colder temperatures were noted for 600–100 BCE; 200–290 CE), oscillating patterns of precipitation (droughts: under Hadrian [ruled 117–138 CE], and 338–377 CE), shifting pressure systems, etc., as teased out by naval archaeology, dendrochronology, palynology, soil studies, and other avenues of examination.[2] Such factors are important in constructing a complete view of Mediterranean antiquity and may have been underpinning determinates of war, expansion, or other human activity.[3] Harper 2017 traces how climactic and environmental stresses contributed to the final chapters of the late Roman Empire. Giraudi et al. 2011 have reconstructed 10,000 years of climate conditions in Central-Southern Italy, utilizing Apennine glacial advances and retreats, together with the δ18O (stable oxygen isotope) composition of carbonate (stalagmite) deposits from caves for estimating precipitation versus evapo-transpiration.[4]

Initiatives and Limitations

Greeks and Romans recognized the perpetual cycle of precipitation and evaporation of the world's water. This cycle has been posited, with varying degrees of currency, to explain tides, annual riverine flooding, and even the very process of change in the world. We turn now to examine how Greek and Roman thinkers perceived and explained water-related atmospheric phenomena. Here we shall focus on a mere slice of what was considered "Meteorology" in antiquity, the study of "things high up," under whose umbrella were also astronomical bodies (comets, meteorites), optical phenomena (sun halos and rainbows), as well as seismic and volcanic activity (both thought to be caused by winds).

The Greeks and Romans lacked the tools to take accurate measurements of temperature and barometric pressure, and gathering data has its own challenges. How far away, for example, are the clouds? Moon-bows, furthermore, are rare.[5] Nor do our sources record efforts at long-term forecasting in the manner of the nightly weather report or the Weather Channel app (in Columella [11.1.30–32] we have an intriguing, solitary reference to long-term weather forecasting). Our sources, nonetheless, are replete with the folk wisdom that reads the signs of impending weather events, often linked to the appearance of the sun, moon, constellations, clouds, and animal behavior. For example, bad weather is portended by dolphins close to shore, birds flying landward, or blooms of jellyfish.[6] Turbidity and color changes in the depths indicate impending storms at Spice Port (*Periplus of the Erythraean Sea* 12).

The ancients were far more receptive than we are to the subtle changes in atmospheric pressure, clouds, and even the winds that shape weather events, like the sailor with the "weather eye" who has honed a sensitivity to data communicated by variable breezes, cloud formations, and ripples on the water.[7] Seasonal weather patterns, moreover, are marked by the stars. The setting of the Pleiades indicates spring; the rising of Sirius, summer; Arcturus rises at dusk in winter. These regular events were well-known and documented in the earliest extant literature.[8]

Aims of Weather Prognostication in Antiquity

Weather prognostication is significant for agriculture, medicine, navigation, warfare, and even the ritual calendar. Hesiod enjoined his indolent brother, Perses, "when the Atlas-born Pleiades rise, begin your reaping, and your ploughing when they set" (*Works* 383–384). For Hesiod, weather prognostication serves a two-fold purpose: as practical advice, and as a means of conveying his didactic message (wealth is acquired only through hard work). For many natural philosophers, meteorological discussions are often vehicles for teaching ethical systems as authors aim to persuade readers that the physical world is governed by orderly, rational, unchanging laws. Phenomena were also explained according to the tenets of natural philosophies. The four-element theory provided the structural principle for Aristotle, Theophrastus, and Seneca.[9] The gods were not viewed as causing rain, thunder, and storms. The world instead behaves in good order according to a rational plan. For the Stoics, the universe is a living creature, permeated by a divine *pneuma*, and sustained by rain, wind, and waterways (Manilius 2.60–104), hearkening back to the earliest theories of the world's creation where moisture nourishes.[10] Weather phenomena thus should not be feared. Capricious gods do not cause bad weather. Physics does.

Sources and Ancient Methodologies

From Homer onward, weather phenomena were observed, described, and explicated both fancifully and rationally by poets, natural philosophers, physicians, astrologers, and farmers. As in other areas of ancient thought, the mythical and the divine were never entirely divorced from meteorological conceptions. Zeus/Jupiter is responsible for thunder and lightning, Poseidon/Neptune causes earthquakes, and weather events indicate divine epiphanies (the messenger goddess, Iris, is a rainbow; St. Elmo's Fire indicates the Dioscuri who protect sailors at sea). Omens also come from weather signs (cf., Chapter 8), and some efforts to control or "bargain with" the weather were strictly magical, including incantations and blood, which could repel clouds according to folk belief. Seneca was skeptical, "how could such a tiny amount of blood contain a force great enough to reach so high and have an effect on the clouds?" (*NQ* 4b6–7).

The Presocratics queried the causes of weather phenomena. But the evidence is spotty, curated first by Aristotle whose watery meteorology was rooted in his theory of Four Causes.[11] Rain does not occur for the purpose of making grain grow but "of necessity;"[12] crop growth is an incidental side-effect of rain, as is the abundance of food for fishes who "thrive best in rainy seasons."[13] The topic was richly treated by Aristotle's peripatetic successors, including his student, Theophrastus. Surviving only in Syriac fragments and two Arabic translations,[14] Theophrastus' *Metarsiology* is characterized by multiple explanations for phenomena, the "hallmark of Theophrastus' meteorology,"[15] supported with empirical evidence and prosaic, terrestrial correlations. For example, seven discrete causes of thunder are listed (or rather "varieties of common phenomena"),[16] distinguished by the noises that they produce: cloud collision (generating a clap),[17] fire (resulting in hisses), and various effects of wind (rumbling noises ensue when the wind twists within a hollow cloud, striking against broad, icy clouds), air pressure (flatulence sounds are emitted when winds split open clouds congested under high pressure), and friction ("rough clouds rub[bing] together" generates a grinding sound) (*Metarsiology* [1] 2–23). Extant in Greek is the peripatetic *On Weather Signs* (written, perhaps, by one of Theophrastus' students),[18] where pride of place is given to celestial signs. Seasonal weather changes are connected with the risings and settings of the major constellations, together with the "circumstances in which they occur" (5–9). Celestial phenomena are indicators of precipitation: asteroids, for example, are harbingers of rain or wind, signifying also from what sector of the sky weather would appear.

For Epicurus, "the exclusion of myth is the sole necessary condition," and phenomena seem to inspire a "multiplicity of explanations," as they had for Theophrastus.[19] Epicurus, furthermore, understood that a variable cocktail shapes the weather:

> Changes of the weather are able to occur both according to the conjuncture of the seasons, just as in the case of animals exhibiting (weather-signs) among us, and also by alterations and changes of air. For both of these explanations are not at odds with phenomena, and it is not possible to ascertain upon which cases the (weather) occurs owing to this or that cause.[20]

In Latin, in the first century BCE, meteorology was treated by the Epicurean poet, Lucretius (book 6), the architectural writer, Vitruvius (8.2.3)—both of whom were highly influenced by Theophrastus' *Metarsiology*—, the Stoic Seneca (*NQ* 4b-6, 2; influenced by Posidonius),[21] and Pliny the Elder.[22] These ancient "farmers' almanacs" share several characteristics, including a reliance on sources that were often foils for the author's own beliefs.[23] Common also were folksy adages ("a north wind rising in the night never sees the third day's light"),[24] autopsy, and, as in other areas of ancient thought, a reliance on analogies with everyday life, rendering explanations explicit and intelligible.

Like his Theophrastean predecessor, Pliny stressed autopsy. Emphasizing the practical value of his work (2.116), he also recognized that local conditions are factors in shifting precipitation at different locations and elevations (e.g., ravines versus mountains). Weather varies according to place, and weather patterns for one region cannot be extrapolated from observations elsewhere.[25] Pliny recognized that Hesiod's date for sowing may work in Boeotia, but it does not necessarily apply to other regions.[26]

Weather and the Stars

We should distinguish between causes (reasons why) and signs (indications) of weather phenomena (Taub 2003: 37). The Presocratics aimed to explain the root causes of phenomena. But what of the compilers of astronomical weather calendars (*parapegmata*) who correlated daily weather phenomena and/or seasonal weather changes with astral signs? Did they understand the links between weather phenomena and celestial signs as concomitant or coincidental?[27] For Pliny, the influence is direct:

> who would doubt that summer and winter and whatever events are perceived with respect to annual change occur because of the motion of the stars?
>
> 2.105

These heavenly bodies include both the planets and the sphere of fixed stars, which in turn are affected by planetary motion. As proof, Pliny linked increasing seasonal temperatures ("heat of the sun") with Sirius' rising. Higher summer temperatures result from the combined forces of Sirius' rising and the seasonal warming of the sun, in contradiction with earlier wisdom. For Geminus of Rhodes (first century BCE), Sirius does not cause the heat of summer, as claimed in folklore, but it rises coincidentally when the summer heat just happens to be the greatest (17.26-30). Concurring with Geminus, Columella at Rome categorically criticized those who believe that the stars determine weather patterns:

But in those debates what was being examined was this, which the Chaldeans (Mesopotamian astrologers) most shamelessly promise, when they assert that changes of the air occur on certain days fixed as if they were boundaries. However, in our agricultural discipline, obsessive compulsiveness of this kind is not demanded, but that which is declared as useful by a "fat-head" (*pingui Minerva*) is appropriate for the farm-overseer with respect to prognostication of future weather, if he holds it as convincing that the force of a star coincides sometimes earlier, sometimes later, sometimes even on the actual day of the rising or setting. For he will be sufficiently prudent who is on guard for anticipated seasonal weather for many days in advance.

11.1.31-32

Accordingly, seasons are not dictated by the human calendar, and the successful farmer must know how to read the weather in order to plant or harvest at the best possible time. Flexibility and sensitivity to the signs from the natural world and its changing rhythms are key.

Red Skies . . .

Even non-sailors are familiar with the ancient aphorism: "red sky at night, sailors' delight; red sky at morning, sailors take warning," a bit of folk wisdom based on the refraction of light against the clouds, assuming that the prevailing winds are from the west, as they generally are in the Mediterranean. Skies appear red in the morning when light from a clear eastern horizon is reflected beneath clouds from the west (weather thus emanates from the west). Red evening skies are caused by the penetration of solar rays through high concentrations of atmospheric dust, usually signifying rising barometric pressure and fair weather.[28] The author of *On Weather Signs* may have been the first to record a version of the adage:

> These signs, therefore, seem to indicate rain: the clearest is thus at dawn when just at dawn before the sun rises the sky appears all red. For it signifies rain either on the same day or within three days more or less. There are other clear signs: if not beforehand, then to be sure within the third day a reddened sky at sunset signifies (rain), but less so than at dawn.[29]

Our author made no attempt to explain the phenomenon, nor did he analyze the causes of rain. Instead, he offered a list of folk signs, many of which have endured. Frogs do indeed croak more loudly, to summon mates, when rain impends, possibly as a breeding instinct: rain provides fresh water in which frogs can lay their eggs (15; cf., Pliny 18.359–360). The Theophrastean material finds expression in the versified account of Aratus who, in a coda to his *Phaenomena* (inspired by Hesiod's *Works*), surveyed the appearance of celestial bodies (moon, sun, stars), weather phenomena (clouds, winds), and the behavior of animals for weather signs. For example, a bright moon indicates fair weather, a broken halo around the moon suggests wind, and rain is presaged when oxen sniff the air and crows caw in imitation of splashing rain.[30] In a similar vein, Vergil, in turn influenced by Aratus, recorded weather prognostication according to animal behavior and the appearance of the sun and moon: a ruddy moon, for example, presages rain and squalls, as do the caws of ravens and the skyward glances of cows.[31] We now turn to survey some ancient theories of precipitation.

"Things High Up"

Clouds

The correlation between cloudy skies and atmospheric conditions is well-known. In Vergil, the return to fair weather after a storm occurs when "delicate fleeces of wool are no longer borne through the sky" (*Georgics* 1.397). Early hypotheses concerning the nature and formation of clouds are in alignment with natural philosophies. The Presocratic, Anaximenes, whose fundamental material principle (*arche*) was air, explained clouds as forming when air congeals. While clouds condense, precipitation develops:

> clouds are formed when air is thickened more, and when it is gathered together still more rain is expressed, snow when the water freezes as it descends, and hail when some gaseous stuff is included with the moisture.
>
> *TEGP* 26

Aristotle accepted the airy nature of clouds because of the reciprocal nature of air and water: each is generated one from the other depending on temperature: air thickens into a denser, moister material as it cools. Aristotle also theorized about cloud behavior (*Meteorology* 1.3.340a19–341a13): clouds "do not form in the upper regions"—although they should—because of the lower temperatures further from the earth. At high altitudes air should condense into water in order to form clouds. The congealment of clouds, according to Aristotle, "happens where the celestial rays already dissipate into the chasm, on account of being separated off by reflection," that is to say, in the zone that does not receive the warmth of the stars or heat reflected from the earth. Calefaction causes clouds to disintegrate. What, then, prevents cold air from condensing into clouds? Arguing that fire cannot exist in that upper-region, Aristotle continued "for all the other things would have been dried out." The solution is twofold: motion (of the atmosphere and celestial spheres) generates heat, and the upper strata are filled not with air but rather with a material that resembles fire. "What we call fire according to custom, but it is not fire." Not explicitly citing *aither*, Aristotle referred cryptically to a "first element" that differs from air and fire, varying in purity, admixtures, and quality, all the more when it is closer to terrestrial airs. Clouds thus do not and cannot occur "in the upper regions."

Theophrastus added a new query to the mix (*Metarsiology* 7.10–27). "Why do clouds float?" Theophrastus recognized that clouds must be denser than air (indeed they are). Because of their watery nature, clouds should be heavier than air, and they should consequently sink. Theophrastus' resolutions are mechanical. Clouds "float" on air, supported by the great quantity of air, relative to the smallness of the suspended clouds, just like wood floats when supported by a sufficient quantity of water ("if there is little water, the wood touches the bottom": this is actually a factor of the difference in specific densities). Additionally, Theophrastus observed that clouds are in constant motion, moving sideways "just as an arrow goes obliquely without falling" (the arrow eventually falls to the ground because of gravity and deceleration owing to wind resistance). Finally, Theophrastus proposed that the continual ascent of vapor from the earth helps support the clouds in the sky. Droplets of moisture, discretely protected within cushions of air, are usually too small to be affected by the force of gravity, and so they "float" like dust particles.

Needless to say, ancient meteorologists recognized different types of clouds, and they were aware of relative distances between clouds and the earth. For Posidonius, reported in Pliny (2.85), the upper limit was 5 Roman miles, with a distance of 250,000 Roman miles (ca. 370,000 km) between cloud-level and the moon, and 625,000 Roman miles (ca. 924,000 km) between the moon and the sun. The author of *On Weather Signs* observed different types of clouds: in particular, dense, nimbus clouds, "like fleeces of wool," which appear at less than 2,000 feet (ca. 610 meters), perhaps inspiring Vergil's "delicate wooly fleeces."

Rainbows

Rainbows are not so much meteorological phenomena as optical, caused by the interplay of light and clouds, as understood in antiquity. For Anaxagoras, rainbows are signs of impending storms: "water dispersed around a cloud produces wind or pours down rain" (*TEGP* 47). Empedocles concurred: "Iris [rainbow] from sea brings wind or great rain" (*TEGP* 102, f60). Anaximenes' explanation of the rainbow is the earliest:

> a rainbow is formed when the rays of the sun fall on thick, dense air. Hence the leading edge of it shines crimson, being burned by the rays of the sun itself, part of it dark, where it is overcome by the moisture. And he [Anaximenes] says rainbows are formed at night, too, by the moon, but not often because the full moon does not appear continuously, and the moon's light is weaker than that of the sun.
>
> *TEGP* 33

Rainbows are thus caused by light filtering through clouds, and the heat of the sun affects the resulting colors of the bows. In Xenophanes, rainbows are clouds "purple and scarlet and green to the view" (*TEGP* 72). Anaxagoras was the first to explain rainbows as an optical phenomenon: "the reflection in clouds back towards the sun" (*TEGP* 47). Aristotle's approach is more rigorous. He had observed double rainbows ("and rainbows do not occur more than two together") and noted that the colors of the rainbow are "primary" (*Meteorology* 3.2.371b26–372a9). Aristotle's geometry of the rainbow is carefully wrought: in essence, rainbows occur when sunlight is reflected at fixed angles from clouds, thus generating an arc (*Meteorology* 3.4.373a33–377a27). These optical wonders can be seen during the day (rainbows) or, rarely, at night under a full moon (moonbows: *Meteorology* 3.2.372a17–28).

Thunder and Lightning

Precipitation can be accompanied by thunder and lightning, weather events that were understood as functions of clouds. Thunder was variously conceived as stemming from motion/collision, wind, water, or friction. The prevailing interpretation is that thunder—a noise emanating from clouds during stormy atmospheric conditions—is caused when winds hit those clouds,[32] a theory that has its roots in the Presocratics. Following his teacher, Anaximander (*TEGP* 30–33), Anaximenes suggested that thunder would occur when winds strike the clouds, as when a red-hot oar dipped in water "is quenched with a loud hiss" (*TEGP* 28). Anaxagoras' explanation is more analytical: thunder is the hiss that results when lightning (fire from the upper atmosphere) is quenched by the cold moisture of the lower atmosphere:

> when the hot air collides with the cold (i.e. the aetherial portion with the airy), by its sound it produces thunder, by its color against the darkness of the cloudy

lightning; by the great quantity of light, a lightning bolt; by having a great multitude of fire particles, a whirlwind; by its mixture with cloud, a fire burst.

TEGP 50–51

Lightning often occurs with thunder (together producing a "thunderbolt," treated separately in the ancient sources). For Xenophanes, lightning is caused by the movement of clouds (*TEGP* 71). In Empedocles, who seems to have understood light as corporeal, we find an interesting theory, that particles of light fall on a cloud and shut in the air, resulting in thunderbolts as the air crashes forcefully out of the cloud (*TEGP* 104). This theory was refined by the atomists: when clouds collide, friction ensues and fire-generating bodies accrete in confined spaces, creating lightning. As these concentrated fiery bodies force their way violently through the clouds, displacing matter, thunder is generated:

> a fire burst occurs when compounds of fire with much void are contained in places and regions with much void, so as to create a body with its own membranes, which by its extreme mixture carries the force to the depths.
>
> *TEGP* 82

It was recognized that lightning is observed before thunder is heard. Empedocles and Anaxagoras posited that lightning, naturally, must occur first, a view staunchly rejected by Aristotle (*Meteorology* 2.9.369b17–18). Noting, however, the delay between the observation of lightning and thunder, Aristotle countered that we see the lightning flash before we hear the thunder that sparked it:

> Lightning, which happens later, falls upon our senses before thunder, since what is heard is outstripped by what is seen. For something seen can be noted from further away, but sound is detectable only when brought close, and especially when the one is the swiftest of all things—fire, I say—but the other is less swift, being airy in nature, and thus hearing occurs on contact.[33]

Aristotle erroneously argued that the thunder occurs first:

> [lightning] happens after the strike and later than the thunder, but it appears to advance on the sense of sight before the sense of hearing. (The reasons) are clear, on analogy with the rowers of triremes. For the first sound of the oars (striking the water) is heard as the oarsmen are already drawing back their oars.
>
> *Meteorology* 2.9.369b7–10

Theophrastus corrected Aristotle's error, suggesting two possibilities: either lightning occurs first (as Epicurus would later argue[34]), or both phenomena occur simultaneously (the latter is the case). Theophrastus, nonetheless, preserved Aristotle's explanation of the time-lapse in our perception of thunder and lightning, hinting at—but not quite grasping—the differences between the speeds of sound and light: "the perception arrives at the eyes quicker than the noise reaches the ears" (*Metarsiology* [5] 7–8). That is to say, the sense of sight is quicker than the sense of hearing.

In the *Meteorology*, Aristotle's explanation of thunder and lightning is grounded within his conception of exhalations and elemental exchange. He synthesized earlier views:

> but whatever of the dry exhalation is trapped in the exchange of air as it cools, the same is expelled from clouds as they condense. Being carried out by force and striking the surrounding clouds, it makes a blow whose racket is called "thunder."
>
> *Meteorology* 2.9.369a26–30

Xenophanes' cloud collision was retained, but rejected were Empedocles' belief of fires pre-existing in clouds, and Anaxagoras' hypothesis of lightning as an optical phenomenon (occurring when ensuing brightness is reflected against the moisture of the clouds). Different sounds and qualities of thunder arise owing to the unevenness of clouds, hollows within clouds, and variant densities.[35] When the ejections are scattered, thunder and lightning follow. Hurricanes occur when dense air is expelled *en bloc* (the celerity of the ejection is particularly savage). If the air is dislodged in a continuous current, then rain ensues.

Posidonius' understanding of thunder was inspired by Aristotle, but his theory is "distinctively individual and personal."[36] On Posidonius' theory (apart from Aristotelian exhalations), whatever is rarefied also becomes dry and hot. As pressure builds up within clouds that contain expanding, attenuated, warming air, the air breaks out with a great noise (thunder). How the air escapes—all at once or gradually—determines the quality of the thunder.

Precipitation

Ever mindful of general patterns, and unconcerned with specific causes, Aristotle recognized various incarnations of precipitation (rain, dew, frost, snow, hail, etc.) as differing only in degree and quantity. He correlated types of precipitation (rain : snow :: dew : frost) based on altitude (upper/lower), temperature (cooler/warmer), and quantity (greater/lesser).[37]

Dew

Although rain, snow, and hail are generated over long periods of time in the upper regions (troposphere), thus accumulating more moisture, dew and frost derive from vapor at the lower ("ground") level:

> but dew is scant with respect to quantity; for its composition is of a day and its area is small as its quick formation and meager magnitude make clear.
> Aristotle, *Meteorology* 1.11.347b21-23

Dew ensues with warming southerly breezes, except, paradoxically, along the banks of the Pontus where the north wind causes it:

> Dew is formed everywhere by southerly breezes, not by northerly winds, except in the Pontus. There the opposite happens, and dew is formed by southerly breezes but not by the northern winds. The reason is the same as that which causes mild weather but not wintry weather. For the south winds produces mild weather; but the north wind yields wintry weather, for it is chilly, so that it quells the warmth of the exhalations But in the Pontus, the south wind does not bring on fair weather such that vapor can occur, but the north wind on account of its chilliness having surrounded the heat coalesces it so that so that it steams more.
> *Meteorology* 347a36-b7

Rain

For the Presocratics, the process of evaporation and condensation accounts for rain: fresh water is separated from the sea and drawn up as mist into clouds, and is then felted into raindrops that finally return to the earth when clouds exhale their winds. For Aristotle, rain is produced according to a hydrological cycle governed by the sun's heat:

> Therefore the efficient, primary and first of the principles is the circle, along which the orbit of the sun—as it approaches and recedes—is obviously the cause of generation and destruction by which condensation and dissipation occur. While earth remains, the steaming moisture around it is carried away by the rays (of the sun) and the other warmth up above. But while the heat lifting it up dissipates and scatters to the upper region, and while it is then quenched on account of being raised up further into the air above the earth, the vapor, which has been cooled on account of the region and the removal of the heat, it then condenses, and it again becomes water from air. Having become so, it is carried back to the earth.
> *Meteorology* 1.9.346b22-33

This cycle depends on the process of elemental change as heat is added or removed: the exhalation of water is vapor, and air condenses into water to form clouds. Aristotle's analogy is poetic:

> It is necessary to consider this just like a river flowing up and down in a circle, composed of air and water. For when the sun is near, the river of vapor flows up, when the sun stands away, the stream of moisture flows down.
>
> *Meteorology* 1.9.347a3–5

Heat, thus, is the efficient cause of precipitation. Aristotle was essentially correct, and his account describes convective rains that occur through the vertical process of evaporation, condensation, and precipitation. He did not, however, consider the effects of topography or air masses on precipitation.

Frost and Snow

For Aristotle, frost occurs when vapor freezes at ground level. Snow is not constituted from rain that freezes as it descends (as Anaximenes believed: *TEGP* 26), but from within clouds as they freeze. Theophrastus concurred:

> snow comes into existence when coldness freezes the clouds before they turn into water and before one part of the water is connected with the other, as long as (the water) is scattered and dispersed in the form of very small drops which are separated by air. For we can see with our own eyes that in snow much air is contained. A proof of this is its softness; when it is compressed by the hand, it becomes little in quantity and when it melts and becomes water, from its large quantity only a small quantity of water comes out.[38]

The admixture of air accounts for the whiteness of both snow and frost (*Metarsiology* [9] 11, [12] 20): a theory that originates with Aristotle: "for snow is foam" (*GA* 2.2.735b21). But, what accounts for the flakes? Do they form before clouds congeal into liquid or after the process has already begun? Aristotle argued for the former: as condensed (icy) clouds break up, snow is produced, and the fracturing that occurs before liquefaction results in snow's "resemblance to foam" and its "intense whiteness" (*Cosmos* 4.394a33–35).

Hail

Icy hail, which paradoxically falls during the warmer months, is altogether another matter, provoking curiosity and demanding explanation. Inspired, perhaps, by Anaximenes' theory that hail is generated "when some gaseous stuff is included with the moisture [snow],"[39] Anaximander explained hail according to the reciprocal effects of cold and hot: when clouds are pushed up to a higher altitude (where rays of the sun—reflected from the earth—do not reach), water is frozen. This often happens during the hot season, thus explaining why hail is more common in hot climates during the torrid, summer months, and why summer rain showers appear "heavier" and "more violent" than winter rains (*TEGP* 52).

Aristotle contentiously disagreed.[40] For Aristotle, hail must be interpreted in light of the qualities that determine the formal differences between rain and other modes of precipitation. Hail is unusual for both its rarity and size, and its efficient cause must therefore also be unusual. Although Anaxagoras had utilized a sound methodology based on scientific principles (vapor, altitude, temperature, season), Aristotle's "main objection to Anaxagoras is that he had attempted to explain the paradox of ice in the warm seasons merely by applying the ordinary principles of condensation in a heightened degree."[41] On the strength of empirical evidence, Aristotle contended that hail does not occur at high altitudes (like snow) but only at low altitudes. Hail is generated only from low-lying clouds as proved by the terrifyingly loud noises emanating from them. Furthermore, the angular and rough shape of large hailstones, so Aristotle observed, is proof that the stones originate close to the earth. Had they fallen from a higher altitude, they would have become spherical and smooth (cf., Seneca, *NQ* 4b.3).

Arguing to the contrary that hailstones are spherical, Theophrastus offered three incompatible explanations (reliant on conflicting theories of cause and formation):

1) they become smooth (or weathered?) by the friction of their descent (they are already ice before falling as hail);
2) the very spherical nature of water is transferred to the watery hailstones (they begin as spherical water droplets which then freeze); or
3) or cold hardens and compacts the ice "from all sides in an equal manner".

Metarsiology [10] 2

Theophrastus here is forcing the meteorological phenomenon to fit his theory. Hailstones, however, are not uniformly spherical, and even small hailstones can have jagged protrusions as lumps of ice fuse together into irregular shapes.

So, then, why does hail occur? Aristotle posited a paradoxical solution to the counter-intuitive problem: hail does not originate when clouds ascend into colder air, as Anaxagoras had suggested, but rather when they descend or sink into warmer air: "it is not possible for hail to freeze before becoming water; nor can water remain in mid-air for any period of time" (*Meteorology* 1.12.348a 6–8). Hail is generated neither in the upper region (like rain and snow) nor at the lower (like dew and frost), but midway. Nor does hail occur during the winter (like snow and frost) or summer (like rain and dew), but in between. "Hailstorms occur generally in spring and autumn and less often in the late summer." To explain paradoxical summer rainstorms in the deserts of Arabia and Aithiopia, Aristotle advocated a hot analog to icy hail:

> around Arabia and Aithiopia rain occurs in the summer and not in the winter, and these (rainstorms) are torrential, and often occur on the same day, for this reason. For cooling occurs quickly in the clouds because of the recoil against the excessively (hot) territory.

Aristotle proposed that hot water freezes more quickly than cold (the so-called Mpemba effect).

We are handicapped by the fragmentary state both of Posidonius' corpus and Seneca's *NQ* 4b ("On Clouds"). Posidonius' explanation was later distilled with approval by his fellow Stoic, Seneca: hail is not produced in the same way as ice "at our level," but from "a cloud that is already watery and turning to liquid."[42] This original theory is preserved nowhere else. Posidonius, nonetheless, seems to have inverted Aristotle's theory. On Posidonius' theory, ice crystals melt partially before falling as hail.[43]

Seneca, however, perpetuated Aristotle's analogy:

> hail is nothing other than ice in suspension; snow is suspended frost. For we have already said the difference between dew and water is the same as the difference between frost and ice, and also as that between snow and hail.
>
> *NQ* 4.3.6

The passage to which Seneca referred has not survived. Seneca also maintained Aristotle's seasonal error, that hail falls only in the spring: Why does it not hail in the winter? Because the cold air condenses not into water but into snow, which is closer to air. As temperatures warm in the spring, droplets of precipitation become larger, and spring-time rains are more violent because "there is a more vigorous transformation of the air" (*NQ* 4b.4). Theophrastus had raised the same question regarding thunder, perpetuating another seasonal error, and offering the same resolution: why does it thunder most frequently in the spring (instead of during the winter or summer)? Thunderbolts require both wind and fire. Because of excessive seasonal coldness, there is insufficient fire in the winter. Owing to excessive heat, summer wind is meager. Spring, however, is a temperate and moderate season characterized by "enough fire, clouds, and winds," the necessary materials and catalysts for thunderbolts (*Metarsiology* [6] 68–74).

Anaxagoras was a little closer to the truth. He believed that hailstorms occur when thunderstorms experience strong updrafts within unstable air masses, wherein the temperature plunges drastically, occurring usually in both spring and summer (especially at higher elevations).

Devastating Weather Events

Earthquakes

Water is also a component in ancient explanations of earthquakes. According to the mythological tradition, Poseidon/Neptune—the Earth-shaker[44]—the god who holds sway over the untamed sea, not only causes storms at sea.[45] He also generates earthquakes as he angrily strikes the earth with his trident (a common fishing tool). Poseidon's violent anger is translated into a rational exegesis by Thales who, we presume, suggested that earthquakes result from disturbances in the waters upon which the earth rests, a theory which Seneca would much later dismiss as foolish (*inepta*), antiquated (*veterem*), and undeveloped (*rudem*), since the earth does not rest on water—so Seneca believed—and earthquakes generally affect one part of the earth, but not its entirety, as would have to occur in Thales' theory.[46] Anaximenes credited the desiccation of the earth as the phenomenon's cause (*TEGP* 34). Quakes occur where the earth loses all its moisture. On Democritus' theory, quakes occur when the earth's hollows become too full (from rain), and water is displaced from saturated places into voids, an effect occasionally exacerbated by wind (*TEGP* 78-79), a hypothesis that aligns elegantly with his atomic theory. Earthquakes are caused sometimes by water, sometimes by wind, sometimes by both.

Aristotle dismissed water as the cause of earthquakes, preferring instead winds, a product of both moisture and heat, as the sun warms the rain that fills the earth with moisture. According to Aristotle, earthquakes occur when dry exhalations are trapped within the moist earth (*Meteorology* 2.8.365b30-366a5).[47] Consequently, earthquakes tend to occur in calm weather, when available wind is depleted. They are most violent, so he thought, where the earth is hollow, and most frequent during the wet spring and dry autumn as well as during rains and droughts, when the greatest quantity of exhalations occur (*Meteorology* 366a23-14; the historical record does not bear this out: earthquakes occur at every season of the year in the Mediterranean). Aristotle also posited that earthquakes accompanied by winds tend to be less violent, since the motive causes (dry and wet exhalations) are more evenly distributed.[48]

Centuries later, Seneca, who had written a book on earthquakes in his youth (now lost), included a fresh treatment in *Natural Questions*, debunking the hypothesis that earthquakes were caused by water and adhering to Aristotle's theory of exhalations. Utilizing the quake that struck Pompeii in 62/63 CE as a case study for the benefits of philosophy (i.e., in deterring fear and anxiety), Seneca correlated the human body with the earth, whose exhalations nourish the universe. It is air, contained within underground spaces, that helps keep the earth and the *cosmos* healthy. Seneca noted that earthquakes do not occur at the surface of the earth, but instead they arise from deep within the land in reaction to obstructions to the normal flow of underground air (comparable to a flowing river).[49] To the Stoic Seneca, the underlying principle of earthquakes is breath (*pneuma*).

Pliny's treatment (2.191–195) is less philosophical but more descriptive. For Pliny, there are several varieties and causes. In some places, walls come crashing down, in others, rock formations are thrust up. Pliny also considered the various noises that accompany earthquakes (rumbling, lowing, human-like shouts, clash of weapons). He considered, furthermore, the effects of quakes on the earth's water supply: rivers emerge during some earthquakes, or fires, or even hot springs; and occasionally earthquakes cause the diversion of rivers.

Some Greek thinkers, furthermore, had supposedly predicted earthquakes: Anaximander had warned the Spartans of an earthquake that did, in fact, in the sixth century BCE, devastate their city, plucking off the top of Mt. Taygetus like a ship's stern (Cicero, *Divination* 1.50.112; Pliny 2.191). Pythagoras' teacher, Pherecydes, presumably also foretold an earthquake, receiving the premonition as he was drawing water from a well.[50]

Tsunamis

Strabo suggested that earthquakes and volcanoes could cause major floods and tsunamis (occurring when earthquakes shift the seabed, displacing several hundred kilometers of water), and he himself had witnessed the aftermath of a tsunami on Egypt's eastern coast (the precise year cannot be pinpointed):

> when I was living in Alexandria next to Egypt, the sea around Pelousion and Mt. Kasion rose and inundated the land, making an island of the mountain, so that the road from Kasion into Phoenicia became navigable.
> 1.3.17

The earliest description appears in Homer, when the Scamander and Simois Rivers angrily emerge from their beds as powerful, destructive, cresting waves in their pursuit of Achilles who was glutting their waters with the corpses of Trojan heroes:

> Nor did the Scamander check his force, but all the more he raged against the son of Peleus, and lifting himself loftily he turned the wave of his stream into a crest, and shouting he called to Simois "my dear brother, let us both hold back the strength of a man, since he will utterly destroy the great city of Lord Priam, nor will the Trojans stand fast against the battle din. But very quickly come to my aid and fill your stream to the full from springs of water, and urge on all the streams, and make your vast waves stand, and rouse a great roaring of tree trunks and boulders so that we can stop this savage man who is indeed now powerful and intends something equal to the gods."
>
> *Iliad* 21.305–315

Several historical tsunamis are attested. Seismic sea waves are not uncommon in the Mediterranean owing to a major fracture zone that runs between the Peloponnese and Crete, through the Aegean, into Turkey, the Black Sea, and the Crimea.[51] This fracture zone, in part, helps to explain Plato's account of Atlantis, the wandering rocks in Homer (*Odyssey* 12.59–72), and Jason's Symplegades, vividly depicted by Apollonius:

> And then the seething brine arose like a cloud, and the sea roared terribly, and all around the vast aither howled. The hollow caverns beneath the jagged rocks— over which the sea dashes—boomed while the salt-water surged within. The white froth of the boiling wave spurted high above. Then the current jostled the ship all around.... Again the rocks were opening up, and a tremor held the rowers as they urged on, until the returning rise of the sea carried them back between the rocks. Then a very grim dread seized them all: for above their heads was destruction. Already, on this side and that, the broad Pontus was visible and without warning a great wave crested in front of them, arched like a sheer cliff.
>
> *Argonautica* 2.565-571, 574-581

The waves were unexpected, and the weather was otherwise fair. Apollonius' detailed description indicates, at the very least, a superficial familiarity with the effects of tsunamis, if not their causes. Generated by earthquakes, tsunamis are characterized by waves with very large crests (and answering troughs) that can travel long distances at great speed.[52] They are not necessarily accompanied by bad weather.[53]

In Herodotus (8.129), we find the earliest historical record of the phenomenon. In 479 BCE, Persian invaders intended to take advantage of an unusually shallow sea caused by an extreme ebb tide to capture Potidea. Mid-transit, the waters refluxed with a huge flood tide that destroyed the Persian forces. According to the locals, such flood tides were frequent, but they had never witnessed one "quite so high as this one." The event was caused by an earthquake in the Aegean, a detail which Herodotus was unable to infer.[54]

Thucydides, however, was able to connect the phenomenon with seismic activity. He described the tsunami of 426 BCE in the Malian Gulf:

> The following summer the Peloponnesians and their allies came as far as the Isthmus in order to invade Attica under the command of Agis, son of Archidamus, king of the Lacedaemonians; but because of a number of earthquakes, they turned back again, and the invasion did not occur. At about the same time as these earthquakes were rife, the sea at Orobiae in Euboea withdrew from the coastline, and then returned and inundated some portion of the city, then it withdrew again, and yet again it is now sea what was once land; and they perished, those inhabitants who were unable to reach higher ground as they ran back.
>
> 3.89.1–2

The tremors and flooding affected other coastal settlements as well. Thucydides correctly inferred the causal connection between tsunamis and earthquakes:

> The cause of this, I think, is insofar as a very strong earthquake happened: that accordingly the earthquake withdrew the sea and suddenly recoiling it rather brutally created a flood. Without the earthquake, it does not seem possible to me that so great a natural disaster could occur.
>
> 3.89.5

Neither Thucydides nor Herodotus noted additional weather anomalies. The weather, we assume, was otherwise clear.

In 373 BCE, a catastrophic tsunami swept through the Peloponnese. Aristotle implied that the tsunami and its attendant earthquake were coincidental: "when a tidal wave happens with an earthquake, the reason is because contrary winds are occurring" as at Achaea (western Peloponnese).[55] Diodorus Siculus (15.48) and Strabo (8.7.2) both attributed the tsunami to an earthquake. The event happened in two stages. An intense night quake, causing devastation, chaos, and fear, was followed by a morning tidal wave. The fact that the quake occurred in the middle of the night rendered the situation all the more terrifying because the cover of darkness made it impossible to discover avenues of escape from the toppled houses. Those who survived the quake were destroyed by the tidal wave the following morning:

> For while the sea soared up to a great height and a high-cresting wave was lifted up, all the inhabitants together with their lands were flooded and obliterated.
>
> Diodorus Siculus 15.48.3

About 500 years later, Pausanias would visit the ruins of Helike, which were "corroded by salt water" (7.24.5–6), and Aelian reported an exodus of wildlife (mice, weasels, serpents, centipedes, beetles, and the rest of the creatures) five days before the tsunami occurred, leaving "in a mass on the road leading to Ceryneia" (Aelian, *NA* 11.19).[56]

Ca. 400 BCE at Pithekoussai, an island west of Naples, the volcano at Epomeus Hill erupted, shaking the island, enflaming the land, and causing the sea to withdraw for three stades (ca. 1,800 feet). The sea quickly refluxed and quenched the fire (Strabo 5.4.9). In 126 BCE, Posidonius witnessed a tidal wave that was caused by the eruption of a volcano between Vulcano (Hiera) and Panarea (Euonymus), north of Sicily (f227). The sea was raised to an extraordinary height and sustained by an upward blast. Eventually it crashed back down, leaving in its wake dead fish, terrible heat, and an unbearable stench (likely sulfur from an erupting volcano). Days later, mud surfaced on the sea, and flames continued to flicker.

Of uncertain date, though perhaps also during Posidonius' lifetime, a major seismic event that originated in the Levant is traceable all the way to northern Euboea (f232Kidd; Strabo 1.3.16). Strabo speculated that movements of the seabed may have caused the tidal wave that destroyed Tryphon's army on the Syrian coast between Ptolemais and Tyre in 144/3 BCE (16.2.26). Posidonius' thoughts on the tsunami's cause have not come down to us (f226Kidd).

Storms and Superstorms

Water could be dangerous, and massive storms pelted both the literary and historical landscapes of the ancient Greco-Roman world. Mythology preserves the memory of a destructive flood, sent by Zeus to punish the *hubris* of mankind, and Seneca vividly imagined the course and effects of such a colossal storm.[57] Cataclysmic events found expression in Presocratic cosmogony. Plato also commented on the effects of cataclysmic flooding:

> Therefore, since many immense floods have occurred in 9,000 years—for such were the years from that time to our own—in all that time and in all those occurrences, the soil washed down from higher ground does not pile up any amount of sediment worthy of mention, as in other regions, but—flowing around in a circle—it always disappears into the deep. Indeed, just as on small

islands, what is left now with regard to what existed at one time is like the bones of a sick body, the soft and rich parts of the earth have flowed away, there remains only the stripped body of the land.

Critias 111a–b

Storm scenes are common in literature, so we shall be selective. In his wanderings, Odysseus survived at least two superstorms at sea: as the crew set out from Helios' island, Thrinakia, the fleet was annihilated (*Odyssey* 12.404–425). The hero himself, later bound for the land of the Phaeacians, was nearly destroyed by Poseidon (5.283–296). Following closely Homer's description in *Odyssey* 5, the storm that scattered Aeneas' fleet off the coast of Carthage is harrowing, histrionic, and technical:

> and the winds, as if marshalled in a battle line, rush out where openings are given and blow over the lands with a whirlpool. They press upon the sea, and the East and South winds together cast up the whole sea from its deepest foundations, and the Southwest wind is thick with storms. The winds roll huge waves towards the shores. There follows the shouts of men and also the creaks of the rigging. Suddenly clouds snatch away both the sky and day from the eyes of the Teucrian refugees. Black night lies over the sea. The poles thunder and aither flashes with thick fires, and all things threaten the men with imminent death.
>
> Vergil, *Aeneid* 1.82–91

Squalls were agitated by the combined efforts of the spiteful East, South, and Southwest Winds; huge waves were churned up from the depths; the sky grew black; and a sheer "mountain of water" towered over the ships, suspending them in the crests and valleys. The North wind twisted ships towards rocky shoals. Swimmers struggled in the storm's whirlpool, and the Trojan treasures that the exiles had so carefully stowed in their ships were scattered over the waves (*Aeneid* 1.81–123).

Seneca's nephew, Lucan (39–65 CE), sketched another picture of a stunning gale that is so vicious that the crew were compelled to abandon ship:

> Just as when the stormy South Wind routs the immense sea from Libyan Syrtes, and the shattered bulks of the mast and sail-bearing yard-arm boom, the skipper leaps down into the waves—the ship's stern deserted—and also each seaman. Since the framework of the hull is not yet scattered, each makes a shipwreck for himself.
>
> 1.498–503

Although poetic descriptions are exaggerated, the poets' broad brush-strokes were shaped by the excessively fierce rains, storm surges, and destructive effects of real storms. They were inspired by historical events, storms so severe that they affected the course of history.

Historical Storms at Sea

Sudden storms are a perennial threat for navigators and those who travel on water. In his epitome of Roman military science, Vegetius recorded the signs of impending storms: a ruddy moon, for example, foretells rain and squalls.[58] Quick-thinking skippers could avoid storms or save their ships. But not all were so skillful or so lucky. The Persian fleet was devastated by a sudden spring storm off Magnesia in 480 BCE. Since the beach was too small to accommodate the fleet, the ships rode at anchor, eight rows deep, their prows pointing out to sea. The sea suddenly began to churn under a clear, windless sky, and a turbulent east-wind ("Hellespontian") dashed some of the vessels against the "Ovens of Pelion," others against the stretch of beach (Herodotus 7.188). In the same year, off Mt. Pelion where thunder was booming, a savage thunderstorm caused heavy losses to the Persian fleet shortly before the battle of Thermopylae (Herodotus 8.12–14).

During the first war between Rome and Carthage, early in the summer of 255 BCE, the Roman generals, Marcus Aemilius Paullus and Servius Fulvius Paetinus Nobilior, had been warned by local pilots not to sail along the outer shorelines of Sicily because of the ongoing shifts in violent weather off the coast ("between the rising of Orion and Sirius"). They unwisely disregarded the warning and lost 284 ships (Polybius 1.37.4–5). The Roman fleet heading to Africa in the summer of 203 BCE was ravaged by a storm: the fleet had nearly made landfall when the winds failed, shifting to the southwest, damaging and scattering the fleet (Livy 30.24.6–7). Caesar complained about the difficulty of sailing in the stormy Atlantic, but he admired the ships of the Veneti (which he described in detail), well-suited to treacherous sea conditions with their flat bottoms and high prows (*BG* 3.12–13). Caesar also recounted the sudden summer storms that damaged his fleet in Britain in 55 and 54 BCE, delaying landfall (*BG* 4.29, 5.5, 10).

In the summer of 36 BCE, Octavian's fleet suffered damage because of unseasonable storms.[59] But in August of 31 BCE, the battle at Actium was delayed by a severe storm caused by a depression over the Ionian Sea and Gulf of Ambrakia. Anchored in Gomaros Bay, Agrippa's fleet fared better than Antony's in the choppier straits of the Gulf. The weather broke after five days, and battle finally ensued on September 2 under gentle breezes on calm seas. Agrippa was further advantaged by the *Iapyx* winds (as Vergil called them), blowing from the west during the late morning, building strength over the next several hours, driving Agrippa's fleet forward and Antony's back (Antony had to rely on his rowers).[60]

Caught in powerful southerly winds on the Amisia River (Ems) in the middle of the summer campaign season of 16 CE, Germanicus' fleet also suffered through hail and fierce squalls from "all directions" (*ubique*). Steering and baling were impossible. The panicked soldiers were "ignorant of the calamities of the sea"; and animals, packs, and weapons were jettisoned to lighten the hull (Tacitus, *Annals* 2.23). The storm was recounted by the early imperial poet, Albinovanus Pedo, in a hexameter extract preserved in Seneca the Elder (*Suasoria* 1.15). With language evoking the epic storm suffered by Aeneas' fleet off Carthage, Pedo emphasized that Germanicus' fleet has traversed into unknown and unpredictable waters replete with sea monsters who prey on the tempest-tossed ships and where different laws of physics seem to prevail. In Pedo, Germanicus' helmsman asked if the storm was a conduit to another, forbidden world, suggesting that the gale has been sent by the gods to punish the *hubris* of Roman incursion into the divine realm at the edge of the world. Two years later, in 18 CE, Germanicus endured another stormy voyage, this one through the Adriatic and Ionian seas. While the fleet was being repaired, Germanicus took the opportunity to visit Actium before resuming course to Syria. North winds, incidentally, prevented him from visiting Samothrace (Tacitus, *Annals* 2.53–54).

In Moesia Superior (101 CE), the emperor, Trajan, commemorated his hydraulic initiatives—a canal at Sip—to render the Danube's cataracts there less perilous (*ob periculum cataractarum*).[61] Arrian, Hadrian's governor of Cappadocia (132–138 CE), described the choppy conditions of the Black Sea where waves washed over the sides of his ships (*Periplus* 3.3–4, 6.1). Sailing from Alexandria along the coast to Cyrene in 407 CE, finally, one experienced skipper sought the open sea as a precaution against a looming storm (as skilled sailors continue to do), but his passengers urged him to turn the ship back to the coast (Synesius, *Letter* 5 Martin).

Floods

Aside from cataclysmic natural disasters (like the eruption of Vesuvius in 79 CE, which resulted in a "modest" tsunami caused by a volcanic earthquake), storms on land were usually recorded only because of their effects on military expeditions. Marching through Armenia in deep snow in 399 BCE, Xenophon's mercenaries, who had supported Cyrus in a Persian civil war, suffered frostbite from the blizzard-like conditions they endured during their homeward trek

(*Anabasis* 4.5). Hannibal's troops struggled against cold, twisting wind and blinding rain as they crossed the Alps in 217 BCE (Livy 21.25). Additionally, Theophanes of Miletus, who participated in Pompey's Armenian expedition (66 BCE), described the winter in the high Caucasus where the summits were impassable. In the summertime, locals would ascend the summits with the aid of spiked, ox-hide snow-shoes, and then sled back down on animal skins (Strabo 11.5.6). Authors tended to focus on the unusual, and such icy conditions, though perfectly normal for Armenia and the Caucasus, were in stark contrast with the wintry conditions at our cultural epicenters: Athens, Alexandria, and Rome.

At Rome, moreover, the Tiber was prone to flooding, especially after winter rainstorms, causing damage to bridges, fields, houses, and other structures and, occasionally, rendering the streets navigable by boats.[62] Attempts to control flooding were a matter of municipal concern, including efforts to raise the ground level (1–4 meters = 3.3–13 feet) in the sixth century BCE,[63] and the periodic dredging and canalizing of the river.[64] Caesar's plans for a relief channel near the Milvian Bridge were aborted by his assassination (Cicero, *To Atticus* #330 [13.33.4]). Claudius' relief canals at Portus, explicitly intended to ameliorate the Tiber's flooding (*CIL* 14.85 [*ILS* 207]), were not altogether successful, and additional canals were cut between the Tiber and the sea under Trajan (*ILS* 5797).

In 60 BCE, a powerful storm shattered houses, sank ships that were moored in the Tiber, uprooted trees, and caused the deaths of "great numbers of people" (Dio 37.58.2–4). Six years later (54 BCE), Rome suffered another catastrophic storm with the onset of the wet, winter season:

> Meanwhile, either because excessive rains had occurred somewhere upstream above the city, or because a powerful sea-wind had driven back its outgoing tide, or still more likely, as assumed, by the act of some god, the Tiber suddenly rose so high as to inundate all the lower levels of the city and to overwhelm even many of the more elevated regions. The houses, therefore, constructed of brick, were soaked through and they collapsed, while all the animals perished in the flood. Of those who did not find refuge in time, all of them were caught either in their dwellings or in the streets, and they lost their lives. The remaining houses furthermore became weakened since the disaster lasted for several days, causing injuries to many, either then or later.
>
> Dio 39.61.1–2

Writing in late October from Tusculum in the Alban Hills east of Rome, Cicero described the same storm for his brother Quintus, recounting which famous monuments suffered from storm damage:

You would be amazed at the storms at Rome, and particularly along the Appian Way up to the Temple of Mars. The promenade of Crassipes[65] was carried away, as were gardens, and many shops. A great force of water came all the way up to the public fishpond.

To Quintus #25.8 [3.5]

In 5 CE, excessive flooding destroyed a bridge and left the city streets under water (and, consequently, navigable) for seven days (Dio 55.22.3). Neither Dio nor Cicero mentioned if any measures were taken for flood relief.[66] Pliny the Elder did, however, note the correlation between deforestation and worsening floods.[67]

Conclusion

Ancient Greek and Roman thinkers sought to understand the causes of precipitation and the differences between various modalities of weather. Thinkers often did violence to the evidence in order to make their interpretations of watery weather fit philosophical theories, as shaped by their *perception* of the visible and invisible forces from which various ancient theories derived. The exercise, however, was not merely abstract. Epic poets, historians, and other writers recorded imaginary and real weather events that shaped the course of events, meting out victory to those who understood how to work within their environments and disaster to those who disregarded the signs from nature. The "weather-wise" learned to interpret clues from water, clouds, air, and animals, a powerful form of divination that could guide farmers, sailors, merchants, and many others in navigating the rhythms of the natural world in the Mediterranean.

5

Water, Health, and Disease

Introduction

Water is a complex component that both nourishes and destroys, feeding crops with gentle rains or destroying communities with the onslaught of tsunamis. Water's ambivalent nature was also recognized as a factor of human health. Water was a *pharmakon*, both supporting and endangering health. Some waters fostered noxious creatures that caused disease. Yet water could also be therapeutic. Here we shall explore Greco-Roman conceptions of water-related illnesses and the healing benefits of thermal springs, understood as shaped by visible and invisible factors.

Water-related Diseases

In 232 CE, Alexander Severus' Illyrian troops fell ill because of the "stifling air" near the Euphrates in northern Syria, but they were revived at Antioch where the air was cooler and the water was "better," resembling the nippier, moister environment of their homeland (Herodian 6.6.2). Like other ancient authors, Herodian of Antioch (170–240 CE) lacked a clear understanding of pathogens and pollutants, and he conceived of disease as occurring because of interactions with the environment.[1] Infectious disease was thus usually attributed to poor food or unusual airs and waters. Nor was sanitation fully comprehended. In antiquity, private facilities frequently lacked running water, and toilet paper was unknown. The ancients instead may have employed community sponges in public latrines,[2] an assumption challenged without citation by Rogers (2018: 45–6). Ancient latrines, consequently, were petri dishes with ample opportunity for dispersing intestinal pathogens.

Any number of infectious diseases are well-documented, especially in crowded military camps where highly contagious eye complaints, in particular, were all too common because of the smoke produced by oil lamps.³ Before the battle of Syracuse in 212 BCE, Carthaginian and Roman troops fell ill at Acragas, and many soldiers died or suffered vague "dislocations" or terrible pains (Diodorus Siculus 13.86.2). The account lacks clinical details. Additionally, in 190 BCE, the Roman army was compelled to weigh anchor, abandoning a Lycian campaign, because so many galley slaves had fallen ill after enduring the unhealthy midsummer season and an unusual stench that gave rise to widespread ailment.⁴ Again, clinical details are omitted. Such conditions, especially for the galley slaves, are "an ideal breeding ground for disease,"⁵ despite concerns to establish camps near adequate water supplies and to provide sufficient latrines.

Some ancient thinkers speculated on the link between disease and waterborne pathogens, generating a nascent theory of germs and microbes. Among them, Lucretius acknowledged that "foreign" (*alienum*) pathogens can infect the water, crops, food supply, or air (6.1119–1130).⁶ Varro referred to the tiny (*minuta*), "invisible" creatures that the "eyes are not able to follow," but which enter the human body through the mouth or nose, causing serious disease (*difficilis morbos*: *Farming* 1.12.2). These creatures and their effects, however, were little understood, and Greco-Roman physicians lacked the tools to investigate such pathogens and how they might spread. Nonetheless, initiatives in bio-archaeology are currently delving into the microbial world in antiquity with attempts at differentiating pathogens from the many forms of mycobacteria that live in soil and water.⁷

Greco-Roman physicians, however, understood that something in "bad" water could make an individual or a population sick, as, for example, poor drinking water from the dirty Tiber River, which Tacitus blamed for an epidemic outbreak in the praetorian camp in 65 CE, together with overcrowding, the insalubrity of the boggy Vatican district, and the general bad health of the soldiers, owing to their inactivity and debauchery (*Histories* 2.93).⁸ Galen, likewise, attributed disregard for basic hygiene, often resulting in "bad waters," as the cause of epidemics in military encampments (*Nature of Man* 119 Kühn). In order to avoid the outbreak of "most destructive disease(s)" (such as ancient "cholera" or "typhoid") from polluted water supplies, Vegetius advocated frequent changes of camp, which would thus move troops from over-stressed water sources, especially during the hot, summer months: one commander had irresponsibly refused to do so until forced by the stench from the latrines (Sallust, *Jugurtha* 44). Withholding details, Vegetius asserted that "a drink of bad water, like poison, generates disease in the drinkers" (3.2). Contaminated water may also have accounted for an elevated mortality rate among recently weaned children.⁹

Before we turn to our survey of water-borne ailments, we must note the ambiguity of ancient medical terminology. It is often impossible to correlate ancient descriptions of ailments to modern diseases, and much ink has been spilt, for example, in trying to determine the pathogens that caused the Athenian plague of 430–427 BCE. The terminology for disease in antiquity is obscure, and ailments are difficult to reconstruct. What is described as "cholera" is not Asiatic cholera but more likely food poisoning (*nostra cholera*), "a common defect for the stomach and intestines"[10] and perhaps impossible to disentangle from water-borne diseases. According to the Hippocratic author of *Affections* (47), "cholera" is caused by rich, fatty foods containing cheese, honey, and sesame, and its symptoms include vomiting and diarrhea, classic indications of food poisoning.[11]

As settled agriculture increased, nonetheless, fields fertilized by pathogen-rich manures were in turn irrigated by neighboring bodies of water. Run-off from irrigation would be returned to those same water-courses, and thus the water supply could easily become tainted.[12] Moreover, during the summer and early autumn, water resources dwindled with increasing heat and diminishing rainfall, and intestinal diseases were circulated by the pathogens in dirty waters. Represented in the bio-archaeological record is amebic dysentery, proliferated by the parasite *Entamoeba hystolytica*, common still in Greece and dispersed through polluted water.[13]

"Typhoid"

Caused by *Salmonella typhi* in contaminated food or water, "typhoid" enters the blood stream through the intestines, multiplying within leukocytes and incubating for about two weeks. The pathogen is resistant to the acidic environment of the stomach in which other micro-organisms usually perish. Since the pathogen hides within the immune system, it can be particularly trenchant and recurring. Once removed by the liver from the bloodstream, the infection is deposited in the bile, thus entering the intestinal tract for excretion. Instead, the pathogen returns to the bloodstream causing re-infection. "Typhoid" was likely escalated by unhygienic conditions, including sewage that corrupted drinking water, public latrines, and spoiled food from street vendors.[14] "Typhoid" is indicated by weakness, abdominal pain, constipation, headaches, delirium, and occasionally rosy, splotchy skin-rashes. The ailment (salmonellosis) was outlined in an elegantly descriptive passage in the Hippocratic *Epidemics* (2.7.11). The patient, the wife of Hermoptolemos, suffered from chills and hot flushes, full body soreness, thirst, and delirium. Her stool was yellow-ochre, a color "characteristic of typhoid and still used as a diagnostic sign nowadays."[15] The five varieties of "typhoid" in the Hippocratic *Internal Affections* 39–43 do not align with typhoid fever as understood by modern medicine, but nonetheless may have been shaped by the observation of actual cases.[16]

The usual treatment for typhoid fever would have been the application of cold packs (still used when the Prince of Wales, the future King Edward VII, became ill with typhoid fever in 1871). Some have speculated that the symptoms that Alexander suffered in 323 BCE align with "typhoid," including an incubation period of nine days from the onset of the fever until death. In June, when he died, Alexander was in Babylon, a city whose water was supplied by the Euphrates through an elaborate network of waterworks and irrigation systems. Alexander was also accustomed to daily baths and may have thus exposed himself to waterborne vectors both through drinking and bathing. We do not know how polluted the river was that June. But we do have a clinical account of Alexander's illness: falling ill on the 18th of the month Daesius (June 2), Alexander ate and slept little, and his fever continued to climb. Throughout the illness, he slept or bathed in the Great Bath at Babylon because of the fever (we are not told if the bathwater was heated or cold). By the 26th, Alexander was "speechless," and, by the evening of the 28th (June 13), he was dead.[17]

In 23 BCE, Augustus fell ill with what Suetonius described as a liver abscess (*iocinere vitiato*). Augustus' physician, Antonius Musa, treated the malady first with the usual course of hot compresses, as prescribed by Celsus for liver complaints.[18] But, when the hot compresses failed to bring the emperor any relief, Musa then switched to cold applications, contrary to accepted medical wisdom (Suetonius considered the treatment an act of desperation: *Augustus* 81). Augustus recovered, and some have suggested that the *princeps* was suffering from "typhoid" fever.[19] He had just returned from Cantabria in northwestern Spain, abutting the Atlantic Ocean. Strabo hinted at pestilential diseases in Cantabria, where mice were recognized as vectors of disease:

> Also not limited to them are the mice, from whom pestilential diseases often result. This happened to the Romans in Cantabria, so that they barely survived, although mouse-catchers were given a bounty proportionate to what they had caught.
> 3.4.18

It is not known in which season of the year the Romans in Cantabria were affected by mouse-borne illness, nor do any of our sources clearly link Augustus' campaign with a virulent outbreak, whether induced by mice or other causes.[20] "Typhoid," however, cannot be ruled out.

Malaria

In antiquity, malaria could be difficult to distinguish from "typhoid."[21] Caused by *Anopheline* mosquitos infected with the parasitic *P. falciparum, P. malariae, P. ovale, P. vivax,* and *P. knowlesi*,[22] malaria was widespread and well-known as a factor of swampy terrain, where the summer sun evaporates stagnant waters and leaves behind thick, putrid pools of colored, harmful, bile-like waters, ascribed as the cause of fevers, vomiting, diarrhea, ear-aches, ulcers, gangrene, and heat spots.[23] Malarial swamps are attested in the literature and recognized by the evidence of town planning.[24] Widespread in Italy from the second century BCE, deforestation, moreover, exacerbated the spread of malaria. As trees are removed, temperatures rise, soil erodes, and alluvial deposition becomes marshy, providing the ideal conditions for *P. falciparum* and its vector mosquitos.[25]

Malarial outbreaks are historically attested. Casually cited by poets, the disease was common.[26] A sanctuary to "Fever" was, moreover, dedicated on the Palatine Hill in Rome.[27] The Hippocratic patient, Philiskos, may have suffered from complications of chronic malaria or "Blackwater fever," where the urine is dark with black sediment.[28] In 413 BCE during the Peloponnesian War, while the Athenian armada attempted a siege of Syracuse, the inhabitants were able to hedge the would-be invaders in the malarial coastlands, thus contributing to the Athenian defeat.[29] Cicero's friend, Atticus, moreover, may have suffered from malaria as perhaps did Julius Caesar in his youth.[30] Malaria may have also weakened Antony's men in advance of the campaign at Actium, where marshes dot the coastline. Antony's troops suffered from illness in winter "but more so during the summer" (Dio 50.12.8).[31] Complications from malaria may also have caused the sudden death of the Visigoth Alaric I who invaded Rome in the late fourth century CE.[32] Finally, endemic malaria would have resulted in very high mortality rates.[33]

The ancients, however, failed to recognize mosquitos as the vectors of the disease, although Varro's microbes seem to come near the mark, as do Columella's thick swarms of contagion-bearing "swimming and creeping things" with their dangerous barbs (1.5.6).[34] It was not until 1880 that the link between mosquitos and malaria was finally made when the French army physician, Charles Louis Alphonse Laveran, noticed parasites in the erythrocytes of infected patients in Algeria. He proposed that malaria was caused by a parasitic organism. In the following year in Havana, Carlos Finlay proved that mosquitos were the vectors of yellow fever.[35]

Most Greco-Roman authors believed that the disease was caused by the bad air emanating from swampy regions (thus, in Latin, eponymously called the disease of "bad air": *malum aer*, coming into English through Medieval Italian *mala aria*). Empedocles had hoped to divert an outbreak at Selinus in Sicily—where the indications were fevers, death, and miscarriage in women—by bringing together two rivers: "by mixing the waters he sweetened them."[36] Attempts to drain lakes and wetlands, moreover, may have resulted in environments that fostered the growth of malarial pathogens, as did urban overcrowding, together with deforestation, erosion, and waterlogging ensuing from efforts to create arable farmland.[37]

In the Hippocratic authors, we have our earliest detailed descriptions. Ancient physicians observed the symptomatic, periodic fevers abating and recurring on a cycle of three or four days (respectively, tertian and quartan fevers).[38] For the Hippocratic author of *Affections*, the initial indication is observed when the body, warmed by the summer sun, grows moist and then becomes ill, either entirely or just "where bile and phlegm are fixed." In its earliest stages, the disease is easily treated, but, if untended, it can be fatal. The Hippocratic physician noted that, although the ailment is most common in the summer, cases occur also during the winter. The author specified different remedies for tertian and quartan fevers. Patients suffering from highly dangerous tertian fevers, when the periodic fever spikes every 48 hours, should be medicated on the fourth day, taking brothy meals on the days when the fever rises, but laxative foods in between. The Hippocratic author advised treating quartan fevers (when the fever spikes every 72 hours and the spleen becomes enlarged) with medications that act upward through the body as the fever spikes, and downward as the fever falls. Patients can also recover from quartan fevers that are diagnosed and treated at the outset.[39]

Celsus' clinical description has been admired for its accuracy since 1889 when Ettore Marchiafava and Angelo Celli first identified *falciparum* parasites under the microscope. Celsus described the disease thusly:[40]

> There are two kinds of tertian fever. One begins and ends in the same way as quartan fevers, but with the distinction that there is complete recovery for one day, but the fever returns on the third. More dangerous by far is the other type which also recurs on the third day. In 48 hours, the paroxysm lasts for nearly 36 hours (sometimes a little less, sometimes a little more). Nor is there ever a complete cessation during the remission, but it is a little lighter (less severe). Most physicians call this type *hemitritaion*.[41]

Like the Hippocratic physicians centuries earlier, Celsus aligned treatment with the stages of the malady, warning that the physician must be careful to avoid mistakes since the outcome of any error (giving food too soon, or letting blood at the wrong time) could result in death.

Aside from diet and blood-letting to restore the humoral balance, magico-medical cures are also attested. One is preserved in the medical poem of Quintus Serenus Sammonicus, court physician to Caracalla (ruled 198–217 CE):[42]

> You will inscribe "Abracadabra" on a piece of paper, and underneath, you will repeat the word, but take out the last letter, and again and again let the elements fall away one by one in succeeding lines, until the letter is reduced into a narrow cone. Be mindful to encircle the neck (of the patient) with this amulet bound up with linen.
>
> ABRACADABRA
> ABRACADABR
> ABRACADAB
> ABRACADA
> ABRACAD
> ABRACA
> ABRAC
> ABRA
> ABR
> AB
> A

The patient was instructed to wear the talisman for nine days, and then throw it over their shoulder into an eastward-running stream. Should the ABRACADABRA treatment fail, Sammonicus prescribed lion's fat, or a cat-skin amulet with yellow coral and green emeralds tied around the neck. The magical word may be meaningless, or it may be a Latinization of the Hebrew *Abrai seda brai* ("let the thing be destroyed").[43]

Epidemics/"Plagues"

It was believed that living in an area with extreme conditions of temperature (cold or heat) or humidity (moisture or dryness) could produce illness, including "plagues" (not to be confused with the Bubonic plague at the root of the "Black Death" in the fourteenth century). Several epidemics are vividly recounted in the literary record.

Most famously at the beginning of the Peloponnesian War, a "plague" was brought to Athens, originating, as Thucydides surmised, from Aithiopia and extending northward through the Mediterranean Basin, baffling the best Hellenic physicians. The disease appeared first in 430 BCE, devastating the Athenians for two years, and it recurred again in 427 BCE, abating after a year.[44] Other cities were affected, including Potidea when Athenian sailors came to the aid of the besieged *polis*.[45] A survivor of the pestilence, Thucydides recorded its symptoms and course with a clinician's eye.[46] The disease, which took the life of the Athenian leader, Pericles, was exacerbated by the influx of people from the countryside into the city. Among the indications are a slowly rising fever (up to 104°F [39–40°C]), chills, slow heart rate, headaches, muscle pain, diarrhea and constipation, stomach aches, lack of appetite, and occasionally rosy-colored splotches on the skin. Victims would try to slake their unquenchable thirst by reeling about the fountains, plunging themselves into water tanks, or flinging themselves naked into ice-cold streams to alleviate their burning limbs (Lucretius 6.1172–1173; cf. Thucydides 2.49.5). Although Thucydides was quick to dismiss the epidemic as having been caused by spies or Peloponnesian sympathizers who allegedly contaminated the water supply, measures were taken afterwards to improve the water supply, perhaps in response to the outbreak.[47] Many of Thucydides' indications are also symptomatic of "typhoid," and some scholars suggest that—although the epidemic may have been caused by multiple pathogens—the principle illness may have been "typhoid" stemming from a sewage-contaminated water supply.[48]

The Carthaginian army, in 396 BCE, suffered from another disease (νόσος: *nosos*). Our source, Diodorus Siculus (14.70.4–71.4, 14.76.2), attributed the disease to the boggy terrain. It was exacerbated by a concentrated population and a particularly sweltering, humid summer. The first indications include cough, swollen throat, gradual sensation of burning, pain in the tendons in the back of the neck, and a heavy feeling in the limbs, followed by dysentery and pustules appearing over the whole body. In some cases, memory loss and violent behavior ensued. The epidemic was highly contagious, and victims died within five or six days. Dionysius of Syracuse (432–367 BCE), consequently, was able to defeat the weakened Carthaginian forces. Quick to point out that the Carthaginians suffered more severely from the epidemic than the Romans, Livy (25.26.7) emphasized the disease's highly contagious nature, the fear experienced by survivors, and the preponderance of corpses. He nonetheless withheld clinical details, aside from the fact that caregivers consequently succumbed and were quickly shunned. Scholarly consensus has identified the outbreak as smallpox, but Littman argues that the Carthaginians suffered from the same malady that struck Athens, as Diodorus himself believed.[49]

Water, Health, and Disease 119

Hydrophobia ("Rabies")

In contrast with the parched victims of Thucydides' plague, victims of animal bites (especially from canines) might suffer from hydrophobia ("fear of water"). Folklore and medical sources were inconsistent regarding whether and how hydrophobia could be treated. In one passage, Pliny declared hydrophobia "incurable" (*insanabile*). But elsewhere he preserved an antidote to the bite of a rabid dog, as decreed by an oracle: the root of a forest-rose (*silvestris rosae*), also called the "dog-rose" (*cynorrhoda*), a magico-medical cure on the strength of its etymology.[50] Caused by *lyssavirus*, rabies (an inflammation of the brain) is indicated by fever, tingling at the wound site, violent thrashing, paralysis, confusion, loss of consciousness, and fear of water.

In Celsus we have a detailed description of the treatments of this ailment transmitted by dogs who are "full of madness" (*rabiosi*). Some physicians, Celsus reported, would send their patients immediately to the baths to sweat out the toxins, keeping the wound open so that the poison could flow out. After receiving such treatment for three days, the patient was deemed to be out of danger and was to be given undiluted wine, "the antidote to all poisons." Celsus remarked that insufficient attention to the wound usually resulted in "hydrophobia" as the Greeks called it, "a very miserable type of disease" whose indications include both severe thirst and fear of water. The only remedy is to restore the balance of humors in a rather shocking way:

> toss the patient into a pool. If the patient does not have the knowledge of swimming, allow him to drink (the pool water) after sinking, and now lift the victim up; if the patient does know how to swim, press him down, now and then, so that—although unwilling—he will be sated by water. For thus at the same time thirst and fear of water are simultaneously removed.
>
> 5.27.2

Patients were compelled to face their fear of water "head-on," literally. We are not told how successful the plunge-pool remedy might have been. Celsus did concede, however, that the treatment might have dangerous side-effects, including muscle spasms and weakness.

Other treatments were also recorded. Pliny prescribed the preserved ashes of fresh river crabs as efficacious against hydrophobia, especially when mixed with gentian and wine and shaped into lozenges (32.53–55). Pliny remarked that the symptoms could be relieved by cauterizing the wound or providing the patient with boiling water to drink (34.151: heat for treating the fever sympathetically). He suggested that the ailment most likely occurred in conjunction with the rise of Sirius (8.152: the dog star, a transparent magico-etymological connection). Also etymologically conspicuous is Dioscorides' prescription of madwort (ἄλυσσον: "lacking rage") whose name suggests the desired outcome.[51] Dioscorides recorded the use of amulets and treatments that are also more magical than rational (2.47): consuming the roasted liver of the rabid dog that bit the victim might cure a current case of hydrophobia (the so-called "hair of the dog"), while wearing the teeth of the canine culprit might protect against future bites. Additional magico-medical treatments include the ash from the head of a burnt dog, eating the head of a dog, or wearing amulets that contain a "worm" from a dead dog (Pliny 29.98–100). Suggestions for preventing dogs from becoming *rabiosi* are, likewise, magico-medical: mixing chicken dung into the dog's food and docking the tail of a puppy (by biting it off) within forty days of birth (Pliny 8.152).

Other Complaints Caused by Water

Local water sources can have precise effects on those who drink from them, including changing the coat-color of cattle (Chapter 3). Some water sources also induced disease. In Nabataea in 26 BCE, Gallus' men suffered gum disease and lameness, "a kind of paralysis, the former of the mouth and the latter around the legs," caused by "the water and plants" in Strabo's assessment (16.4.24). Germanicus' troops endured similar ailments while campaigning on the Rhine. The army could find only one source of potable water, which caused teeth to fall out (*stomace*: "disease of the mouth" [scurvy?]) and the destruction of knee joints (*skelotyrbe*). The only remedy was "britannica," the antiscorbutic *radix britannica* whose properties were likely learned by Germanicus' army doctors from the local Frisians.[52]

In general, latrines, sewer systems, and the baths were hatcheries for the pathogens that cause disease.[53] At Rome, sewage was dumped into the Tiber, feeding and infecting the fish, which would then be consumed by residents in the city.[54] The baths themselves were hardly clean. We do not know how often— or if—water in the plunge pools was changed (once a day?). The bacteria in excrement would remain warm and continue to grow.[55] But defecating in the baths, which did occur, was considered inauspicious.[56] Celsus advised those with fresh, open wounds against visiting the baths owing to the threat of gangrene (5.26.28), and Marcus Aurelius complained about the slimy, disgusting bath water: "Such as bathing seems to you—oil, sweat, dirt, slimy water, all vile things" (*Meditations* 8.24).

Illness and Travel on Water

It is well known that during the Age of Exploration (fifteenth to seventeenth centuries), European sailors to the Caribbean and Americas imported the pathogens that caused chickenpox, diphtheria, measles, mumps, smallpox, and other infectious diseases against which the native populations had no immunity. The results were devastating: for example, the Taino population of San Salvador and Hispaniola was all but wiped out. Diseases were likewise imported from the Americas to Europe, such as the "Great Pox," venereal syphilis, with a devastating outbreak in the early 1500s.[57] Disease was also dispersed by travelers in the ancient Mediterranean: the Antonine "plague" (165–180 CE), for example, was brought to Rome by soldiers returning from the east, and the pathogens were dispersed further afield to the Gallic and German provinces.[58] Generally speaking, however, since the Europeans had been a migratory people as early as the Neolithic Era, the Mediterranean Basin quickly came to share "a pool of infectious diseases."[59] As trade and travel increased (ca. 200 BCE–200 CE), the spread of parasites that caused malaria and other water-borne ailments nonetheless intensified.[60] Initiatives in paleopathology seek to understand the transmission of diseases, including evidence from skeletal and dental remains, decomposed soft tissue, mummified corpses, and the excavation of latrines.[61]

Sailor Diseases

Those who travel on water have their own medical complaints, namely scurvy and seasickness. Most Greco-Roman sailing was coastal, and even Hanno's expedition to circumnavigate Africa was rarely, if ever, out of sight of land. Most expeditions of exploration, furthermore, frequently put in to shore. Scurvy was, thus, not the same concern for the ancient sailor that it was for trans-oceanic mariners during the European Age of Exploration.[62] Sailors would, however, be prone to the pathologies endemic in coastal areas, including water-borne diseases and insects,[63] but, as Lucretius observed, some diseases affect some peoples but not all (6.959–978, 1110–1117).

Seasickness, however, was common, cited casually in literary sources, especially in the context of sudden storms at sea.[64] The malady is caused by the misalignment of sensory information: e.g., we perceive motion with one sense (hearing) but not another (sight) when we focus on an object on the boat rather than on the land (seasick passengers should gaze at the horizon). The Greek word, ναυτία (*nautia*) is etymologically related to ναῦς (*naus*), the word for ship, and thus refers fundamentally to an adverse reaction to motion aboard ship (cp., Latin *navis* [ship], *nausea* [seasickness, nausea]: Plutarch *Table Talk* 694b). But the term quickly came to refer more generally to "nausea."[65] Plutarch explained why people become more seasick on calm seawater than when traveling on a rough river as a factor of fear and the sense of smell. The sea and its scents are "unfamiliar" and often disagreeable, and the weather, as everyone who has ever traveled by sea knows, can change quickly, thus exacerbating fear in anticipation of violent seas:

> there is no help in the stillness of the sea; but the soul fearing and imagining the pitch and roll is stirred up and fills the body with confusion.
>
> *Natural Phenomena* 11 914f[66]

Plutarch also indicated that the seasick traveler craves salty foods (*Precepts of Statecraft* 801b), while Celsus (1.3.11) advised the seasick sailor (*qui navigavit* ["those who sail"]) to abstain from food or to eat very lightly. In Athenaeus (15.675f–676c), the scent of myrtle proved efficacious against debilitating nausea that was caused by seasickness (cf., Chapter 9).

Healing Waters, Bathing, and Hot Springs

We turn now from the incapacitating effects of water to its healthful uses. "It is a healthful thing to have bathed," so claims a mosaic inscription attached to a public bath house in Sabratha, Libya. The mosaic features a pair of sandals and three strigils, commonly used for scraping oils from the skin.[67] Galen discussed the therapeutic benefits of bathing in warm and cold potable waters, especially after strenuous exercise where toxins have been evacuated through sweat (which should be scraped away with a strigil).[68] Sweet, warm baths return moisture to the body. Cold baths contract the pores, preventing both the absorption of external moisture and the loss of internal toxins, according to Galen. Cold baths, however, were thought to strengthen the entire body and toughen up the skin. But such therapies, on Galen's theory, impede growth and should be avoided by younger athletes.

Both cold and thermal springs, moreover, have been used for the curative properties of their mineral-rich waters to treat many ailments. In Strabo, hot springs are suitable for curing unspecified diseases at Baiae (5.4.5), and kidney stones at Pithekoussai (5.4.9).[69] The waters of the Alpheus River were also thought to cure leprosy, perhaps on the strength of the etymology (ἀλφός: *alphos* ("dull white leprosy"]: Strabo 8.3.19).[70] In Vitruvius (8.3.4–5), we find descriptions of the medicinal properties of different types of hot springs: sulphurous, aluminous, bituminous, alkaline. Celsus, Pliny, and others provided instructions for thermo-mineral healing of various complaints: barrenness, dislocations, fractures, gout, foot conditions, headaches, psoriasis, diseases of the eyes and the ears, and mental insanity.[71] Pliny (31.10) especially recommended the sulphurous springs at Aquae Albulae, between Rome and Tivoli, for treating wounds,[72] and he praised the salubrious, hot springs at Tabariah, near the Jordan (5.71). Seneca reported (and dismissed) the belief that the Nile's waters could cure infertility in women and that water from Lycia could protect a foetus during a difficult pregnancy (NQ 3.25.11).

Vitruvius (8.3.1) theorized that springs become hot by virtue of underground alum, bitumen (asphalt/viscous petroleum), or sulfur (a product of volcanic activity) warming the soil above, minerals that are associated with fiery properties (Strabo 16.1.15; Pliny 35.178).[73] These sites and others were valued for their curative properties: heated in vitiated soils, their waters are rendered medicinal. Sulfuric waters restore muscular weakness by "burning away" poisonous humors; alum-springs warm the body and overcome the coldness that sets into paralyzed or diseased body parts; bitumen-springs are purgative, according to Vitruvius (8.3.4–5).

Thermal and mineral springs were also exploited. Patients took their cures both by soaking/swimming or imbibing: thermo-mineral water was prescribed to relieve internal pain and bladder stones (Seneca, NQ 3.1.2), but Pliny warned against excessive consumption of mineral water (31.59–61). With imperial expansion, the Romans quickly appropriated many of the curative springs in western Europe for medicinal and recreational use. Interestingly, under Hadrian, some spas were reserved exclusively for the ill during the morning hours (Augustan History, Hadrian 22.7–8).

Many of these springs, and other bodies of water (rivers, wells), were centers of healing cults dedicated to a variety of gods with healing associations: Apollo and Diana, Aesculapius and Hygieia, Jupiter, Vulcan, Mars, Minerva, Venus, Dionysus, Silvanus, and others.[74] Pre-eminent was Herakles (Hercules) who presided over thermal springs at Thermopylae, which Athena created for the hero (Herodotus 7.176.3), and at Trachiniae where he died,[75] as well as at spas in Italy, Sicily, and Dacia.[76] Priests and physicians attended the sick at the Fontes Sequana, where dwelt the water spirit who personified the river Seine, and Aquae Sulis (Bath, England) where Sulis Minerva presided over a thermal spring (104°F = 40°C), a popular healing sanctuary from pre-Roman times into the Georgian era. Small finds there include votive body parts and an oculist's collyrium stamp.[77] Soldiers were sent to spas to seek cures and convalescence or for rest and recreation.[78] The presence of the Roman army encouraged economic growth and prosperity,[79] but soldiers occasionally came into conflict with local residents over spa sites. At Scaptopara in Thrace, in 238 CE, residents complained to Gordianus III that many Roman officials, including soldiers, would descend upon the hot springs, demanding lodging and other services without offering payment.[80]

In contrast, other patrons respected the divine nature of healing shrines. In the early second century CE, the Ninth Legion *Hispania* came to Aachen (Aquae Granni) to recuperate, for which convalescence its prefect and senior centurion (*primus pilus*), Latinius Macer, a native of Verona, dedicated an altar in fulfilment of his vow to Apollo, who had been syncretised with Grannus, the eponymous local patron of the waters. Human health here connects with the land through its divine patron.[81] The altar shows an enthroned Apollo Grannus holding his lyre, a quiver on his right shoulder. Baden (Aquae Helveticae), near the legionary headquarters of the Eighth Legion Augusta (Vindonissa), was the site of a military hospital and a healing spa to Mercury.[82] Water from the hot springs was piped to therapeutic basins, one of which could accommodate nearly one hundred bathers.

Conclusion

Water was recognized as both a source of disease and a cure for it. Medical writers described the course of water-borne diseases—e.g., typhoid, malaria, the "plague"—offering theories of their causes and suggesting treatments. Varro and Columella even seem to have developed a rudimentary "germ theory." But water is medically ambivalent. Owing to the effects of visible and invisible forces, the wrong waters could cause the proliferation of intestinal pathogens, while the right waters, either consumed or soaked in, could induce healing, both physical and spiritual.

Part Three

Imagining the Watery World

6

Biological Creatures of the Sea

Introduction

In 15 CE, Germanicus' fleet encountered a virulent storm on the Amisia River (Ems). In a hexameter extract preserved by the elder Seneca, Albinovanus Pedo histrionically described the episode, focusing as much on the terrors of the huge sea monsters that he imagined would savage the boats as on the storm:

> Now they see that Ocean—who bears huge monsters everywhere beneath its sluggish waves as well as savage whales and watery dogs—rises up and snatches the ships (the very din increases the sailors' fears). Now they believe that ships sink in the mire and the fleet is deserted by the breeze, and that they themselves are abandoned by helpless fate to be pulled to pieces by the sea-beasts. Unlucky lot!
>
> *Suasoria* 1.15

Germanicus' fleet was far from Rome and the protected waters of the Mediterranean Basin. The Romans moreover never fully came to understand the North Atlantic, much less the mysterious fauna that populated its depths. Vergil had also envisioned monsters lurking beneath the marble-smooth Mediterranean (*Aeneid* 6.729). In the fourth century CE, Avienus would describe Ocean as monster filled (*beluosi*: *Ora Maritima* 102). Even with modern technology, many deepwater creatures remain poorly understood. They inhabit a realm that was all but closed off to ancient thinkers who could scrutinize only the occasional blue-water fauna that washed up on shore. According to Aelian, people have explored the seas down to a depth of 300 *orguias* deep (ca. 2,000 feet [550 meters]). He observed, nonetheless, that the characteristics of deeper waters remain obscure:

> Nor does anyone else tell whether fishes or even other animals swim beneath (that limit) or if even the depths are untrodden by them.
>
> *NA* 9.35

Plutarch moreover remarked that the sea metes out only scant glimpses of its environment, remaining mostly hidden (*Cleverness* 975e).

Despite (or perhaps because of) the inaccessibility of the deep sea, marine animals were a source of fascination, fueling myth and imagery, and receiving attention from many writers, especially Aristotle, Pliny, Aelian, and Plutarch. It was in water that life forms arose, nurtured in egg-like shells or even within the bodies of dogfish before maturing and coming to land.[1] The very creation of animal life was thus connected with the sea/water. But the nature of those life forms remained mysterious. Pliny was surprised that some folk still believed that marine creatures were devoid of sensibility and feeling.[2] But Plutarch, who generally advocated for the humane treatment of animals, considered most deep-sea creatures unworthy of justice since they lack natural affection and are utterly devoid of any pleasantness or sweetness (*Cleverness* 970b). Such attitudes shaped the ancient conception of (terrestrial and) marine animals.

Some animals are ambiguously amphibious, occupying both land and sea. Seals bask daily on sunbaked rocks,[3] and near the dwelling of the Turtle-eaters was a "seal-island" (Strabo 16.4.14). Seals come to shore to give birth; and dolphins leap from the water. Scuttling between the sand and the edges of the water, crabs moreover have "many legs and many possessions on both land and sea," like Petronius' Trimalchio who was born under the sign of the amphibious crustacean (*Satyricon* 39). From this amphibious nature hybridity and metamorphosis follow.

Metamorphosis

Like sea gods and the shifting nature of water that erodes and distorts, marine animals were creatures of transformation. Dolphins came into existence through an act of metamorphosis. The nerites, a beautiful, small shellfish with a spiral shell, was also generated from an act of transfiguration. Loved by Aphrodite but preferring the company of his ocean family, Nerites, the gorgeous son of Nereus and Doris, rejected Aphrodite's offer of Olympian status. Aphrodite turned the godling into the shellfish (Aelian, *NA* 14.28). In addition, it was thought that some sea creatures could transform themselves. Known to avoid fishermen's lures, the fox shark could turn itself inside out so that the hook would fall away (Plutarch, *Cleverness* 977b). Like snakes, both crabs and langoustes renew their shells and consequently "slough off old age" (Pliny 9.95). The octopus was the consummate aqueous creature of metamorphosis. Aristotle and his successors remarked on the octopod's fear-inspired ability to match its coloration to the surroundings in order to hide from enemies (Aristotle, *HA* 3.1.661b32) or as camouflage when hunting.[4] The octopus became a metaphor for social and political flexibility (Athenaeus 12.513d; Scholiast on Pindar, *Nemean* 9.30). For this reason, the octopus was regarded as "clever" by Theognis (πολυπλόκος: *poluplokos*: Plutarch, *Natural Phenomena* 916b–f), despite Aristotle's accusations of wickedness (for using its ink to trick prey) and stupidity (for approaching human hands submerged in the water).[5]

Anthropomorphism

Animals, moreover, were utilized to probe human character and morality. In the richly illustrated medieval bestiaries, the acme of a long Greco-Roman tradition that started with Aristotle, real (and imaginary) animals were employed as Christianizing metaphors to interrogate human society and foibles.[6] This harmonizes with the pronouncement of French anthropologist, Claude Levi Strauss, who famously said that "animals are good to think [with]."[7] Anthropomorphism is key to the interpretation of sea creatures. Mating, even among fish, was characterized in terms of marriage, and some male fish even cohabit with the females, "taking care of them as though wives" (Aelian, *NA* 9.63). Some fish travel in formation "like a battle-array of hoplites" (*NA* 9.53). Whelks obediently submit to the rule of their king, following along if the king migrates (*NA* 7.32). Fishermen have discovered that crabs could be lured by music. "As if by a spell" they are mesmerized by flute music and would follow musicians, retreating inland Pied-Piper style.[8] Stingrays also "rejoice beyond measure" at flute music and dancing fishermen, "for they have ears, so they say, and eyes." Crabs, furthermore, would rise gently to the surface where—lured by the beguiling music—they would be wrenched up by fishermen (*NA* 1.39, 17.18). At Epidamnus (Roman Dyrrhachium in Illyria), fishermen were said to have negotiated peace treaties with mackerel: men fed the fish, and the mackerel helped wrangle other fishes for the blue-water catch (Aelian, *NA* 14.1). The sacred fish of Hierapolis were "always under truce," owing perhaps to the influence of the goddess, Astarte, who inspired unity and concord (ἔνσπονδα: *ensponda*: Aelian, *NA* 12.2). Other tame fishes include Crassus' pet moray eel whom the triumvir bedecked with jeweled necklaces (for the bejeweled fish of Labraunda, Chapter 8). When the eel died, it was mourned and buried (Aelian, *NA* 8.4; cf., 12.29–30). Like Crassus' eel, other fish also came when called. In the "Old Men's Harbor" at Chios, fish provided comfort to the elderly. It was also recognized that fish could express emotions, including jealousy and fear: flying fish and squid, for example, leap out of the water when they are frightened (Aelian, NA 9.52, 9.63). Other sea creatures were aligned with human inventions, especially ships. The nautilus employs its arms like oars. Resembling a ship under sail, the nauplius even has a keel on its shell, as explained by Pliny (9.88, 94).

Despite the anthropomorphic conception of non-human animals, compassion and sympathy were rare in ancient authors. Although Aristotle conceded that animals deserve praise or censure for their voluntary acts, he maintained that we owe them no justice because they lack reason and speech (consequently, they also lack souls, thus influencing the Medieval Catholic view). Others, including the eponymous founder of the Epicurean school and the early Stoic philosopher, Chrysippus, maintained that non-human animals—incapable of memory, emotion, forethought, intention, or even voluntary action—were inherently irrational.[9] Plutarch, nonetheless, ascribed reason (and a rational soul) to many animals, and he advocated for humane treatment: whereas children might throw stones at frogs or "in sport" (παίζοντα: *paizonta*) as they play around, the frogs do not die "in sport" (*Cleverness* 965b). There are, however, examples of compassion. Empathy for beached dolphins was expressed poignantly in the *Greek Anthology* (7.214–216). Fishermen, moreover, would only gather crabs on land because they did not wish to appear "more brutal than the waves" that battered the helpless crustaceans violently about the headlands (Aelian, NA 7.24).

Creatures of the Deep

Most fauna in the Mediterranean Sea are harmless and non-poisonous. Pliny, for example, knew that sharks could be frightened away by divers swimming aggressively towards them (9.152–153). The perils of the submarine environment, nonetheless, were exaggerated, as, for example, the whales and watery dogs of the North Atlantic who threatened Germanicus' fleet (above). In Oppian (*Fishing* 2.434–453), moreover, *iulides* would swarm by the thousands and nip at sponge divers. Oppian's *iulides* might have been rainbow wrasses, abundant but harmless, colorful carnivores that feed on small crustaceans and mollusks. In Pliny (9.151), *nubes* (*nubila*) would obstruct divers from returning to the surface. Pliny's *nubila* ("cloud-like" creatures) might have been benign, curious manta rays who often swim around divers. Terminology is ambiguous, and it is not always possible to correlate ancient nomenclature with modern species.

Aristotle was the first thinker who considered the animal world systematically. He classified creatures into two *genera* according to similar characteristics: blooded (roughly corresponding with "vertebrate" creatures) and bloodless (invertebrates). Within these broad *genera*, Aristotle further distinguished animals by types (εἶδος: *eidos*).¹⁰ His marine types include: fishes with scales; scaleless fish (murenas, congers, ox-rays, stingrays, electric rays, and dogfish); viviparous cetaceans (dolphins, whales, and seals); boneless animals lacking blood and intestines (octopods, cuttlefish, squids); crustaceans (soft-shelled lobsters, prawns, and sea- and river-crabs; hard-shelled oysters, murex, whelks, trumpet-shells, sea urchins, and spiny lobsters); and "plant-animals" (zoophytes that resemble plants, such as jellyfish, anemones).¹¹ Relying on second-hand reports from travelers, fishermen, and hunters, together with his own autopsy of dissected specimens, Aristotle's descriptions are meticulous and largely accurate. He noted, for example, that the octopus employs a hectocotyl arm for mating (*HA* 4.1.524a2-9), that catfish eggs are protected by the males (*HA* 9.37.621a20-b2), and that dogfish (a species of shark) are viviparous, with umbilical cords linking the embryo to a placenta-like membrane.¹² Aristotle's categories were adopted by Aelian (*NA* 11.37).

We focus on marine species who figured prominently in the fabric of Greco-Roman culture (dolphins), and those that provided material for the development of fantastical, marine creatures and polymorphic, shape-shifting, theriomorphic, sea deities (seals, whales, sharks, octopods).

Dolphins

The elegant dolphin was lauded and esteemed. Bottlenose dolphins were as plentiful in the myth and art of the ancient world as they are in the Mediterranean waters, although they are now listed as vulnerable by the IUCN (International Union for Conservation of Nature). Most common is the striped dolphin (*Stenella coeruleoalba*) with its blue-gray dorsal, characteristic striped, white underside, and conspicuous beak. Also known in Mediterranean waters are the large, narrow-beaked, bulbous-headed Risso's dolphin (*Grampus griseus*), the short-beaked common dolphin (*Delphinus delphis*), and the common bottlenose dolphin (*Tursiops truncatus*), which frequently gambol in shallow waters near the shore and are today trained for entertainment.¹³ It is difficult to distinguish dolphin species from the iconography. Beyond the waters of the Mediterranean, Aelian recognized two types in the Erythraean Sea: the fierce, sharp-toothed, hostile dolphin (river porpoise?) and the gentle (Indo-Pacific bottlenose) dolphin, which submits to being petted.¹⁴

Pods of dolphins porpoise in Minoan artwork at Knossos, Agia Triadha in Crete, and Agia Irini in Kea, rendered in blue and yellow with wavy reddish-orange stripes to emphasize their lively motion, partially in the water and partially above.[15] Humans ride dolphins on the coinage of Taras (cf., *UCWW*: Chapter 8), and dolphin-mounted cupids frolic across the frescoes in the House of the Vettii at Pompeii and the mosaic floor of Cogidubnus' Roman palace at Fishbourne.[16] Dolphin figurines mark the laps in the Roman circus, a focus of Neptune's cult in Rome. In relief and on pots, dolphins escort ships into harbor (as they continue to do),[17] they announce storms, and sailors still consider them auspicious.[18] Dolphins also accompany the deceased to the underworld (the Isles of the Blessed) on numerous funerary monuments,[19] and they adorn many a Roman fountain. Dolphins attend Dionysus, Aphrodite, and Apollo whose cult epithets include "delphinios" (dolphin-like). Apollo sent a dolphin to guide his Cretan worshippers to Delphi in order to establish his cult there (Plutarch, *Cleverness* 984a). Beloved by Poseidon, the creatures rejoice in water (Oppian, *Fishing* 1.383–385). Like crabs, they reputedly appreciate music, including the water organ (Pliny 9.24) and diaulos,[20] and they befriend musicians such as the shepherd boy who played his pipes along Libyan shores (Oppian, *Fishing* 5.466).

Celebrated for their anthropomorphic morality, they were friendly to humans because they once were human. "For requiring nothing from any man, it is kind to all and has been helpful to many" (Plutarch, *Cleverness* 984d). Dionysus, the god who preferred to transform than to destroy, converted his kidnapping pirates into benevolent dolphins who later came to regret their miscrearce.[21] A vase from Vulci, by the Micali painter (seventh century BCE), shows the pirates at the moment of their transformation into dolphins[22] (Fig. 6.1). Because of their sympathy towards humankind, it was thus considered a sacrilege to hunt them, and slaying a dolphin was "equal to killing a human" because delphinian "understanding was equal to human thinking."[23] The Thracians, consequently, were "wicked" for their cruel method of hunting dolphins. They targeted the young animals still under the protection of their mothers who would also perish trying to defend their pups (Oppian, *Fishing* 5.519–588). In Oppian, dolphins were "kingly" and "harmless." But, responding to calls of *Simones* ("snub-nose") and deploying into delphinian military formations, dolphins were known to have assisted fishermen, wrangling catches of fish into the shoals at Narbonensis (Nîmes), waiting for their rewards, a feed of bread and wine.[24] If trapped as bycatch, the wily dolphin would bide its time feasting on fish in the net until the fishing boats came close to shore. The dolphin would then bite through the nets, making an escape (Plutarch, *Cleverness* 978a). Fishermen usually allowed miscreant dolphins to go free, sewing rushes onto their dorsal fins so they could recognize repeat offenders whom they would "chastise with blows."

Fig. 6.1 Pirates at the moment of transformation into dolphins: Vulci, Micali painter (seventh century BCE).

Dolphins are effective hunters, but Seneca's description of dolphins that defeated crocodiles in the Nile is not rooted in biological reality but instead reflects the anthropomorphizing tendency to heroize human-like animals as vanquishers of dangerous enemies (*NQ* 4b.2.13).

Of all marine animals, dolphins were the most thoroughly anthropomorphized, as we see in Lucian's *Dialogues of the Sea Gods* (8) where Poseidon and his dolphins chat about sailors saved from drowning by dolphins. Dolphins were thought to remember kindnesses, and they were celebrated for saving men from drowning (e.g., Arion [Chapter 8]; however fanciful such a rescue might be, according to Plato, *Republic* 5.453d). Coeranus of Paros (in the central Aegean) was befriended by a pod of dolphins (Aelian, *NA* 8.3; Plutarch, *Cleverness* 985a–b) that he had once ransomed from slaughter. Later shipwrecked off Naxos (in the Cyclades), he was rescued by the grateful cetaceans who deposited him safely ashore. Years later when their benefactor died, the dolphins returned, assembling off the shores at Coeranus Point, poignantly attending the funeral "as one faithful friend to another." Out of respect they did not depart until the funeral pyre was extinguished. There was, furthermore, a lingering tradition that Hesiod's body was tossed into the sea by murderers but returned to the beach at Rhion by a pod of dolphins.[25] In the *Greek Anthology* (12.52), Meleager lamented the death at sea of his soul mate ("the other half of my soul"), Andragathus. Meleager wished to become a dolphin so he could carry his friend back to Rhodes.

In another account recorded by Aelian, a tame dolphin gamboled in the harbor at Poroselene near Lesbos, accepting treats from the residents, and coming to the calls of a young boy who played by the water. Boy and dolphin would swim and race, and sometimes the dolphin—the superior athlete who competed against swift marine creatures and oared ships—allowed the lad to win. The dolphin was given a name, which, unfortunately, was not preserved (Aelian, *NA* 2.6. Pliny 9.26 recounted a similar tale at Hippo Diarrhytus in Africa). In Aelian we also hear of a dolphin who fell in love with a beautiful boy at Iasus on the coast of Caria. Mourning the boy's death (accidentally caused by the sea creature), the dolphin committed suicide, and the pair were thus commemorated on Carian coinage.[26] A gentle dolphin in Libya would convey a shepherd boy on its back, and he wasted away from grief over the boy's unexplained death (Oppian, *Fishing* 5.459–518).[27]

According to Aristotle, dolphins are also protective. Mature dolphins, for example, tend to supervise the younger ones from a safe distance, close enough to offer aid as needed, but far enough away to allow the juveniles to find their sea legs, as it were, on their own. A pod once sought medical help for a wounded companion whom they transported to a harbor. A pod of dolphins also entered the harbor at Aenus in Thrace seeking the release of a wounded, captured comrade.[28] Dolphins have also been seen supporting deceased compatriots on their backs in order to prevent other creatures from devouring them.[29] Although Aristotle did not speculate why dolphins might come to dry land, epigrammists imagined that ill or dying dolphins would beach themselves out of a desire for the dignity of a proper burial. Owing to their compelling "humanity," some dolphins were duly buried and eulogized with epitaphs and memorials.[30]

Aristotle recognized that dolphins are air-breathing, viviparous mammals with life expectancies of up to thirty years.[31] With a gestation period of ten months, dolphins produce one or two calves at a time, suckling their young "with abundant milk." They are, furthermore, attentive parents.[32] Aristotle noted additionally the ambivalent nature of dolphins (and whales) who take in air through their lungs (like land creatures) but dwell in water which, like air, they also ingest but eject through blowholes. Aristotle did not fully understand the function of the blowhole, which he thought served to expire ingested water.[33] Aristotle, however, did know that cetaceans could suffocate when deprived of air, and that they could survive out of water, although with much difficulty, "gasping and groaning" for a long time.[34] Aristotle thus expanded his classifications to include composite, aquatic mammals who breathe in order to cool their blood and who also take in water as incidental to feeding. The cetacean blowhole, thus, correlates to the piscine gill. Pliny added that a tube connects the "mouth" (*ora*: blowhole in the forehead) to the lungs, allowing dolphins and whales to breathe (9.16–19). It was thought, furthermore, that, like sharks, dolphins had to swim constantly, even when asleep. Rising to the water's surface just before going to sleep, they would sink to the bottom, and then be awakened into a twilight state by the impact of hitting the sea bottom, rising and sinking again and again until they completed their sleep cycle, all the while remaining in perpetual motion.[35]

Dolphins were distinguished as the speediest swimmers,[36] and they were renowned for their acrobatic skill: "of all fish and creatures on dry land, they are the best at leaping." Aristotle explained their acrobatic ability according to the principles of mechanics. Their ability to hold their breath under water, "like divers," creates a bowstring-like strain. When air compressed within the body is finally released, the dolphin is shot upwards like an arrow.[37] Because of their physical prowess, they were naturally recognized as skillful hunters.[38]

On the ancient vision, dolphins uniquely resemble humans in their sense of community (they are usually observed in pods), and, exceptionally, they interact with humans as benefactors, saving those who fall into the sea and colluding with fishermen to improve the catch or simply for the sport of it. Despite this perceived affinity with human beings, dolphins provided protein and fat for hunters along the Black Sea.[39] Delphinian body parts were also used in a number of medicinal applications including the treatment of skin ulcers (dolphin ash), fevers (delphinian liver), and dropsy (melted fat) (Pliny 32.83, 113, 117, 129).

Seals

Featured on the coinage of Phocaea ("Seal"-Island), monk seals were once plentiful on Mediterranean beaches, with a range that extended from the Black Sea throughout the Mediterranean into the Atlantic as far west as the Canaries and Azores. The population began to decrease in the early years of the Roman Empire because of the stresses of commercial fishing, and they are now critically endangered.[40] Because of this dual existence, belonging to both water and land (or to neither region, thus considered "stunted" by Aristotle), seals were difficult to categorize and only partially understood.[41] The seal's earlessness, large penis, forked tongue, and compact kidneys were a source of fascination to Aristotle.

Aristotle and others recognized that, like other mammals, seals are quadrupeds, but they slip instead of walk because they "cannot steady themselves on their feet." Seals are also viviparous. Like humans, they give birth on land at any time of the year to "no more than three pups" at a time (seals usually bear single pups), and they nurse their young. But seals spend most of their time in the water where they also feed. They must, consequently, be classed among marine animals. Aristotle correctly noted that seals acclimate their young to the water at about twelve days of age,[42] and the pups quickly become excellent swimmers (cf., Aelian, NA 9.9). Curious and friendly like dolphins, seals make devoted mothers who nurse their pups and cuddle them until they are strong enough to explore the "wonders of the deep," the watery world that Oppian considered their natural habitat (Fishing 1.686–701; cf., Pliny 9.41). They were, moreover, deemed capable of emotion. Seals were observed grieving over the death of their young (Philostratus, Apollonius of Tyana 2.14.54), and—like dolphins—they could fall in love. According to Aelian (NA 4.56), one seal reputedly fell in love with a very ugly sponge diver, commensurate with the animal's unsavory reputation as a noisy, redolent, misshapen creature (cf., Aristotle, HA 2.1.498a31–b4). Menelaus, for example, had complained of their stench (Homer, Odyssey 4.398–455; Chapter 7).

Seals were not eaten and therefore not deliberately hunted. To Oppian, they were destructive, sharp-clawed, and violent, and he recommended killing them by bludgeoning them on the head.[43] Nonetheless, their vomited rennet was considered a remedy for epilepsy, tetanus, and swooning women.[44] Difficult to penetrate, their thick hides were regarded as suitable for tents (Pliny 2.146) and as amulets against lightning (Augustus reputedly kept a sealskin for such a purpose: Suetonius, Augustus 90). According to Pliny (9.42), sealskins would bristle during ebb tides because they retain a connection to the ocean.

Seals are rare in ancient art, but not altogether unattested. On a Caeretan hydria (530–500 BCE), "a very alert seal watching among the waves" is featured swimming behind a sea monster.[45] From Cos, the "fair-swimming" seal, Euploia, is shown on a mosaic floor.[46] They are trainable, as Pliny recognized, and may have performed in shows. They were also included in the arena, kept in artificial pools, and forced to fight bears,[47] owing to their reputation as vicious man-eaters.[48] Seal-lore likely contributed to the legend of larger and more dangerous sea monsters.

Biological Creatures of the Sea 137

Giant Sea Creatures

The largest creatures, so Pliny (9.2) surmised, dwell in the sea because of the sheer quantity and extent of nutrient-rich water (contradicting other views where Ocean is "barren"). The ocean is thus infused with generative principles "from above," and consequently always fosters the creation of living creatures. These lifeforms arose because of the mixture of seeds that were affected by first principles and shaped by wind and wave, producing a wealth of animals who reflect terrestrial fauna and flora (the sea cucumber, for example, resembles the eponymous terrestrial plant in both color and scent).

Whales

Even more mysterious than dolphins and seals, and less understood, were whales who rarely come near the shores but dwell only in the deep (together with other large marine creatures: Aelian, *NA* 9.49). Consequently, they were impossible to study in their native blue-water habitat. "Whales existed at the margins of belief in antiquity,"[49] as creatures more mythical than biological (e.g., the huge fish that swallowed Jonah or the ship-swallowing whale in Lucian).[50] In artistic representations, whales are sinuous, serpentine, scaly creatures who sport horns or tusks.[51] Aristotle recognized their similarity to dolphins: both are viviparous; both expel air (and water) through blowholes on their foreheads. But Aelian imagined whales emerging from the sea and, like seals, basking on rocks to sun themselves (*NA* 9.50).

Although whales (i.e., large marine creatures) were roughly lumped together as *kete* ("sea monsters:" Chapter 7), several species were recognized: the common phallaina/baleina (fin whale [*Balaenoptera physalus*]), the toothed orca (*Orcinus orca*; Aelian's "sea-ram" with a white band on its forehead: *NA* 15.2), and the large toothed, spouting, blue-water *physeter* ("whale" with a blow hole: sperm whale? whose range includes all the earth's oceans [*Physeter macrocephalus*]). From examination of beached specimens, Aristotle may also have observed the differences between toothed and baleen whales. His "moustache-whale," for instance, has pig-like bristles in its mouth instead of teeth.[52] But there is no indication that he understood the function of baleen.

Cetaceans were observed both in the Atlantic Ocean and the Erythraean Sea. Oppian compared the whales of the Atlantic with ships powered by twenty oars (*Fishing* 5.59). In Pliny, the whales in the Gallic Ocean are the largest, "rearing up higher than a ship's rigging and disgorging a flood" (9.8). Strabo described the spray of whales off the coast of Tourdetania as a "cloud-like pillar" (3.2.7). In the Erythraean Sea ("the outer ocean"), very large marine animals (25 *orguiae* [150 feet = 46 meters]) troubled the crew of Alexander's admiral, Nearchus, off Cyzica in 325 BCE. The spraying whales gave off "great streams and a large body of mist from their eruptions." The crew were so startled that they dropped their oars. But Nearchus encouraged his fleet to enjoin battle, raise a battle-cry, splash their oars, and make a great deal of noise with their trumpets, thus sending the pod back to the depths. The crew praised their skipper for his courage and wisdom.[53] Nearchus' whales were possibly blue whales (*Balaenoptera musculus*, ca. 80 feet [25 meters]) or sperm whales (ca. 65 feet [20 meters]): both species are native to the Erythraean Sea. Whales were also observed off Britain.[54]

In the ancient sources, whales were generally characterized as insatiable, gluttonous, slow, and lumbering (owing to their extreme size). Because of their monstrosity, they were thought to threaten both ships and weaker, smaller specimens of their own kind, including calves, pregnant cows, and other cetaceans, "as if with the rams of warships" (Aelian, *NA* 15.2). On account of their immense size and power, they were considered fearless, routed only by violent collisions (Oppian, *Fishing* 5.50–51; Pliny 9.7). Orcas, in particular, were demonized, construed by Pliny as "huge with flesh and ferocious with teeth." One orca, sighted in the Ostian harbor during Claudius' reign (41–54 CE), was trapped by its gluttony and driven to shore by the currents. Ordering the praetorian guard to block the harbor entrance with nets and hunt the creature with javelins, Claudius thus provided the spectators with an entertainment.[55]

Pliny (9.7) reported that the Gedrosi on the northern coast of the Arabian Sea used whale bones, as long as 40 cubits (60 feet [18 meters]), for doorways and roofing beams.[56] The Gedrosi and others would scavenge whale bones that had washed up with the tide on the beaches of Cyzica (Arrian, *Indika* 30.8–9). Plutarch blamed an epidemic on the decomposing body of a beached whale at Boulis on the Corinthian Gulf (*Cleverness* 981b).

Like seals, whales were not usually deliberately hunted for food, but they were consumed opportunistically. For example, the notorious "Porphyrios" (probably an orca) enjoyed a long career terrorizing the waters near Byzantium. For decades, he sank ships, forcing travelers to seek safety by sailing well off the usual courses.[57] In 548 CE, perplexed and in an effort to keep the sea roads safe for travel and commerce, Justinian hoped to capture the beast. Porphyrios' demise, however, occurred serendipitously when the whale was lured close to shore by a pod of dolphins. Like Claudius' orca, "driven on either by hunger or simply by love of slaughter," the demonized creature eventually became stuck in

Biological Creatures of the Sea 139

the shoals, and a citizen army assaulted the 30-cubit long (45 feet [14 meters]) *ketos* with axes. Dragging Porphyrios ashore and dividing the flesh for food, they took delight in their retribution (Procopius, *History of the Wars* 7.29.9–16). The whale's slaughter was thus glorified as human conquest over raw nature (cf., Chapter 7).

Sharks

Pliny included sharks among the largest animals of the Erythraean Sea. Three types of sharks (*kuones*: sea hounds) were recognized among the fifty or so species of the carnivorous cartilaginous fish that hunt in packs in the Mediterranean: the huge man-eating *zugaena*, the "most warlike" of the deep-sea creatures (hammerhead shark [*Sphyrna zugaena*] and blue shark [*Prionace glauca*] known to attack humans); the small (18 inch [45 cm]) but fierce bottom-dwelling speckled *galeotes* (spotted dogfish [*Scyliorhinus canicula*]); and the bottom-dwelling *kentrinas* with hard, poisonous black spines on its crest and tail (spiny dogfish [*Squalus acanthias*]).[58] Sharks (especially the spotted dogfish) are frequently represented in Roman mosaic seascapes.[59] Aristotle knew that some sharks are viviparous while others lay eggs in cases, wisdom that was repeated by Pliny.[60] Aristotle's observation of the placenta-like membrane in sharks was remarkable, and forgotten until Guillaume Rondelet repeated Aristotle's description in 1554 (*Libri de piscibus marinis in quibus verae piscium effigies expressae sunt* ["Books about marine Fish in which the true Images of Fish are expressed"]). The Danish anatomist, Nicolas Steno, finally confirmed the shark's placenta in 1673. The long-held belief that some shark species, including the oviparous dogfish, give birth through their mouths, may derive from confusion between dogfish and other species: freshwater cichlidae do incubate their eggs and protect their fry in their mouths, as do saltwater apogonidae.[61] The sources are surprisingly silent on the shark's multiple rows of teeth (two or three rows for the spotted dogfish: see Chapter 7 for Scylla's three rows of teeth in Homer).

Fishermen would lure sharks with baited hooks: when one shark was captured, the others in the shiver would rush and even leap into the boat hoping to snatch up free food. They were consequently caught "of their own free will" (Aelian, *NA* 1.55). Because of their perceived ferocity and power, their fossilized teeth were worn as amulets,[62] and their rough skins were used for polishing wood and ivory (Pliny 9.40, 32.108). Because they were thought to be notorious man-eaters, their flesh was not popular among gourmands, but in Athenaeus (7.310c–e), we have

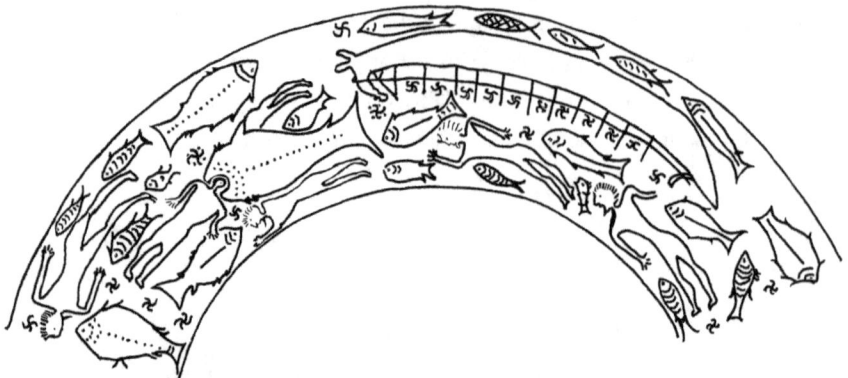

Fig. 6.2 Shipwrecked sailors. Geometric Krater, Ischia: Museo Archeologico, Pithekoussai 618813.

Archestratus' "divine" recipe for saw-toothed shark prepared with cumin, salt, and olive oil.

Sharks are deep-sea creatures who could endanger sponge divers, attacking their heels, loins, and "every white part of their bodies." But dogfish were as afraid of divers as the divers were of the sharks, and their only safety was an aggressive, underwater attack (Pliny 9.152–153). Leonidas of Tarentum nonetheless described the horrible fate of Tharsys, son of Kharmides, the diver sent to loosen a fouled anchor. As Tharsys ascended, just reaching the surface, a shark "swallowed me just up to my navel." Tharsys' fellow sailors retrieved and buried his torso in the sand (*Greek Anthology* 7.506). In another epigram, a shark ate one of two shipwrecked men as they were fighting over the same floating plank (Antipater of Thessalonica, *Greek Anthology* 9.269). Such scenes are also recorded artistically. A Geometric krater shows six shipwrecked sailors at the mercy of a horde of predatory fishes: one victim has lost an arm, another is swallowed head-first, and others try to escape[63] (Fig. 6.2). Plutarch also reported an unusual (and inauspicious) shark attack close to shore. As an Eleusinian initiate was washing his sacrificial piglet in the Kantharos harbor at Athens, he was snatched up by a shark and swallowed, like Tharsys, up to his belly.[64] The event was particularly concerning because it occurred, unusually, so close to shore.

Cephalopods

Also known to ancient folk are a variety of octopods, the largest of which dwell in deep water. Aristotle's account of these fascinating, shy creatures was

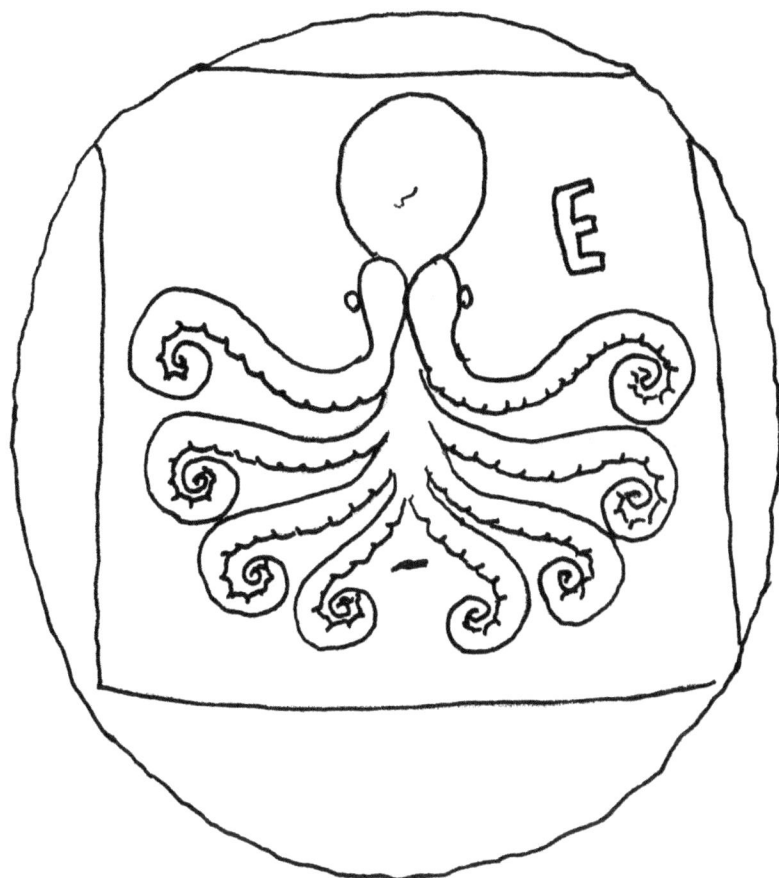

Fig. 6.3 Octopus in incuse square on silver didrachma of Eretria, 500–480 BCE. Athens, Numismatic Museum, Collection Gr. Empedocles, "Coins of the Aegean" #39–156.

replicated in Pliny and Aelian. Because of its tenacity, flexibility, and sinuousness, the octopus was popular in art, adorning coins (cf., *UCWW*: Chapter 8) (Fig. 6.3), Minoan amphorae,[65] Greek vases,[66] shields,[67] and mosaics well into the Roman era.[68]

The cephalopod's defensive ink sac was well-known, situated in the upper body in octopods and near the belly in cuttlefish. Since cuttlefish live close to shore and lack the octopod's defensive ability to change color, it was believed that they produce ink in greater abundance, and that the lower placement of their ink sacs allows them to project ink over a greater distance. Aristotle portrayed cephalopod ink as earthy, resembling the chalky residue in bird excrement because—so it was thought—both birds and cephalopods lack bladders. Earthy excrement thus settles into the ink sac and from there it is excreted. Unlike other cephalopods, octopods have the advantage of being able to change color defensively in order to blend in with the surrounding rocks, appearing "similar to stone," a trait that privileges them as hunters (Lucian, *Dialogues of the Sea Gods* 4: *Proteus and Menelaus*). Neither Aristotle nor any other ancient author speculated on how the octopus is able to do so. The multivalent cephalopod tentacles are effective for swimming, strangling prey, engaging in sexual intercourse, and clinging to rocky surfaces.[69] Octopods are also able to discharge any sea water that they have consumed through a tube in the back of their heads. The heads of living octopods, as observed by Aristotle (*HA* 4.1.524a10–13) and Pliny (9.85), are "hard" and seemingly "inflated."

Aristotle was fascinated by the unusual mating behavior of the octopus and other cephalopods, via the "mouth" or "nostril."[70] The male inserts his hectocotyl tentacle—the forked and pointed tentacle that contains the sexual organ and has the largest suckers —into the female's "nostril" (a muscular mantle cavity that appears to be situated in the "head," as are all octopod organs). When the eggs are fertilized, the female discharges into her lair a tendril-like structure around which she clusters her cache of white eggs (whose volume exceeds her own, so the sources claimed). She tends them for a period of fifty days until the tiny, helpless, spider-like hatchlings emerge. But their mortality rate is high. According to Pliny (9.93), polypods have a life expectancy of three years. In Aelian (*NA* 6.28), they were expected to live only a year. Aelian rightly reports that the female's life expectancy is even shorter because her strength is drained by the act of mating (or at least of incubating her eggs).[71]

Aristotle observed that these unusual creatures are capable of both swimming and crawling because of their eight, suckered tentacles, which they utilize like hands or feet. Aristotle's description is detailed and sensitive. Unlike other cephalopods who can only swim and whose outer two tentacles are larger, octopods have four large middle tentacles. In comparison with other cephalopods, octopods have proportionally longer tentacles, allowing them to cling to rocks for safety during storms. Arranged in a double line in most species, suckers are comprised of interlaced fibers that arc over surfaces when slack but adhere tightly when stretched, yielding an unbreakable seal,[72] comparable with the tools that surgeons employ for resetting dislocated fingers, on Aristotle's simile. The octopod swims "obliquely," extending its head to one side, feet to the other.[73] Octopods have also been seen on land, seeking out rough terrain ("they hate softness") or sunning themselves on sea rocks (Aelian, *NA* 7.11). The octopus can survive on land for a short period of time owing to the fact that oxygen is absorbed through their skin, not through a gill, provided that the skin remains moist.

These creatures were also fond of shellfish whose remains would indicate the location of octopod lairs. Pliny (9.87) denied the long-held assertion that an octopod would eat its own tentacles, as did Plutarch (*Cleverness* 978f). A century after Pliny, Aelian suggested that the voracious creatures would eat their own tentacles: thus they always have access to a ready meal (octopods practice autophagy if under high stress). Both Pliny (9.90) and Aelian (*NA* 1.27), however, correctly noted that a severed tentacle would grow back, like a lizard's tail. Pliny also reported that octopods cleverly use tools to protect themselves from the destructive snap of mollusk shells, which could fracture the tentacles: when the prey opens its "mouth," the octopus drops in a stone that wedges the mollusk open, and then extracts the fleshy morsel.

In Pliny (9.93), we also have the tale of an octopus who persistently looted salted fish from the fish-salting factory at Carteia (on the Bay of Gibraltar). The angry guards there first tried fencing in the tanks, but the clever cephalopod climbed a tree to approach the tempting salt fish from above. Guard dogs were put off by the odd creature with its horrible breath, huge size, and unusual color. The octopus fought back, using its tentacles like clubs. The team finally brought the creature down with tridents. Pliny's source, Trebius Niger, who wrote a treatise on ichthyology (ca. 150 BCE), depicted the 700-pound beast as having a head the size of a 15-amphora barrel, suckers the size of basins, and 34-foot-long "arms."

In Aelian (*NA* 13.6), another mischievously monstrous octopus scuttled through a sewer to burgle pickled fish from Iberian cargo ships at Dichaearchia (Puteoli). Amphorae were in shards on the floor, but the thief left no other incriminating evidence (doors and roofs were untouched). The merchants left their "boldest" slave, fully armed, to lie in ambush. Instead of engaging the athletic beast, the fearful man deemed the polypod too powerful. On the following night, the poor beast was dispatched by a band of vengeful merchants who trapped the creature and hacked it to pieces. The many compelling parallels (larcenous octopods, timid guards, heroic slaughter by a band of men) suggest that Aelian's tale may be adapted from Niger's.[74] Polypods are known to come ashore for better hunting. They do not, however, climb trees.[75]

It was also believed that octopods in the Atlantic grew so large that they were unable to slip through the Pillars of Herakles (Pliny 9.8). Trebius Niger further reported that polypods, like sharks, would attack divers and shipwrecked men, as illustrated on a black-figure *dinos* where a large octopus rises to the surface as if to threaten the two ships on the rim.[76]

Conclusion

The nature of water shapes how the ancients envisioned creatures who inhabit the murky depths, and this conception is ambiguous. On the one hand, marine creatures were closely observed, and astoundingly accurate details are preserved in Aristotle and his intellectual descendants. On the other hand, marine animals, like other non-human creatures, served as mirrors of human character and, like other wild animals, they were deemed examples of raw, untamed, uncultivated nature that was to be vanquished by human superiority. Although there was much sympathy for the playful, winsome, friendly dolphin, most marine animals were imagined as reflective of their mysterious, metamorphic environment, thus inspiring the sea monsters that terrified sailors and harassed folk heroes.

7

Mythical Creatures of the Sea: Sea Monsters and Sea Gods

Introduction

The watery depths were populated by little understood creatures that were either endearing (dolphins, seahorses) or terrifying (most other marine animals). Their range is fluid, amorphic, and metamorphic. Water changes color as the light shifts, and its appearance is affected by fluctuating winds and currents. Objects underwater are distorted by light refraction, and the submarine environment fosters the growth of barnacles that cling to underwater structures, such as piers and even marine animals, including whales.

The watery setting is by its very physics and optics one of change and mystery, eliciting fear. Compounded by tales of ship-wrecking whales, man-strangling octopods, and man-eating sharks, it is no surprise that the ancient imagination created even more fanciful and frightening sea beasts.[1] It was widely believed that the largest sea beasts dwelt in the Erythraean Sea, where they would appear at the solstices when they were churned up from the depths by violent storms and whirlwinds (Pliny 9.4–5). The Erythraean Sea sustains whales (*ballaenae*) of 4 *iugera* (about 3 acres), sharks (*pristes*) extending 100 cubits in length (150 feet [46 meters]), and lobsters (*locustae*) of 4 cubits (6 feet [about 2 meters]).[2] Such accounts fueled the creation of "sea monsters," including hydras and *skolopendrae* ("millipedes").

Imagined Sea Creatures

Pliny declares, "whatever is born in any part of nature exists also in the sea, and many things in addition that do not exist elsewhere" (9.2). Vernacular taxonomy would agree: seahorses, sea lions, sea robins, Steller's sea cows (now extinct), elephant seals, catfish, pigfish, lionfish, lizard fish, frogfish, zebra sharks, eagle rays, sea spiders, sea anemones, sea grass, and sea cucumbers resembling land cucumbers in appearance and scent. Greek and Roman thinkers categorized the unusual in familiar terms, thus making exotic creatures more accessible and comprehendible. Greek and Roman art is replete with references to, and images of, such crossovers. Pliny described sheep-like creatures that would come to shore to graze on roots before returning to the water. Other composite marine creatures—with the heads of horses, donkeys, and bulls—would also eat crops (Pliny 9.7). In Aelian (*NA* 9.49), submarine threats include the invincible *maltha*, the inauspicious sea-hyena, and the ram-fish (orca?), whose splashes create large waves that swirl up the depths. Skippers of twenty-oared galleys reputedly honored sea gods with rib bones of 10,000-foot-long (3 km) *skolopendra* that washed up on shore (Theodoridas, *Greek Anthology* 6.222). Coming across a 4-*orguia*-long (24.5 feet [7.5 meters]) *skolopendra*, a fisherman returned it as an offering to the sea gods (Antipater, *Greek Anthology* 6.223). Fishermen reported seeing such monsters (marine polypods), the size of a trireme, floating in the water, with flat tails and very long hairs growing from their nostrils. As they swam, their countless feet resembled the tholepins of ships under oar (Aelian, *NA* 13.23, 16.12). In Pliny (9.8) and Aelian (*NA* 13.20), "wheels" (*trochoi*) came close to shore in large numbers, timid but crested with four radiating spines of enormous length.

Eliciting less terror, composite man-fish tritons were also attested: one was sighted playing a shell in Lusitania during Tiberius' reign (Pliny 9.9; Suetonius, *Tiberius* 9.9–10; Aelian, *NA* 13.21). In Pausanias (9.21.1), tritons have fine, shark-like scales, gills under their ears, a human nose, but a broader mouth and beast-like teeth. Their eyes are blue-gray (γλαυκά: *glauka*), their fingernails resemble murex shells, and they have dolphin tails in lieu of feet. Nereid sightings were also noted. Their human parts bristled with hair, and the mournful song of a dying nereid could be heard from a distance (Pliny 9.9).

The artistic record emphasizes the whimsical fusion of land and water fauna, and such mythical hybrids may have inspired Empedocles' theory of "evolution" (*TEGP* 118–122), where early creatures were outlandish assemblages of disjointed body parts (neckless heads, shoulderless arms, unconnected eyes) or bizarre amalgams (ox-headed men). The earliest sea monsters are leonine hybrids: with three heads either in a vertical or horizontal row attached to fishy back sides[3] (Fig. 7.1). A particularly fine mosaic shows a sea lion, with a full feline mane and long, sinuous trifurcated fish tail, swatting at the water with his feline forepaws[4] (Fig. 7.2). Miniature *kete*, kept as pets by Nereus' daughters,

Mythical Creatures of the Sea: Sea Monsters and Sea Gods   147

Fig. 7.1 Three-headed leonine sea monster, Boston Museum of Fine Arts 6.67 (Archaic, Proto-Attic vase).

Fig. 7.2 Composite lionfish, Ostia Antica, Baths of Neptune, second century CE.

took many forms (leonine, canine, ovine), rendered with flippers, horse legs, or spiked ruffs, snouts (furrowed and sometimes fitted with goatees or tusks), and delphinian or piscine tails.[5] Like seals, the Nereids and their mother, Doris, are imagined sunning themselves on rocks as they dry their green hair, or swimming and riding fishes or sea monsters with scaly, serpentine bodies (Ovid, *Metamorphoses* 2.8–14). We now turn to consider hybrid mythical creatures of the deep, including the capricorn and hippocamp—popular images in private and civic art—and their more horrifying marine cousins who threatened the heroes of myth.

Capricorn

The composite capricorn emerged from an act of self-metamorphosis by the good-natured hybrid, Pan (aegipan, goat-man). During the battle between Zeus and Typhoeus, Pan had suggested that the Olympians assume animal disguises to deceive the monster.[6] Before hiding in the Nile, Pan disguised himself as an amalgamated creature that synthesized the characteristics of both god (Pan the goat) and his means of evasion (fish in the Nile: cf., Manilius 2.167–72). Pan's transformation into the capricorn—a hybrid amphibious creature—was key in preserving Olympian rule and thwarting Typhon's tyranny. Pan had saved the Olympians by his use of water. In Augustan Rome, Octavian presented himself as having preserved the Roman state in a decisive naval battle at Actium, and the capricorn, his particular emblem and moon sign, which held sway over the sea, came to represent that maritime victory as emphasized on coins and in public art.[7]

Hippocamp[8]

In ancient art, hippocamps (seahorses) commonly adorn ships or draw Poseidon's chariot in braces or quartets, but the creature is little mentioned in the literary record. Homer depicted Poseidon's hippocamps as "swift-springing," "bronze-hoofed" creatures with long, golden manes that fly in the wind (*Iliad* 13.23–31).[9] In Philostratus (*Imagines* 1.8: describing a painting), these skillful swimmers were altogether delphinian, with their watery hooves and gray eyes. Most of our evidence comes from the artwork where the form is largely standard. A fully formed equine torso (usually with two legs, sometimes four) is attached to a sinuous, flowing fish tail. The Romano-British hippocamps on the Mildenhall Dish (Suffolk)[10] and the mosaic floor at Aquae Sulis (Bath) are finless, two-legged creatures. Elsewhere, Poseidon rides a serpentine, black-figure seahorse with a complete set of carefully wrought fins (dorsal, adipose, pelvic, and anal). Scales extend up the neck.[11] One splendid winged hippocamp is encircled by waves in ringlets on a red-figure plate. The elegant wings (with indications of covert feathers) are balanced by a brace of feathery pectoral fins, and a webbed crest runs down the fishy spine (whose stripes repeat the black lines that separate the primary and greater coverts from the median coverts)[12] (Fig. 7.3). In the hands of artists, thus, the tiny, delicate dorsal fins have evolved into Pegasus wings.

Fig. 7.3 Winged hippocamp, State Hermitage Museum, Saint Petersburg, ca. 320–310 BCE.

Sea Monsters

There is little doubt that those who vanquished biological sea monsters (larcenous octopods and ship-wrecking whales) considered their acts as heroic conquests of human civilization over the forces of raw nature. Aelian's band of octopus hunters were "eager to risk the danger" of encountering the hated beast, and others joined the fray thirsting to be a part of this "strange and incredulous spectacle" (*NA* 13.6; Chapter 6). Sea monsters also populate the mythic landscape, challenging both gods and heroes. The creation of Pandora is all the more devastating to man because she sports a crown featuring "terrible monsters" nourished by "land and sea" (Hesiod, *Theogony* 582). The conquest of such beasts represents victory over chaos on a cosmogonic level, extending back to the Ugaritic tradition (where sea monsters were among the enemies of Baal and

Anat) and Hebrew lore (where YHWHs conquest of the Leviathan was an allegory of the battle that ushered in the new age).[13]

Typhoeus

The sea remained a symbol of chaos and a boundary between the human and divine realms, between life and death.[14] Like Marduk, Yahweh, and Baal, Zeus must overcome a powerful marine creature. The monstrous son of Gaia (Earth) and Tartarus, Typhoeus was a flame-breathing, serpentine monster with one hundred snaky heads emerging from his shoulders, uttering every "indescribable sound" (Hesiod, *Theogony* 829). In Hesiod (*Theogony* 820–880), the battle is a violent tempest of fire and wind over the dark sea. Long waves crash back and forth, as land and sky and underworld seethe. Zeus' strategy anticipates Herakles' battle against the Lernean hydra: Zeus must scorch each of the snaky heads. Typhoeus then becomes a maritime entity posing a navigational threat, the source of wet winds that scatter ships and drown sailors. In artistic renditions, moreover, sucker-like details appear on the underside of Typhoeus' tail, resembling Minoan decorative octopods and anticipating the sea monsters of medieval maps.[15] The gentle octopus was thus re-imagined into the fearful sea monster that threatened the settled rule of law overseen by Zeus and the safe transit of Greco-Roman sailors.

The polypod may also have inspired the polycephalous, serpentine, Lernean Hydra, child of Typhoeus and Echidna, one of the monsters of Herakles' canonical twelve labors (Hesiod, *Theogony* 313–315). The tentacles were recast as heads. Literary and artistic sources vary on the number of heads, from six to fifty.[16] But the Hydra's heads, like octopus tentacles, were regenerative.[17] In Aelian (*NA* 9.23), Herakles' Hydra was fictive, but the image of the invincible, reviving water snake became a metaphor of tenacity.[18]

Scylla

Forged from an act of metamorphosis, Scylla was often pictured as a human-canine-piscine composite. In Ovid, Scylla was once a beautiful maiden whom Glaucus wooed. Jealous of a rival for Glaucus' affections, Circe poisoned the spring where Scylla bathed, thus turning the girl into a hideous sea monster.[19] Homer's cave-dwelling Scylla bears a superficial resemblance to the polycephalous Hydra. Stealthy and quiet, Scylla has twelve misshapen legs and six very long necks, each of which sports a fearful head. In each there are "teeth in three rows"

(*Odyssey* 12.85-100). It is uncertain whether the detail is merely a dramatic flourish or if Homer knew of the shark's conveyor-belt rows of rootless teeth.[20] In Vergil, Scylla was a maiden "with a lovely chest" whose fishy bottom half was monstrously large, fitted with delphinian tailfins and wolf bellies (*Aeneid* 3.426–428), but there is no mention of the teeth. Scylla was represented frequently in artwork as a lovely woman with a dragon tail and canine heads emanating from her belly.[21]

Kete

Herakles and Perseus also faced sea monsters of the man-eating variety. For Cassiopeia's *hubris*—claiming that she, the Aithiopian queen, was more beautiful than the Nereids—Poseidon sent a *ketos* to terrorize the Aithiopian coast. An oracle revealed that the gods could be appeased only by the sacrifice of Cassiopeia's lovely daughter, Andromeda. The hero, Perseus, however, saved the princess by petrifying the monster with the head of the Medusa (which he had recently slain).[22] Ovid's description of the beast is vague. A serpentine dragon (*draco*) with deadly, greedy fangs, the creature reacts like a wild boar held at bay by a pack of hunting dogs. In the artwork, the monster takes many forms: for example, a pig-snouted creature with a long, flickering tongue and crested forehead;[23] or a winged serpentine creature with a long saw-toothed snout, spiky crest down its neck, and dorsal fin. On one depiction, scales are rendered with white dots, and the coiling, serpentine tail ends in a bifurcated caudal fin.[24] According to Pliny (9.11), Marcus Scaurus claimed to have brought the monster's bones from Jaffa to Rome.

The story was duplicated in one of Herakles' side-labors. At Troy, whose king, Laomedon, had refused to pay the gods after they had built the impenetrable city walls, Poseidon and Apollo punished Troy with a *ketos* and a plague, respectively. Oracular pronouncements offered the same advice given to the Aithiopian royals, that a virgin princess (Hesione) must be sacrificed to the monster. She would be saved by Herakles who—after being swallowed by the monster— hacked his way out, thus killing the beast.[25] This scene was also rendered artistically. One depiction shows a nude Herakles threatening the Trojan *ketos* with a large hook. The shark-like *ketos* has a mouth full of sharp teeth, dorsal fins, gills, flippers, a spiny back, and sinuous tail with a bifurcated caudal fin. A seal, two dolphins, and a cephalopod swim around the *ketos*. Herakles' sea monster is typically shown with clearly-rendered scales, dorsal fins, often in triplicate, and bifurcated caudal fins.[26]

We also have vivid literary accounts of the sea bull sent by Poseidon to destroy Theseus' son, Hippolytus.[27] Over time, the image of Poseidon's sea bull evolved from Euripides' broad brush-strokes to Seneca the Younger's hyper-photographic realism. In Euripides, the bull emerges from the seafoam, an image that evokes Aphrodite rising from the spray that forms around Ouranos' severed testicles.[28] In Ovid (*Metamorphoses* 15.11–13), the bull snorts "a portion of the sea from his nostrils and wide-open mouth." In Seneca, the details are meticulous: a *capax* (spacious) *physeter*, the snorting beast has a blue neck, crest on its green forehead, bristling ears, and flame-darting eyes. Moss (or seaweed) drips from its dewlaps and chest, and the scaly creature drags a huge "part (tail) such as could smash boats in the distant sea."[29] Euripides' bull appears from the foam, Ovid's merely swims (and spits up water like one who has been submerged for too long). In Seneca, the bull was fully envisioned, with piscine characteristics and, like Scylla and Typhoeus, a legitimate threat to maritime traffic.

Skiron's Turtle

Theseus also encountered a monstrous, man-eating sea turtle who swam beneath Skiron's lair and consumed the corpses of the brigand's victims (Skiron would kick travelers into the sea after robbing them). Theseus dispatched Skiron by pushing him off the same cliff. The bandit was presumably consumed by his own turtle.[30] Pausanias explicitly distinguished Skiron's turtle as a sea turtle: sea tortoises (αἱ θαλάσσιαι [χελῶναι]: *hai thalassiae chelonai*), he said, resemble land tortoises except in size and their "seal-like" feet (1.44.8–12).

Hybrid Gods of the Deep

Marine gods, likewise, share the characteristics of fluidity, metamorphosis, and the terror that we have observed in the ancient conception of biological and mythical marine animals. No wonder, then, that Greek and Roman storytellers would understand such deities as they did their theriomorphic avatars. Water gods could inspire and soothe (like dryads), but they could also be violent and unpredictable. We consider elsewhere the polyvalent Olympian deities whose functions include, but are not limited to, the sea (Chapter 9). We now turn to composite and metamorphic marine gods—cerulean (dark-hued) and strong—that are associated exclusively with the watery realm.

Mermen

The man-fish Triton was anticipated in the early eighth-century BCE Assyrian "Frieze of the Transportation of Timber."[31] Sporting a distinctive Assyrian beard and headdress, his human torso is completed by a scaly fish tail. The merman swims in front of a horse-beaked vessel with a fish tail stern from which three men offload a log. Like his Assyrian predecessor, Triton was a mature "broad-bearded" man and half-green fish ("half-finished"). Covered with barnacles, he emerges from the water with a dripping beard (Ovid, *Metamorphoses* 1.330–347; Nonnus, *Dionysiaca* 36.93–94, 43.205). The son of Poseidon and Amphitrite, he often accompanied and assisted his father.[32] In Hesiod, Triton is "a dreadful god" (*Theogony* 930–933). But elsewhere he is the "tuneful" sea deity who is often shown blowing on his hollow, spiraling conch shell filling the shores with his song (Nonnus, *Dionysiaca* 1.59).

Triton was not unique. In Aeschylus we meet another merman, Glaucus, a "humanoid" beast, neither fully human nor fully fishy, whose mortality was constantly cleansed by running river water (*Glaucus of the Sea* fragment 26Radt). In Ovid (*Metamorphoses* 13.898–965), the tale is fully developed. Glaucus was once a fisherman whose catch one day began leaping from the fishing net onto the grass. The fish then returned to the deep. Perplexed, Glaucus wondered if the odd behavior was owing to divine intervention or some magical property in the grass' sap. He consequently decided to chew some of it. Glaucus' conversion from fisherman to merman is vividly wrought: the beard turns dark blue, the hair begins to flow like water, his shoulders and arms become blue, and the legs taper into a fishy tail fin. Glaucus would be worshipped as a prophetic "old man of the sea" in Spain (Scholiast, Apollonius, *Argonautica* 2.767).

Old Men of the Sea

Although most gods were metamorphic, transformation was the hallmark of gods who dwelt primarily in the water. Born from Gaia and Ouranos, Okeanos contained the essence of all matter, thus accounting for his transfigurative ability, a trait that conveys to other marine gods and creatures.[33] The prescient, seal-loving Proteus was the cranky old man of the sea who guarded his knowledge jealously, transforming himself in order to avoid answering the queries of mortals. Only Proteus knew how Menelaus could return safely to Sparta from Troy. In order to learn this wisdom, Menelaus was instructed to undergo his own metamorphosis, "becoming" one of the seals in Proteus' beloved pod who sunned themselves on the shores of Pharos island where they beached at the same time every day (*Odyssey* 4.398–455).[34] The metamorphosis had to be convincing. Disguising himself in a sealskin, Menelaus took his place among the brood and waited. When Proteus arrived for his nap, he was ambushed by Menelaus, who locked his arms on the old man. Proteus transmuted himself into lion, serpent, leopard, boar, flowing water, and finally a tree. Weary and annoyed, Proteus eventually answered Menelaus' queries. The metamorphic Proteus was himself seal-like: like his noisy herd, the old man beached daily at the same time when he "counts his 'harem' of seals and sleeps in their midst."[35]

Other old men of the sea, Nereus and Phorcys, were also immortal, omniscient, and prophetic. They existed at the threshold between immortality and death. Like Proteus, they guarded their privileged knowledge jealously. Anyone who desired to learn proprietary data from these sea gods must, like Menelaus, cling to the transfiguring deity. From the "gentle" Nereus, Herakles learned how to find the island of the Hesperides, but only after the god mutated into every form (Hesiod, *Theogony* 235; Apollodorus, *Library* 2.114). The scene is illustrated on the Friedlaender hydria where Herakles, under the gaze of Poseidon (and possibly Hermes and Athena), wrestles a coiling, serpentine merman wearing a bathing cap. Herakles is entwined within the creature's first helix, and the hero is "holding the merman in an apparently painless version of the half-nelson." Herakles' fingers are intertwined, and the sea beast struggles to escape.[36] These gods could advise those who sought guidance for journeys on the pathless and changing sea. The old men of the sea also communicated divine knowledge that linked the worlds of mortal men and deathless gods.[37] Proteus, for example, revealed that Menelaus would go to the Isles of the Blessed when he died. From Nereus, Herakles discovered how he could attain immortality.

Like her father Nereus, Thetis, too, was a prophetic metamorph who straddled the space between divine and mortal.[38] She was fated to bear a son more powerful than his father, and she was consequently affianced to a mortal king, Peleus, whose proposal she rejected. Like her father, she eschewed cooperating with mortals. Peleus was then compelled to cling to Thetis as she transformed herself into a lion, a snake, and other creatures, until her resolve was weakened. The scene is rendered famously on an Attic red-figure kylix from Vulci.[39]

Still other marine gods changed the shapes of their worshippers or rivals, motivated often by clemency. Okeanos was generally viewed as a benevolent force, and he listened with a sympathetic ear to Prometheus (the Titan had challenged the authority of Zeus), bound to a mountain in the Caucasus.[40] Dionysus' pirates became dolphins, and his aunt, the jealous Ino, became

Leukothea, the white-armed goddess who would rescue Odysseus from a storm at sea (Chapter 9). After the suicide of queen Dido, Aeneas' ill-fated Carthaginian lover, her sister, Anna, sought asylum in Latium, but Aeneas' jealous wife, Lavinia, became wrathful. The River Numicus rescued Anna by changing her into the nymph, Anna Perenna (the perpetual year), the patroness of an eponymous annual festival, one of Rome's most popular, when, on the Ides of March, on the banks of the Tiber, revelers would imbibe as many cups of wine as years they hoped yet to live (Ovid, *Fasti* 3.531–532).[41] Ruined by the fires of her sister's unhappy pier, Anna was later saved by water.

Conclusion

The ever-changing appearance of water influenced the conception of imaginary sea creatures, both the benign (the conch-playing Triton, the capricorn, the winsome seahorse) and the menacing (Typhoeus, Scylla, *Kete*), envisioned as composite creatures built from different biological animals. These hybrids reflect an essential characteristic of the metamorphic sea god. They were creatures between forms, they were animals in a process of metamorphosis, but they were neither one nor the other, and yet they were both. The gods who possessed "the vast sea where there is nothing but the abode of monsters" were theriomorphic, hybrid, changing, unpredictable, and very sea-like.[42]

8

Water and the Divine: Unseen and Magical Forces of the Spiritual World

Introduction

For the ancient Mediterranean peoples, there was little distinction between the spiritual and physical worlds. Nature could only be understood in conjunction with the divine. The physical world was envisioned as a living, anthropomorphic, dynamic entity that undergoes change (either in elemental cycles[1] or anthropomorphic terms through "generations" of humanity [e.g., Five Ages]) and where gods are connected with elemental forces and components of the material world.[2] In both the mythological and "rational" systems, the world was subject to oversight, either by fate or by the laws of physics: Fire in Heraclitus, Love and Strife in Empedocles, Plato's Demiurge, Aristotle's Unmoved Mover, or the all-pervasive Stoic *pneuma*.[3] This culminated in the Neoplatonic tenet that nature was nothing but an expression of the divine. Water was a powerful tool of unseen forces that connected the human and divine realms. Here we explore watery aspects of that liminal zone where human and divine meet: miasma, the communication between the spiritual and the mundane, and the conduits between the living and the dead.

Miasma and Ritual Purity

Water in the form of cataclysmic flooding was a powerful tool of cosmogony and purification. The Near Eastern accounts of floods sent by the gods to punish humankind for their wickedness or *hubris* (setting oneself as equal to the gods) are preserved in the *Atrahasis*, *Gilgamesh*, and *Genesis*. Greek and Roman writers adapted these tales for their own ends. In Ovid (*Metamorphoses* 1.262–312), we have the elegant account of a flood sent by Jupiter to cleanse the world of those people who had refused to worship the gods, the ultimate act of *hubris*. As the winds and rivers are unleashed, all living creatures are completely displaced: marine dolphins invade the forests; birds cannot find places on which to alight; predatory wolves and their sheep-prey swim together; the swift deer cannot outrun the torrents; and the wild boar, renowned for his strength, finds his muscles useless.

Hubris often results in miasma, imprecisely understood as "ritual pollution." Miasma refers to transgressions against institutional laws, ethics, and (for our purposes) the inappropriate shedding of bodily fluids (blood, menses, urine) on sacred ground or in sacred waters.[4] The epigrammatist, Apollonides of Smyrna (first century CE), imagined a sacred spring bemoaning its victimization by an act of miasma:

> Pure (for the Nymphs have gifted the name to my spring, pre-eminent over all dripping things), when a pirate had killed some men reclining nearby and washed his bloody hands in my hallowed waters, having reversed my sweet flow, and it no longer bubbles up for travelers. For who will still call me "Pure?"
>
> *Greek Anthology* 9.257

Here the miasma has stopped the flow of water altogether. Antiphanes of Megalopolis recounted a similar lament:

> Having streamed once upon a time with well-watered flows, I am now deprived of all my nymphs, even to the droplets. Defiling me with gore, a man-killer has washed his hands in my streams. From that time my maidens have fled the sunlight, saying, "we are nymphs who mingle only with Bacchus, and not with Ares."
>
> *Greek Anthology* 9.258

In both epigrams, miasma culminates in the complete stoppage of the water. Shedding blood into the water supply, a crime with deep ramifications beyond the act of murder that precipitated the miasma, represents both physical and ritual contamination (cf., *UCWW*: Chapter 1). The water is no longer "pure," and it has been rendered entirely unsafe for human use. Historical examples are also noted. The emperor Nero had angered the gods by swimming in the pure, potable waters of the Aqua Marcia. The waters were spared, but Nero—so it was believed—was punished with declining health (Tacitus, *Annals* 14.22.4).

Acts of miasma that could result in the pollution of the entire community (giving birth, having sex, or dying within a sacred precinct) must be absolved,[5] since their effects could be dire. Not burying a corpse, for example, has far-reaching hygienic repercussions. In Sophocles' *Antigone* (1005–1022), Creon's refusal to bury a man whom he considered a traitor, resulted in the pollution of the entire community, creating a rift in the delicate balance between the human and divine worlds. Nature no longer behaved predictably. Fires did not burn as they should. Sacrifices were rejected. The gods no longer communicated with human beings by sending signs through their animal avatars. Miasma must be reconciled, and prescriptions for expiation are detailed in the documentary record. For example, according to the purity regulations of Cyrene, men must wash after sex, and women must visit Artemis' Nympheion before giving birth.[6] According to the *Lex Sacra* of Selinunte, blood sacrifices and libations must be offered to atone for transgressions, and some residents are excommunicated from public sacrifices for a period of time.[7]

It is best to avoid miasma altogether in order to prevent the spread of ritual pollution. Measures were consequently taken before entering sacred spaces and after contact with miasma. Purification, the antidote to miasma, occurs most commonly by means of water (sprinkling) or fire (fumigation). Pliny observed that such cleansing behavior was universal among intelligent animals. Mauretanian elephants, for example, would spray themselves with water from the Amilo River before worshipping the moon (Pliny 8.2). Participants would wash their hands before praying, entering sacred spaces, or making sacrifices (Ovid, *Fasti* 4.778), and many sacred spaces featured washbasins for this very purpose from the Minoan Era onward.[8] Aeneas, for example, dusted himself with water before entering the Elysian fields in the underworld (Vergil, *Aeneid* 6.628–36). Sacrificial animals were also strewn with water before blood sacrifices.[9] These acts of purification by water are analogs to baptism and purification by water in mystery rites, e.g., in the cults of Isis, Mithras, and Christ.

Purification is just as important after contact with death. Corpses were washed in warm water and then perfumed to prevent decomposition. At Troy, Thetis diffused ambrosia and red nectar through the nostrils of Patroclus, slain in battle by Achilles' nemesis, Hector, so that his skin (Patroclus') might remain firm (*Iliad* 19.38–39). We also note that Achilles refused to bathe (and thus purify himself) until he had buried his friend (23.40–41). When the Trojan refugees landed in Italy, they purified the corpse of their bugler, Misenus, with fire, wine, and water sprinkled from an olive branch (Misenus was a son of Aeolus, whom Triton had drowned for the *hubris* of challenging the gods: *Aeneid* 6.164–174).[10] These actions may have further ensured that the body was in fact deceased.[11] In Rome, those returning from funerals (i.e., after contact with a corpse), underwent purification by means of fire and water: as they passed under the smoke of a fire, participants were peppered with water from a laurel branch.[12]

It was, moreover, important to purify ships that had been contaminated by battle in foreign waters.[13]

At Rome, the Vestal Virgins were held to the highest standards of moral conduct. Before entering the priesthood, the inductees swore oaths of chastity, and infractions were punishable by being buried alive.[14] Among other things, these six women were responsible for collecting water for state sacrifices. Ritual water could thus only be drawn by the purest women in Rome. We also note that Rhea Silvia, the mother of Romulus and Remus, was fetching water and washing holy implements when Mars raped her (Ovid, *Fasti* 3.11–12). Additionally, under Romulus, Tarpeia had betrayed Rome to the Sabines as she was fetching water for rituals (Livy 1.11), making her act of treason all the more abhorrent.

In contrast, the historical Tuccia, who had been accused of breaking her Vestal vow of chastity, proved her innocence by performing an impossible task— carrying water in a sieve—according to the legend.[15] During the second Punic War, Claudia Quinta also proved her innocence with the ritual use of water. In 204 BCE, the Roman senate was advised to import the cult image of the Magna Mater (the Great Mother Goddess Cybele) from Phrygia to Rome (to improve the morale of a war-weary populace).[16] After a long voyage, the goddess' barge, finally reaching Italy, was to be towed from the port at Ostia to the city by the men who "wearied their arms" under the weight of the "foreign ship," which "proceeded with difficulty" against the waters. The men's labor was exacerbated by a drought that had left the land dry and thirsty, the grass burnt, and the Tiber a muddy wash in which the barge was stuck. Among the spectators was Claudia Quinta who had falsely been accused of promiscuity. Three times Claudia dipped her hands into the dirty Tiber. Three times she dusted herself with the river water. Three times she lifted her hands to the sky. On bent knee she supplicated the cult image on the barge, appealing to the goddess' sense of justice on the strength of her own innocence (*Fasti* 4.319–324). Claudia's plea—and her chastity—were acknowledged by the goddess, and the unblemished Roman matron succeeded in leading the barge to Rome "with a slight exertion." Like Tuccia, Claudia was able to prove her innocence and purity by the agency of lustral waters.

The ritual calendar also included periodic rites of communal cleansing. In Rome, the month of March was dedicated to opening the campaign season. A series of rituals prepared the army and their tools (armor, weapons, trumpets) for battle. The campaign season was closed in October with an answering set of rituals that ritually decontaminated the army after contact with bloodshed and foreigners.[17] Ovid's account of the March festivals opens with the image of Rhea Silvia collecting water for washing the sacred implements (*Fasti* 3.11-12). Rome's annual birthday celebration (Parilia, April 21), furthermore, featured the purification of the people by fire and water sprinkled from laurel branches (*Fasti* 4.27-728). Fire and water were the two fundamental, binary elements that nourished, sustained, and destroyed life (Jupiter had considered destroying the world by holocaust: *Metamorphoses* 1.253-255). In Rome, fire and water symbolized the shelter offered by a husband to his bride, and the protection of the state over its citizens.[18] The same juxtaposition was applied to the purification of the souls of the deceased. Anchises described the process to his son in the underworld:

> (the souls of the deceased) were spread out suspended towards the empty winds; from others the imbued crime is washed away under a vast whirlpool or is burnt out by fire.
>
> Vergil, *Aeneid* 6.740-742

A soul must be cleansed of the transgressions of one life before it can be reborn into another body. Additionally, before the fisherman, Glaucus, could become a merman, he had to be purified by water.[19] The common methods of purification, flood (water) and fumigation (fire), could be brought to bear microscopically on an individual soul or macroscopically on the entire world.

Reading the Will of the Gods: Divination by Weather Signs and Water

> In the Britanic sea facing the coasts of the Ossismici the (island of) Sena[20] is remarkable for an oracle of a Gallic divinity, whose priestesses—sacred because of their perpetual virginity—are said to be nine in number. The people call them "Gallizenae," and they think that the priestesses, gifted with unique talents, rouse the seas and winds with chants, and that they turn themselves into whatever animals they wish, they heal whatever is incurable among other peoples, they know what will come and they predict these things, revealed to sailors and only to those who have set out in order to consult them.
>
> Mela 3.48

Mela's maritime priestesses are otherwise unattested,[21] but their number and chastity evoke the tradition of Apollo, the poetic god closely associated with the nine muses who accompany him at Mt. Helicon and the oracular deity who spoke through undefiled women who had received the god's gift of prophecy by promising him sex but, in the end, safeguarding their virginity.[22] The transformative nature of the Muses, furthermore, aligns with the skills of Greek sea deities, including the Nereids and the old men of the sea. It is impossible to flesh out the cult of this Gallic cult beyond what is recorded in Mela. The deity held sway over the sea, and he communicated proprietary information through his priestesses, data which could only be shared with pilgrims to his sacred precinct.[23] It is tempting to correlate the god and his cult to the Dioscuri or other protectors of sailors at sea, especially the *Theoi Megaloi* who were worshipped by pilgrims seeking initiation at Samothrace (Chapter 9).

Through prayer, sacrifice, votive offerings, and divination, individuals and communities made requests of the gods and sought signs sent by them. We shall only focus here on water-intensive methods. On the Stoic system, everything in the cosmos was interlinked by *pneuma* and governed by divine providence, thus allowing for the revelation of divine will as interpreted by a rational understanding of signs from the gods (Posidonius f7Kidd). Quite apart from weather prognostication, where the behavior of animals, for example, often aids in weather forecasting (cf., Chapter 4), some signs were sent via meteorological phenomena: e.g., an eclipse after Julius Caesar's death.[24] Eclipses were interpreted as indicating many things, including terrestrial and human fertility and the rising or sinking of major bodies of water, as we read in Petosiris (fl. 150–100 BCE), a priest of the Egyptian Thoth (god of magic and the moon) who recorded celestial omens in the Babylonian tradition:

> A solar eclipse especially in Water-Pourer (Aquarius) and Lion indicates barrenness and the desiccation of rivers and other bodies of water: in Water-Pourer, water withdraws in the North; in Lion, the waters in Egypt and the South dry up.[25]

Some thinkers, moreover, considered astrometeorology a viable method of ascertaining the will of the gods. Quintus Cicero, for example, defended the scientific underpinnings of divination according to Stoic doctrine, arguments refuted by his brother Marcus.[26] Manilius (1.867–873) and Seneca (*NQ* 2.32.2) also considered astrometeorology an empirical science, where meteorological phenomena were products of the natural world, inextricably linked with the four elements, but divine guidance was among their complex causes. As such, natural events could portend the future, including war and insurrection.[27] The practice has its roots in Babylonian astrology, codified in the seventy-or-so fragmentary cuneiform tablets of the *Enuma Anu Enlil* (seventh century BCE), six tablets of which outline signs from thunderstorms, rain, and winds.[28] For example:

If Adad (the Mesopotamian deity of weather) thunders like a lion, the king will flee.

44.1.11

If Adad shouts fifteen times, the harvest of the land will prosper for three years.

44.1.37

If Adad thunders at sunrise, on that day showers will come and (then) cease, there will be sick people in the land.

45.1

Future disaster and success could thus be foretold by the timing and quality of weather events. Much later, the Justinian-era scholar, John Lydus, listed meteorological omens and divination by means of thunder and thunderbolts in *On Signs* (*de Ostentis*): e.g., an invasion of wolves would destroy livestock (39.9), joy and relief from illness (39.23), or war in the east (40.16).

Signs were also sought from terrestrial waters. Popular in antiquity and beyond, but associated with witchcraft, early forms of scrying were also practiced: hydromancy and lecanomancy (divination by bowl of water, among the magical charges against Apuleius: *Apology* 42.3). Originating in Babylon and entering the Greco-Roman world through Egypt in the first century BCE,[29] scrying involves looking at a reflective or bright surface (like a crystal ball) to foretell the future or garner information from the spirit world. In Patrai (in the western Peloponnese), the ill sought cures from Demeter by praying and burning incense, while a mirror, attached to a string, was lowered into her well, just deeply enough to touch the water. An image of the patient—either dead or alive—would be imprinted on the mirror by the smoke from the incense (Pausanias 7.21.12–13). The supplicant, consequently, would know whether the goddess granted healing or not.[30] In Lycia, near Cyaneae, worshippers, who peered into a spring that was sacred to Apollo Thyrxuscan, could see "all the things that they want to behold" (Pausanias 7.21.14). Another spring at Poseidon's sanctuary at Taenaron in the southern Peloponnese was also divinatory. Visitors who looked into it could see harbors and ships at sea, until a woman polluted the spring by washing her dirty clothes in it, destroying its divine nature (Pausanias 3.25.8). From London, a magical papyrus preserves the ritual of lecanomancy for invoking Serapis:[31] water is poured into a bowl, and a lamp is lit while a medium (a young boy), glancing into the water, searches for visions. The priest would utter a chant to dismiss the god. He would then recite a charm to protect his medium. The practice is described in other magical papyri where lecanomancy is prescribed for obtaining power and information or equal status with the gods in a "face-to-face vision" (Apuleius, *Apology* 4.162–167, 220–232).

Different types of water were appropriate for invoking different types of deities: rainwater for Olympian deities, seawater for gods of the earth, river water for Serapis or Osiris, spring water—emerging from deep within the earth—for deities of the dead (Apuleius, *Apology* 4.220–232). Coming from the sky, rainwater is transparently called celestial water (*aqua caelestis*) in Latin sources. Seawater was thought to contain earthy particulates and was thus, like spring water, an appropriate mechanism for communication with chthonic earth deities. The Egyptian gods, moreover, were closely associated with the Nile River. River water and springs were believed to issue from deep beneath the earth where dwelt the gods of the underworld and death. The process for invoking a deity by water is preserved in the *Greek Magical Papyri*: hold a phiale or *patera* of water on the knees, pour a libation of green olive oil, bend over the vessel, and then address "whatever god you want and ask about whatever you wish, and he will reply to you and tell you about anything." Formulaically, the god who has been invoked should be dismissed with "the spell of dismissal." The scribe assures his reader that whoever uses the spell "will be amazed."[32]

The ritual is attested artistically as early as the late fourth century BCE. An Etruscan bronze mirror from Volterra shows a seated goddess nursing a grown Hercules as she enumerates his future achievements. In the upper zone, Silenus gazes into (and drinks from) a phiale.[33] Augustine of Hippo (354–430 CE), the Christian bishop who reviled wasting his youth on "pagan" literature, recounted its practice in Italy, to where it had been introduced from Persia (according to Varro). Rome's legendary second king, Numa Pompilius (ruled 715–637 BCE), was significant for introducing important religious rites, much of which he had established by hydromancy, according to Augustine. Numa peeked into a spring and observed the forms of gods "or rather, the mocking illusions of demons" (in Augustine's words), who taught Numa the rites for his fledgling nation. Characteristically hostile to all non-Christian practices, Augustine equated hydromancy with necromancy: "it is the same thing, when the dead seem to make prophecies." Criticizing Numa's lack of transparency, Augustine accused the king of recording the causes of the rites in a secret book that he buried. The book was eventually discovered and later destroyed by the Roman senate who "preferred to burn what Pompilius had hidden, rather than to fear that which he [Numa]—who did not dare to incinerate the collection—dreaded" (*City of God* 7.35).

Centuries before Augustine, Ovid also understood Numa's ritual calendar as having a watery origin, but the poet's more innocent explanation eschewed Augustine's black magic (*Fasti* 3.275-348). Numa's interactions with the gods were shaped by his wife, the water nymph, Egeria, whose name ultimately derives from a sprite who presided over a spring of Diana at Aricia (south east of Rome). Near Porta Capena at Rome's southern gate, the Camenae Spring was the source for ritual water drawn by the Vestal Virgins, and the spring assumed Egeria's name, since the Vestals "carried out" (*egerere*) the water from there. Rome's festival calendar was thus linked with the sacred waters that supported those rites.

Finally, augury was the means of prognosticating the will of the gods from the evidence of animal behavior, such as by the flight patterns of birds (we eschew comment on haruspicium, the Etruscan art of ascertaining divine will by examining the livers of sacrificial animals). Portents came also from marine animals. Priests at Rome interpreted the strange event of fish leaping from the sea at Octavian's feet as an indication that those "who at that time held the seas (i.e., Sextus Pompey) would in future be under Caesar's feet" (Pliny 9.55). Varro also mentioned that sacred fish—confident in their safety—would be lured by the sound of flutes to altars at the edge of the shore during sacrifices.[34] In the Nile, the Oxyrhynchus fish, born from the wounds of Osiris, was sacred. Egyptians, consequently, never fished with hooks in fear of accidentally spearing one, preferring to catch nothing over a success rooted in an act of *hubris* (Aelian, *NA* 10.46). In addition, "tame" crocodiles presaged the demise of an unspecified Ptolemy by refusing to eat the king's food offerings (Aelian, *NA* 8.4). Furthermore, the art of piscine divination was perfected in Lycia where priests attached meaning to whether fish came when called, or ignored a summons, when fish arrived in large numbers, if they leapt or floated, or if they accepted or rejected food in response to priestly hails.[35] Finally, at Zeus' shrine at Labraunda in Caria, a clear spring was populated with fishes bedecked with golden collars and "earrings" ("jewels in their lobes"). Aelian (*NA* 12.30) connected the cult epithet (*Labrandeos*) with Zeus' function as a god of storms that were furious (λάβρος: *labros*). Aelian did not suggest why the fish were so honored. In Strabo, there is mention of an ancient shrine of Zeus, *Stratios* ("Warlike"), at Labraunda, but the jeweled fish are omitted.[36] Fish in Venus' pond at Hierapolis in Syria were also bedecked with gold and seemed to enjoy human contact, "fawning to be scritched" (Pliny 32.17).

Magical Effects of Sea Creatures

Sea creatures, moreover, were deemed to have magical associations. The hyena fish and other marine creatures were believed to possess magical properties. For example, if the right fin of a hyena fish was placed under a pillow, the sleeper would suffer from agitation and nightmares "not at all gracious or welcome." The severed tail fin of a living "rough-tail" (horse mackerel: τράχουρος: *trachouros*) could cause miscarriages in mares, provided that the fish has been returned to the sea (Aelian, *NA* 13.27). The fishermen of Cyprus noticed the sympathetic magical effects of the beautiful moonfish on local trees and springs. Trees would expand if they were fitted with a moonfish that was caught during a full moon (when the fish was also at its fullest half-moon shape). The reverse would also occur—trees would wither—if the fish was taken during a waning moon. Similar results ensued when moonfish were tossed into Cypriot springs, causing the waters to flourish ("flowing continuously and never failing") if the fish were collected under a waxing moon, but a bubbling spring would dry up if moonfish were taken under a waning moon (Aelian, *NA* 15.4).[37] The outcome correlates sympathetically with the condition of the moon: waxing and fullness brings about growth; waning renders desiccation or death. Stingrays were thought to have a similar sympathetic influence on trees, able to destroy tall, flourishing trees without any delay (just as they were thought to kill people: *NA* 2.36, 8.26).

Water and Death

Finally, the Greeks and Romans, like other Indo-Europeans, saw the sea and other waterways as liminal barriers between the worlds of the living and the dead. Those who could cross between these regions could also mediate between humans and gods, between life and death. Gilgamesh must cross the sea—a difficult and treacherous journey that no mortal had yet achieved—to seek his ancestor, Utnapishtim, and learn the secret to immortality (tablet 10). In the *Egyptian Book of the Dead*, souls crossed the Lily Lake to reach the Field of Reeds where they would enjoy eternal life. Herakles acquired Olympian status by reaching Ocean, on whose shores he obtained apples from the Garden of the Hesperides (this fruit could be plucked only by the gods) and by entering Hades through a seaside cave (either at Taenaron or Heraklea on the Acherusian peninsula) in order to kidnap the three-headed guard dog, Cerberus. Herakles was worshipped widely as a savior-god, overseeing curative hot springs.[38]

Although the Greeks called it a "barren expanse," the sea, nonetheless, sustained them economically and culturally. All Greek thought emerges from the outer Ocean.[39] Ocean, the marginal, inaccessible limit of the Greco-Roman world, was even an ingredient of the ambrosia, which assured eternal life for the gods. This boundary is permeable, and entering it constitutes a rite of initiation. Theseus, for example, endured a passage into manhood by diving into the Ocean to retrieve the ring tossed in by his adversary, Minos. Ino and Glaucus acquired divine status by immersion into Ocean.[40] In order to consult the soul of the deceased Theban prophet, Tiresias, Odysseus must sail north until his ship crosses Ocean's stream, and there he would find the groves of Persephone, thickly wooded with black poplars and fruit-perishing willows.[41] In Homer, thus, the world of the dead exists beyond the expanse of Ocean. In a comedic scene, Dionysus reaches the underworld through a marsh (Aristophanes, *Frogs* 352). The Hyperboreans, who dwelt beyond the North Wind and "under the very pole of the stars," lived lives of perfect ease. Absent from labor and warfare, they spent their days banqueting. When they were finally sated with their reveling, they met death by diving into the ocean from "a certain rock" that carried them directly to the underworld in the "most blissful type of funeral."[42] The Hyperborean homeland stood at the crossroads between earth and sky, life and death, and the rock from which they dived "may very well be the white rock that stands at the entrance of Hades."[43] Ocean, consequently, was a bounding principle not only of the earth's geography but also between different realms of existence.

Although all living creatures die and make the transit over the waterways that delimit the world of the living from the realm of the deceased (and there they must stay), some extraordinary mortals ventured over those same waters and then returned to the world of the living. Aeneas was rowed over the River Styx. Herakles, Perseus, and Jason travelled over Ocean (as had Gilgamesh). In the half-light of the Garden of the Hesperides on the western Ocean, Herakles met Atlas who stood precisely where day and night met (Hesiod, *Theogony* 746–750). Perseus also made the treacherous journey over Ocean, which could not be crossed in ordinary ways by walking or sailing. Perseus must fly.[44] By making these round-trip journeys to and from Hades, the hero conquered not only Ocean, but also death.

Death and the sea are also symbolically linked. Blakely notes the chthonic resonance of the layout of the sanctuary at Samothrace. Plato, furthermore, presented his theories of hydrology as a *katabasis* (Chapter 2).[45] Fallen Homeric heroes, moreover, were likened to cliff divers.[46] In Homer, Penelope wished to meet her own death, hoping for a gust of wind to carry her over the "misty ways" to Ocean (*Odyssey* 20.61–66). In Pindar (*Nemean* 7.30–31), death is "a wave" that comes upon everyone in common. Marine imagery is furthermore commonplace in funerary poetry.[47] For Beaulieu (2016: 120–134), the Arion episode is a *katabasis*. Arion leapt into the sea playing his lyre, dressed in funerary attire. A delphinian psychopomp then collected the dead man, depositing him at Poseidon's oracular shrine at Taenaron, where Arion dedicated a small dolphin-rider statue in thanksgiving for his own delphinian rescue.[48] Dolphins had performed similar services for Hesiod (Chapter 6), and Ino's son Melicertes, murdered by his mother in her Hera-sparked insanity (Pausanias 2.1.3). There is also a tradition that dolphins saved Odysseus' son, Telemachus, from drowning in his youth (Plutarch, *Cleverness* 985b).

Additionally, Persephone, the bride and queen of Hades, was closely associated with springs and streams: the cave and spring at Eleusis were sacred to her,[49] and central to her worship there.[50] Her handmaidens were the daughters of Okeanos, with whom she enjoyed picking flowers on a spring day, while the earth and the "salty sea" smiled (vv. 5, 14). At Eleusis, the fountains that greeted the initiates intensified a sense of the extraordinary and fantastic. The reflection of the *mystae's* torches in the pools may have underscored the festival's associations with death and afterlife (Longfellow 2012).

In Sicily, moreover, Herakles was reputed to have instituted a festival to Kore (Persephone) at the Kyane Spring, where worshippers made offerings of animals that were cast into the spring (Diodorus Siculus 4.23.4). According to Diodorus (5.4.1–2), the spring came into existence after Hades, who had just snatched up Persephone, struck the earth in order to return to his underworld kingdom. In another tradition, Kyane was a nymph who tried to protect Persephone from Hades. She was turned into water and her tears created the eponymous spring (Ovid, *Metamorphoses* 5.427). In Pindar (*Nemean* 1.13 –14), Persephone received Sicily, an island that was "hollow underground and full of streams of fire," as a wedding gift from Zeus.[51]

Conclusion

Water was a central component of the spiritual beliefs of Greco-Roman antiquity. Water could purify on both the macroscopic and localized levels, cleansing individuals or entire communities, and water could become ritually polluted and thus unpotable and unusable. Water facilitated communication with the spiritual world, either through weather signs, the behavior of aquatic animals, or various modalities of hydromancy. Water is also a powerful metaphor for death. Ocean separates realms of existence: the human world from the divine; the world of the living from the region of the dead. Water was thus a conduit between the human mind and the unseen, imagined forces of the divine realm.

9

Sailor Cults and Cults of Sea Gods

Introduction

Water gods were abundant and polymorphic, and many deities naturally have maritime associations. Poseidon's wife, Amphitrite, protected her worshippers from piracy at Tenos.[1] Menelaus beseeched the Nereids before escaping from Egypt with Helen (Euripides, *Helen* 1584–1587), and Sappho invoked the Nereids for the safe arrival of her brother (f5). The old men of the sea, Glaucus and Nereus, together with Melicertes would also be propitiated as saviors for those in peril at sea (*Greek Anthology* 6.164, 349; cf., Chapter 7). In 480 BCE, an Athenian sacrifice to Boreas seemed to trigger a three-day storm resulting in the destruction of many Persian ships. To counter the threat, the magi offered sacrifices on behalf of the Persian fleet to Thetis and the other Nereids who responded by ending the storm on the third day; or, in an alternate interpretation, "the storm abated of its own accord."[2] In Ovid, Thetis' sister, Psamanthe, was a "deity of the sea," worshipped at an unadorned seaside temple that was sacred to Nereus and his daughters (*Metamorphoses* 11.359–362, 392). Thetis' son, Achilles, was also honored with a sailor cult at a goat-inhabited island (cleansed daily by sea birds) in the Black Sea where storm-tossed ships could find respite. The hero himself appeared to his pilgrims in distress, revealing the island's best moorage in dreams, occasionally alighting on sails or prows, like the Dioscuri (Arrian, *Periplus* 21–23).

Winds were supplicated by prayer, propitiated with blood sacrifice, or controlled with amulets (Empedocles, *TEGP* 173). Breezes were also contained in skins, as when Aeolus, their divine king, gifted Odysseus with a bag of winds that would send the hero homeward. But Odysseus foolishly napped, and his greedy, untrusting crew opened the bag, releasing the breezes and throwing the ship off-course just within sight of their home island, Ithaca (*Odyssey* 10.1–79). Odysseus would eventually be saved from Poseidon's wrath by Dionysus' aunt,

Ino. According to the myth, Hera had driven Ino insane as punishment for nursing Zeus' illegitimate son, Dionysus. Ino's madness compelled her to leap into the sea, but she was instead transformed into the beneficent "white-armed" Leukothea, the shearwater.[3] In a vividly recounted storm, Leukothea saved Odysseus from drowning with her divine veil. Her instructions were precise: strip the clothing off, abandon the raft, and return her veil "to the wine-colored sea far away from land" (*Odyssey* 5.343–350). The veil served as a safety line, and Odysseus was purified by means of the Poseidon-sent storm. Propertius (first century CE) would later appeal to Leukothea, together with the Dioscuri and Neptune after dreaming that his beloved Cynthia had been shipwrecked (2.26A.9–10).

Here our focus is on ancient attempts to control the imaginary realm of the divine, including the worship of gods who could protect travelers on water and the rituals observed by sailors for safety at sea.

Gods Who Protect Travelers on Water

Poseidon

Poseidon was an ancient god, attested in Linear B sources as the principal deity of Pylos.[4] The Ionians of Asia Minor claimed descent from Nestor, king of Pylos, on the strength of the sanctuary dedicated to Poseidon at Mt. Mykale where young men honored the god with bulls.[5] The importance of his widespread cult is further underscored by the prevalence of toponyms that evoke his name: e.g., Potidea and Poseidonium/Paestum in southern Italy. According to tradition, Poseidon obtained his watery realm by lot.[6]

Poseidon was "the father of horses" (he sired the winged horse Pegasus as well as Arion, the swift steed who carried Adrastus, king of Argos[7]), and the cult of Poseidon Hippios was widespread.[8] As such, Poseidon was recognized as both the tamer of horses and the rescuer of ships.[9] Like his brother, Zeus, Poseidon was honored with bull sacrifices, but he was also propitiated with horse sacrifices, including by drowning. Drowning rituals in Poseidon's honor were by no means uncommon. The Trojans sacrificed both bulls[10] and horses by drowning them in rivers (*Iliad* 21.131–132).[11] Evidence also attests the sacrifice of horses by drowning at the whirlpool at Argos (Pausanias 8.7.2). Mithridates VI sought Poseidon's divine favor for his navy "having deposited a team of white horses into the sea" in 74 BCE (Appian, *Mithridatic Wars* 70). Wearing dark-blue clothing in imitation of the deep, Pompey's youngest son, Sextus, cultivated Neptune and is reputed to have made offerings of live horses to his divine patron.[12]

The practice was, perhaps, metonymic human sacrifice where the sacrificed animals were identified as representing either the god or the worshipper.[13] Poseidon's mythology embraces a tradition of metamorphoses into horses. Poseidon took the form of a horse to pursue Demeter Erinys ("furious") at Telephusa in Arcadia (she had foolishly tried to avoid his "human" advances by turning herself into a mare: Pausanias 8.25.4-5). An apocryphal tradition recounts Odysseus' transfiguration into a horse,[14] in punishment for blinding Poseidon's son, the Cyclops Polyphemus.[15] According to Servius, the equine Odysseus was stabbed to death "with the spine of a sea-beast" by his own son, Telegonos. As a horse, Odysseus was himself the ritual sacrifice that would finally serve as atonement to Poseidon.[16]

Watermen in distress summoned Poseidon, and they celebrated him in their successes. Fishermen would appeal to Poseidon, the averter of disasters (Aelian, NA 15.6). In thanksgiving for a successful tuna haul, the people of Corfu dedicated a bronze bull to Poseidon at Delphi. The fishermen had been unable to catch—in any satisfying quantity—the large numbers of tuna that swam in local waters, despite the assistance of their θυννόσκοπος (thunnoskopos: tuna-scout), a clever bull who bellowed when he saw the schools. After sacrificing their diligent thunnoskopos to Poseidon, the watermen were then successful. The statue base was still on display at Delphi in Pausanias' day. Pausanias, however, questioned the wisdom of killing their talented bovine thunnoskopos (10.9.2).

Poseidon was the savior of ships, his divine privilege and responsibility, and he was called upon to aid those who sailed (HH 22). Sailors would propitiate him mid-sail from the stern.[17] Returning homeward by water, Nestor offered sacrifices to the sea god (Odyssey 3.178-179), and Athena implored her brother for the safe return of Odysseus' son, Telemachus, in his black ship (Homer, Odyssey 3.55-61). Menelaus furthermore called upon Poseidon for a safe voyage before sailing from Egypt where he was reunited with a faithful Helen (Euripides, Helen 1584-1588). Lucian even included a temple to Poseidon among the amenities in the belly of his very large ship-swallowing whale (True Story 1.32). In the precinct, the undigested captives found graves and a spring of clear water.

Poseidon's authority extended beyond epic. Alpheios of Mitylene beseeched the god for fair passage from Rome to Syria (Greek Anthology 9.90). Theognis hoped that the god would return his friend, Chaeon, safely (691-92). In Oppian, fishermen prayed to Poseidon Asphalios ("steadfast") for gentle winds and rich harvests of fish.[18] The Greeks supplicated Poseidon Soter ("savior") to protect their ships during a storm in 480 BCE (Herodotus 7.192). Traveling between Constantinople and Athens in the winter of 336 CE, Libanius "found Poseidon favorable," although he had expected stormy sailing conditions (*Autobiography* 15–16).

Poseidon's Roman counterpart, Neptune, was also invoked in prayer and stone by Latin-speaking sailors and others who traveled by water (at Corinth, Delos, and elsewhere). Augustus honored Neptune with a "small offering" at Carthage.[19] Vows to Neptune were fulfilled at Dougga,[20] Numidia,[21] Capua,[22] and Rome.[23] At Noviomagus Reginorum (Chichester, England), a dedicatory plaque suggests that a guild of smiths erected a temple to Neptune and Minerva in honor of the imperial family. No trace of the temple survives, but Chichester is near Bosham Quay and the Roman naval base at Fishbourne. Large-scale iron working facilities on the Weald were also associated with the fleet.[24] The combination of deities and location suggests a guild (*collegium fabrorum*) of shipbuilders. Cults of Neptune and Ocean, furthermore, were significant: Britain is an island, approachable only by water, and Neptune is attested in dedications near the sea and at forts on both the Hadrianic and Antonine walls.[25]

Poseidon's Temple at Sounion

The striking temple to Poseidon at Sounion, on Attica's southern tip, standing 60 meters (200 feet) above the water, was an important beacon for sailors at sea, like many headland shrines dedicated to maritime deities (including the mountain of Samothrace). The site has long been sacred, and cultic activity can be dated at least to the second half of the seventh century.[26] The Cape was first mentioned by Homer, where Phoebus Apollo, supporting the Trojan cause and consequently hostile to the Greeks, had taken the life of Phrontis, Menelaus' exceptional helmsman (*Odyssey* 3.278–285). Menelaus put in at the Cape to bury his friend, perhaps providing the etiology of the Panathenaic quadrennial games, which the Athenians and others observed there.[27] The poros limestone temple was begun soon after 490 BCE. Never completed, it was destroyed by Xerxes in 480 BCE, when he also razed the sacred structures on the Athenian acropolis.[28] After their victory at Salamis, the Athenians dedicated a captured Phoenician trireme in thanksgiving to Poseidon, a war trophy that Herodotus had seen first-hand.[29] The new Agrileza marble temple at Sounion was erected on the site as part of the building program overseen by Pericles in the 440s as the Athenian navy was expanded.[30] The frieze featured scenes that glorified human and divine conquest over nature and dangerous beasts (the Centauromachy, Gigantomachy), and the deeds of Poseidon's heroic son, Theseus. In the town below the sanctuary, fortifications consisting of defensive walls and small towers 66 feet apart (20 meters) were built during the Peloponnesian War (as the Spartans disrupted seaborne supplies). Two roofed shipsheds, rising sharply from the water, were added probably in the third century BCE.

Apollo

As a god of "colonization," Apollo was also an important maritime deity. As Delphinios (the dolphin god: dolphins are among his sacred animals), Apollo guided worshippers from Crete to Delphi (*HH* 8 *Apollo*). In Apollonius' *Argonautica* (1.359–361, 404–425), before embarking, Jason invoked Apollo for nautical aid, reminding the god of his Pythian promise: that Apollo would assure the voyage's success by pointing out the passages of the sea. Apollo is here the "Lord of Shores" (Aktios), guardian of Iolkos and Pagasai, and "Lord of Embarkations" (Embasios) who guarantees safe voyages. Jason was explicit: "may a gentle breeze blow, by which we might sail on following seas" (1.359, 422–424). As the Argonauts prepared to set sail, they erected a shore-side altar and made sacrifices to Apollo "of Landings" (Ekbasios: 1.966, 1186). Then, on the lonely beach beyond Cape Acherousia, the Argonauts erected an altar to Apollo "the Protector of Ships" (Neosoos) on which they offered burnt thigh bones (2.927). Additionally, as the *Argo* sailed through a "dark shrouding" that produced a "deadly night," which neither the stars nor the twinkling moon could penetrate, Jason called upon Apollo for salvation with the promise of countless gifts.[31] In Manilius, furthermore, Apollo was the "victor of the sea" (*ponti victor*: 5.32).

In addition, material remains include maritime votives to Apollo. At Delphi, the Stoa of the Athenians bears an inscription advertising the dedication of "the equipment and stern ornaments seized from the enemy,"[32] likely rigging from captured Persian ships. Antigonus Gonatas had also dedicated his "sacred trireme" to Apollo, after his victory at Cos over Ptolemy II (262–245 BCE: Athenaeus 5.209e). We are not told where the ship was consecrated, perhaps at Delos[33] or Samothrace where an early-third-century shipshed could accommodate a small vessel, such as a bridal ship, refuge vessel, or pirate boat.[34] Parts and models of ships were common gifts, entire ships were rare, and rarer still were shipsheds that were purpose-built to house such extravagant *ex votos*, such as those at Delos, Samothrace, and Actium (cf., *UCWW*: Chapter 9–10).[35]

The maritime cult of Apollo at Delos—strategically situated at the Aegean's center and a focus of Pan-Hellenic obeisance—is furthermore borne out by nautical *ex votos*, including rudders and anchors.[36] Symbolic remains excavated at Isthmia include ships etched into *pinakes*, terracotta and metal ship models (mostly from Poseidon's Isthmian temple), as well as depictions of Poseidon and tridents.[37] Such dedications may have occurred in imitation of the consecration of the *Argo* to Poseidon at Isthmus when she (the *Argo*) won the boat race at the Isthmian games.[38] The ship was eventually placed among the constellations as if still on the water.[39] Apollo was invoked as a deity of "safe arrival" (Apobaterios) in an inscription from Cyrene erected on behalf of Nero, possibly in 67 CE when the emperor was nearly shipwrecked.[40] Zeus was also honored as such by Alexander of Macedon when he landed on Asian soil in 334 BCE, erecting an altar to Zeus Apobaterios, Athena, and Herakles (Arrian, *Anabasis* 1.11.7). The epithet, one of thanksgiving after a challenging journey, has also been ascribed to Apollo's twin sister, Artemis,[41] and his healer son, Asklepios.[42]

Aphrodite, Isis, and Venus

Aphrodite was another important deity of water in general and the sea in particular. "Sea-born" Aphrodite emerged from the foam of Ouranos' genitals (Sky) in the waters off Salamis in Cyprus, her sacred island.[43] Aphrodite was worshipped as Euploia, a goddess of "fair sailing" and protectress of sailors at sea.[44] Aphrodite's maritime cult was significant and widespread. Solon (640–558 BCE) had solicited Aphrodite's favor before setting out from Cyprus.[45] The goddess was honored at Aegina with a votive anchor (fifth century BCE).[46] Nearly a dozen votive anchors were recovered from Aphrodite's shrine at Gravisca in Etruria.[47]

A patroness of the sea and guarantor of safe sailing, the "guardian of all navigation" (*Greek Anthology* 9.601), Aphrodite, was celebrated as a maritime deity by the epigrammatists.[48] Mnasalkes referred to a "sanctuary of Cypris of the sea" (Einalia), emphasizing its seaside setting (*Greek Anthology* 9.333, third century BCE). An anonymous epigrammatist described a lovely statue of Aphrodite that was erected on a purple shore next to gentle waves (*Greek Anthology* 16.249). In the late 300s, Anyte proclaimed that the goddess' statue enjoyed gazing at her "beloved" (φίλον: *philon*) sea, so that she, in turn, might make sea voyages "pleasant (φίλον) for sailors" (*Greek Anthology* 9.144). It was believed that the goddess could calm the waves as "the one who stills the sea" (Galenaia).[49] As Epilimenia ("along the harbor"), she was considered a patroness of harbors at Corinth,[50] and she was widely attested as Pontia, the "lady (of the sea)."[51] She protected sailors and the navy, and she enriched traders.[52]

Anecdotes also attest Aphrodite's miracles at sea. During the 23rd Olympiad (688–685 BCE), the merchant, Herostratos, had purchased a statuette of the goddess from her temple at Paphos (Cyprus), before putting out for Naukratis. Caught in a storm, the sailors prayed. The goddess immediately imbued the ship with myrtle that eased the severe seasickness suffered by crew and passengers. The healing scent consequently enabled the ship to reach shore. Herostratos dedicated the myrtle branches and the statuette in the goddess' shrine at Naukratis.[53] Athenaeus' description of the mega-yacht *Syrakousia/Alexandris* suggests, furthermore, that Hellenistic royal ships incorporated the shrines of Aphrodite.[54] The opulent state boat constructed by Ptolemy IV (ruled 221–204 BCE), also included shrines to Aphrodite, together with altars to Dionysus and the royal family.[55]

Like her Greek counterpart, the Roman Venus was a significant maritime deity. Venus' patronage of nautical enterprises is apparent on the Torlonia relief (Fig. 9.1). Dominating the relief's center is a larger-than-life Neptune with his trident. Above Neptune's right shoulder, a partially nude Venus holds a wreath above the fire atop the harbor's *pharos* lighthouse. Under full sail, a ship comes into port as if welcomed by the two gods. Venus' descendants, Romulus and Remus, are also shown suckling the she-wolf (a powerful symbol of the founding

Fig. 9.1 Torlonia relief, Museo Torlonia #430.

of Rome) as a duplicated decorative element on the sail. The two upper corners feature other deities: Liber (Dionysus) and Ceres (Demeter). Both were especially important to maritime commerce, wine, and grain, respectively. In the upper-right corner of the relief stands Liber pouring a libation of his sacred wine, holding his thyrsus, accompanied by his familiar panther. To his right, an elephant-drawn cart—a statue group atop a building—evokes Liber's exotic, eastern origin. Between Neptune and Liber, under the yard arm of a moored ship with a furled sail, appears a large apotropaic eye. The relief's upper-left corner features Ceres with her cornucopia, brimming with grain. Jupiter's eagle flutters in front of Ceres. The sails of the ship putting in are inscribed VL ("a vow to Liber:" in fulfilment of a promise for some unrecorded divine favor, from the temple of Liber near the port).

Both Greek and Roman ships were frequently equipped with an *akrostolion* (fitted to the prow) in the shape of a helmet[56] or animal, often a goose or Aphrodite's sacred swan,[57] whose wings resemble the fanning aplustre that adorned the sterns of many ancient ships. The artwork often depicts Aphrodite riding her swan, as on an Attic red-figure pelike (ca. 350 BCE), where the goddess stands on the back of a large, white swan skimming across the waves; a dolphin swims underneath.[58]

Also closely associated with Aphrodite is the highly Hellenized Egyptian goddess, Isis. Throughout the Mediterranean, Isis' epithets included Euploia, Pelagia ("of the sea" as at Akrocorinth: Pausanias 2.4.7), and Pharia, metonymy for Alexandria and its famous lighthouse. The three epithets were largely synonymous, but Isis' particular identification as Pharia distinguishes her as protectress of the Roman grain supply.[59] As Pharia, Isis was also attested epigraphically at Ostia,[60] invoked for the health of Antoninus Pius (ruled 138–161 CE), whose coinage celebrated the goddess and the lighthouse.[61]

Maritime Ruler Cult: Aphrodite and Arsinoë II

In the third century BCE, the cults of Aphrodite and Isis Euploia,[62] symbolic of Ptolemaic naval power, were aligned with a cult of Arsinoë II (316–270 BCE), the sister-wife of Ptolemy II Philadelphos (ruled 283 to 246 BCE; they were the children of Ptolemy I Soter and Berenike I). "Beloved of the west wind" (Athenaeus 11.497d), the queen was honored as Arsinoë Euploia-Zephyritis with sand altars at Alexandria.[63] Arsinoë was also worshipped with a cult at Cape Zephyrion (near Alexandria), established by Ptolemy's admiral, Kallikrates of Samos,[64] where a devotee named Selanaia, who had survived a storm at sea, dedicated a shell with an inscription to Arsinoë Zephyritis (of the gentle, westerly wind).[65] The sanctuary at Samos was intended as a refuge "from every wave" for the men who worked on the salt sea, together with the chaste daughters of the island. Promises of safe voyages were granted by Arsinoë Euploia to her worshippers: "she will grant a fair sailing, and even in the middle of a storm she will make the level sea as smooth as oil for those entreating her."[66]

The Ptolemies and their agents actively encouraged this appealing cult, and they advanced it as the divine authority for their maritime hegemony. The goddess and her human avatar were long considered the patronesses of the "Ptolemaic Sea."[67] Cults of Arsinoë as Aphrodite Zephyritis were established outside Egypt at Cicilia and Delos where games were instituted by the nauarch, Hermias, and where shells were offered.[68]

Maritime Ruler Cult: Zeus/Jupiter and the Pharos Lighthouse

Sailor cults were also aligned with the imperial cult at Rome, and Roman emperors were invoked with epithets that accord with the maritime assimilation of Aphrodite and Arsinoë Euploia. Octavian was addressed as Epibaterios ("setting out"), especially at Alexandria where he was also assimilated with Zeus Soter,[69] likely in association with the Pharos lighthouse that was dedicated to the "Savior Gods" (Lucian, *How to Write History* 62) for the "safety of sailors" (Strabo 17.1.6). The tower was topped with a statue of Zeus (Diodorus Siculus 18.14.1) or Poseidon (*Greek Anthology* 9.674), and the seaward-facing eastern side featured an inscription invoking Zeus. Newly inaugurated, it was praised by the epigrammist, Poseidippos, who emphasized its divine purpose:

> O Lord Proteus, Sostratos of Knidos, son of Dexiphanes, established this savior of the Greeks, guardian of the Pharos. For in Egypt there are no hills for beacons as on the islands, but a sea-level breakwater stretches out, affording a safe anchorage. For the sake of this now a straight and upright tower cutting through the aither shines by day from an unapproachable distance, and all night long a sailor, hastening on the waves, will see the great flame blazing from the summit. Even if he were to run towards the horn of Taurus, he would not miss the mark of Zeus the Savior, Proteus, sailing by this guide.
> 115

In Poseidippos, the tower was sacred to Zeus Soter whose favor underscores this remarkable accomplishment of engineering that guided navigators safely through dangerous shoals into a harbor that was, at long last, clearly marked. In general, lighthouses served as "practical and divine guarantors of passage into port," and they were often built (or re-dedicated) to commemorate triumphs and epiphanies.[70] Sanctuaries of Zeus Soter also stood in the harbor at the Piraeus (Strabo 9.1.15), at Epidauros in Lakonia (Pausanias 3.23.10), and the important port town of Corinth (Pausanias 2.22.6).

Trajan, moreover, was assimilated to Zeus Embaterios ("setting out") at Hermionis, a small port town in the southeastern Argolid on the Aegean (*IG* 4.701), and Hadrian was honored as Apobaterios (" safe arrival") in a birthday dedication at Olympia,[71] an epithet also attested for Zeus on Hadrianic coins.[72]

The Dioscuri and St. Elmo's Fire

> I boarded the king's ship; now on the beak,
> Now in the waist, the deck, in every cabin,
> I flamed amazement: sometime I'd divide,
> And burn in many places; on the topmast,
> The yards and bowsprit, would I flame distinctly,
> Then meet and join.
>
> *The Tempest* 1.2.196–201

Thus does Shakespeare's Ariel describe herself as imitating the mystical phenomenon known as St. Elmo's Fire, which sailors at sea took as the apotheosis of Castor and Pollux, the twin Dioscuri, the "guardians of the sea and saviors of sailors" (Strabo 1.3.2). Described by Xenophanes as a star-like apparition (*TEGP* 73), the faint blue or purple glow, sometimes accompanied by a buzzing or hissing sound, is caused by the ionization of air molecules along the masts of ships or other poles (including the noses of aircraft). They occur also on sails during bad weather, as well as on animals and trees (Seneca, *NQ* 1.1). Caesar observed the phenomenon on spear points in Africa (Aulus Hirtius [?], *African War* 47), as did other generals. A detailed description comes from Pliny:

> These stars occur both at sea and on the lands. During night watches of soldiers in the field, I have seen them, lightening-like in appearance, cling to javelins in front of the entrenchments. They also tread on the yardarms and other parts of ships under sail with a certain tuneful sound like birds darting about. When they occur singly, they are oppressive, and they sink vessels. If they fall on the lowest part of the keel, they set the ship on fire. But if they appear as twins, they are beneficial and presaging a favorable voyage, by whose advent that dire and threatening star, called Helena, is put to flight. On account of this, sailors

assign the names of Castor and Pollux to the phenomenon. They invoke those gods at sea.

2.101

According to legend, such an electrical discharge "shin[ing] round the heads of men" occurred both at Troy to Aeneas' toddler son, Ascanius (Vergil, *Aeneid* 2.632–635), and to Servius Tullius as a child (Livy 1.39), predicting the future greatness of both boys. Distinct from lightning, which is often very dangerous, these electrical discharges seem to float down to whatever they have alighted on, and, once they have settled, seldom cause damage, which may explain why sailors in a storm considered the discharge as divinely protective. Rejecting this optimistic interpretation as irrational, Seneca instead explained the phenomenon as a sign that the winds and storm were losing power "otherwise the fires would be moving, not resting" (*NQ* 1.13).

Sons of Leda, the Dioscuri were half-brothers: Pollux was fathered by Zeus, Castor by Leda's mortal husband, Tyndareus. As teenagers, they besieged Athens whose king, Theseus, had kidnapped their lovely young sister.[73] The twin brothers of Helen of Troy, Castor and Pollux, were among the Argo's crew (Apollonius, *Argonautica* 1.146–150), and they participated in the Caledonian Boar hunt.[74] The brothers sought marriage with the daughters of Leukippos ("White Horse"), Phoebe and Hilaeira, who were already betrothed to other men, the Dioscuri's Theban cousins, Lynkeus and Idas. A feud was sparked, ending in Castor's death (in another tradition, the brothers married the girls). Pollux, however, as a son of Zeus, was semi-divine, and he had the choice of joining the Olympians or sharing his divinity with his brother. Pollux chose the latter, and the brothers spent their days between Olympus and Hades, either together (Ovid, *Fasti* 5.715–720) or alternately (Pseudo-Hyginus, *Astronomika* 2.22). Consequently, they became symbolic of immortality, with a cult that spread from their native Sparta to the rest of the Greek-speaking world (Pindar, *Pythian* 10.51–52; Pausanias 3.16).

Like Poseidon, the Dioscuri were associated with horses and the sea. The husbands of "White Horse's" daughters, they were the "tamers of horses," and their steeds were a gift from Poseidon (Pseudo-Hyginus, *Astronomika* 2.22). They also come to the aid of travelers, especially sailors in distress who would stand at the prow in a storm to vow white lambs for salvation (*HH* 33 *Dioscuri*). Their epiphany was a welcome sight during a storm at sea:

> but a great wind and the waves of the sea have pushed the ship under water. Suddenly they appeared glancing through the air on nimble wings. Straightaway, they check the vexing winds and still the white salt waves in the seas. Beautiful

signs, they have cast the toil asunder. Seeing the brothers, sailors rejoice and they abate from their miserable labor.

HH 33.11–17; cf., Plato *Euthydemus* 293a

Because they were favored by the gods, they were able to protect mariners at sea, as attested in literary and inscriptional evidence. In the quarries south of Dyrrhacium, where Pompey defeated Caesar in battle in 48 BCE, an inscription invokes the Dioscuri as the "protectors of sailors" (*SEG* 49.653). According to pseudo-Hyginus (first century BCE, *Astronomika* 2.22), Neptune gave them the power to aid shipwrecked men, and some ships were fitted with figure-heads of the twins, as, for example, on the ship that conveyed the apostle Paul from Malta to Puteoli (Acts 28:11). Together with the other constellations of the zodiac, Gemini—representing the Dioscuri—is a valuable navigational tool. The moon is another navigational aid associated with the brothers. Their wives were called by epithets of the moon: bright (Phoebe) and serene (Hilaeira). Furthermore, it was believed that the Dioscuri could counteract the power of their sister, Helen, "destroyer of ships" (Aeschylus, *Agamemnon* 689), whose star was thought to endanger maritime traffic in Pliny (above) and Statius: "[Oebelian brothers] put your Ilian sister's cloudy stars to flight far away and shut them out from the entire sky" (*Silvae* 3.2.11–12).

Gods of Samothrace

Samothrace was a natural point of anchorage between Asia and Europe, where Ovid changed ships on his way to Tomi (*Tristia* 1.10.19–22) and where Paul stopped on his way to Macedonia from the Troad (Acts 16:11). In Diodorus Siculus, the Dioscuri were also favored by the sailor gods of Samothrace, who alight as stars on the heads of the brothers during storms at sea. These mysterious "great" sailor-protecting deities in the northern Aegean were frequently confused with the twin, Dioscuri,[75] and their rites in principle paralleled those of the sailor-protecting Gallic deities of Sena (Chapter 8).

The cultural and linguistic ties with Thrace remained strong, and the use of exotic non-Greek cultic language endured down to the first century BCE (Diodorus Siculus 5.47.14–16), enhancing the ritual mystery and evocative of a distant, pre-Greek origin. More than seventy inscriptions on ceramic were written in a Thracian dialect.[76] Attracting foreign dynasts who made increasingly elaborate votives, the cult grew in importance. Famous mythic initiates into the Samothracian rites include the Argonauts (Apollonius, *Argonautica* 1.915–921) and Cadmus, founder of Thebes, whose bride, Harmonia, was a Samothracian princess.[77] Philip of Macedon allegedly met his bride, Olympias, when they were initiated there (Plutarch, *Alexander* 2.2). Arsinoë Philadelphos and her brother, Ptolemy, were benefactors. Other historical initiates include Herodotus (2.51), governors of Macedon,[78] and Roman officials who had fought against the Cilician pirates.[79] In 18 CE, traveling from Lesbos to Thrace, Germanicus intended to be initiated, but adverse winds drove him off course (Tacitus, *Annals* 2.54). The cult attracted pilgrims from across the Mediterranean. *Ex votos* at the site attest initiates from seventy cities, mostly in the Aegean,[80] including men and women, wealthy and humble. The primary motivation for initiation seems to have been for the *mystae* to become "both more dutiful and more just and better in all things" (Diodorus Siculus 5.49.6). According to Livy (45.5.4), the initiates must, furthermore, be ritually clean.

The cult was popular with mariners, and sailors would pray to the Samothracian gods during storms (above) as they did to the Dioscuri. Seafarers, including captains[81] and pilots,[82] came to the sanctuary for the initiation. Entire crews were initiated *en bloc*: one dedication records more than thirty names.[83] It was believed that the Samothracian gods were more likely to save their own initiates at sea than non-initiates. It was also thought (or hoped) that they might protect their own worshippers from shipwreck altogether.[84]

Little is known about the gods or their rites. Even the names have remained secret, and worshippers invoked them obliquely as "gods" (*theoi*) or "great gods" (*theoi megaloi*). A list of possible divine names survives from a late-third-century-BCE source: Axieros, Kasmilos, Axiokersos, and Axiokersa.[85] Not even their number or gender is certain. Sometimes the deities were called upon as a pair or a triad. They were often—but not universally—male.

As with other mystery rites, initiates underwent a common experience, or they learned some shared (and secret) knowledge, which perhaps entitled them to prosperity in life or special treatment after death.[86] One *mystos* claims to have reached old age (eighty years) without pain or trouble owing to his initiations at Samothrace and Eleusis.[87] The penalty for revealing the secrets of the Eleusinian Mysteries (in honor of Demeter and Persephone) was death,[88] and similar proscriptions may have been in place for the Samothracian Mysteries. Herodotus winked at fellow initiates:

> whoever has been initiated into the secret rites of the Kabeiroi, which—after receiving them from the Pelasgians—the Samothracians oversee, this man knows what I say.
>
> 2.51.2

Apollonius was a little more helpful:

> In the evening, on Orpheus' commands, they put in at the island of Elektra, daughter of Atlas, in order that—by learning the unspoken rites by means of gentle initiations—they could sail more safely across the icy cold sea. I will say no more on these matters. But the island herself rejoices as do the deities who have by lot obtained these mysteries of which it is not right for me to sing.
>
> *Argonautica* 1.915–921

Initiates thus guarded their proprietary knowledge, perhaps under threat of losing its privileges.

As tokens of their initiation, *mystae* received iron rings, which might have represented the initiatory rites in one of two ways.[89] Magnetized when touched to a lodestone, the rings could "recall the moment of ritual action and stand in metonymous relationship with the island itself." The magnetized rings might then serve as an indicator of the bond between gods, *mystae*, and the island.[90] Cole suggests that the stones to which libations were poured during the rite might have been lodestones, that were used to magnetize the iron rings; a demonstration of magnetism may have been included in the event.[91]

The magnetized iron ring may provide a link between the mystery rites at Samothrace and variant cults of Kabeiroi-craftsmen at Lemnos overseen by Hephaestus, the smith god, whose medium was metal (iron) and whose agency was the forge. Moreover, the gods at Samothrace were associated with the Idaean Daktyloi ("Fingers"), renowned for discovering iron and teaching the mysteries to Orpheus (Ephorus, *FGrHist* 70 F 104). Their names are "overtly metallurgical": Damnameneus ("hammerer"), Akmon ("anvil"), and Kelmis, a triad whose punishment for offences against the Idaean Mother Goddess was transformation into iron and imprisonment within Mt. Ida.[92] The nautical and aqueous implications are profound: boats were designed and built by master craftsmen who used a variety of materials, including wood and iron, and tradition ascribes the invention of both watercraft and navigation to gods of technology: Athena (Apollonius, *Argonautica* 1.110–112) or Prometheus (Aeschylus, *Prometheus Bound* 467–468).

The lodestone would later be developed into an essential nautical instrument, the magnetic, marine compass. Magnesia in Anatolia was a rich source of lodestones, whose behavior came to be described as "magnetic."[93] But long before such advances in nautical technology were achieved, the properties of the lodestone were known to the Chinese by the third millennium BCE under Hoang-ti,[94] and perhaps even earlier to the Olmec.[95] When suspended on a thread, or placed on a piece of floating wood, one side of the stone would always face the pole star. Consisting of a magnetized iron needle placed on a piece of straw floating in a bowl of water, the magnetic compass was developed by the Arabs, and used by European navigators by at least 1,000 CE.[96]

Ploiaphesia: Rituals of the Sailing Season

The rhythm of life in the ancient Mediterranean, including the nautical, followed the festival calendar. During the Roman imperial period, a sailing festival, *ploiaphesia*, was celebrated in early January in conjunction with the annual oaths of loyalty to the emperor and the public vows that were intended to ensure the happiness and good luck of the state.[97] The mythic etiology for the winter festival is provided by Isis' return to Egypt on January 2 (she had sailed to recover the corpse of her murdered brother-husband, Osiris).[98]

A grander "Sailing of Isis" (*Navigium Isidis* or *Ploiaphesia*) was celebrated on March 5, observed as the "birthday of sailing,"[99] a festival that launched the sailing season. In Apuleius (*Metamorphoses* 11.8–17), we have a vivid depiction of a colorful and raucous parade of richly bedecked revelers, musicians, performing animals, and priests. The procession culminated on the shore at Kenchreai where an elaborately decorated ship ("an untried vessel," perhaps a miniature) was launched while the waves received milk libations and grain, a first-fruits offering of the trading season. The ship was loaded with spices, and its sails proclaimed a prayer for prosperous voyaging through the new season. Isis herself professed authority over the sailing season, commanding "wholesome ocean breezes." Her cult, moreover, was inextricably linked with calming stormy winter waves and with the formal inauguration of the sailing season.[100]

Evidence attests that *Ploiaphesiae* were celebrated widely at port towns throughout the Mediterranean well into late antiquity, e.g., Eretria and Ephesus,[101] Rome,[102] and Lugudunum (Lyon) in honor of Venus.[103] The festival was one of many that informed medieval Carnival galas (ship carriages were featured prominently in Italian and Rhineland Carnival parades[104]). Its nautical aspects endure in "Blessings of the Fleet," which are still observed as community celebrations in port and fishing towns throughout the world.

In antiquity, other deities were also honored with ship processions, including Athena (for the Panathenaic ship, see *UCWW*: Chapter 9) and Dionysus who travelled by ship, vanquishing a pirate band, transforming their ship, and transfiguring the crew into dolphins.¹⁰⁵ In Philostratus of Athens (late second century CE), we have the only textual reference to the Dionysus festival:

> In the month Anthesterion, a trireme is sent, raised up, to the agora, which the priest of Dionysus guides from the sea like a helmsman, and loosing its cables.
> *Lives of the Sophists* 1.25.1

Artistic evidence shows festival ships carried by men or rolled on wheels. An archaic Clazomenean vase found in Egypt depicts a boat carried in procession by at least three men and accompanied by satyrs playing *diauloi* (double flutes: 550–540 BCE). On three late-sixth-century Attic skyphoi, Dionysus is conveyed in a ship on wheels. On the reconstruction of one example,¹⁰⁶ looking front, the god is framed by two satyrs, each playing a *diaulos* and facing the god. A swan's neck aplustre ornaments the float's stern. Another swan with outstretched wings leads the procession. Wheels are detectable on the original.

Sailing Rituals and Superstitions

The seas were fraught with danger, from tempests to sea monsters, and those who traveled on water often sought divine protection. Before setting sail, the skipper would look for auspicious signs, such as the swallows that nested in Cleopatra's flagship (Dio 50.15.2). Inauspicious signs would delay the launch. Bad omens could include sneezing while going up the gangplank, cawing crows or magpies in the rigging, sighting wreckage on shore, and dreams (of goats, boars, or owls presaged storms or pirate attack).¹⁰⁷ If the weather turned foul, hair and nail clippings could be tossed overboard to propitiate the gods. To this end, it was important to interpret the will of the gods correctly.

Rituals for Safety at Sea

Offerings to the gods were made when boats were launched or landed, and many ships were equipped with altars. The Torlonia relief shows a blazing altar on the quarterdeck of the ship coming into port.¹⁰⁸ (Fig. 9.1) Behind the altar stand three passengers: in the center a woman holds an incense box, to her right a man sprinkles incense on the fire, to her left, a man with a *patera* prepares to make a libation of wine. Such altars might be portable,¹⁰⁹ but stone altars may also have been incorporated structurally. At Pompeii, an altar built into a niche is indicated on a graffito depicting the cargo ship, *Europa*.¹¹⁰ Finds from the

Spargi wreck include a stone altar that seems to have been part of a shrine.[111] Rock-cut shrines, as at Kastro on Lemnos, represent epigraphic ships "positioned in view of the sea."[112]

The diligent skipper would ensure that launching and landing rituals occurred in a timely manner, just as Alexander's general, Nearchus, had done, honoring Zeus Soter before exploring the Indus River in 325 BCE (Arrian, *Indika* 21.2). Homer is a crucial source for sailing rituals, and in the *Odyssey* we have detailed accounts of embarkation rites. While Telemachus' ship was made ready to leave Ithaca, the crew prepared mixing bowls for pouring wine libations to the "deathless, eternal gods," but especially to Odysseus' divine patroness, Athena, who would likewise protect the boat at sea (cf., *Odyssey* 2.430–433). While Telemachus' crew prepared to leave Pylos where Nestor had entertained and advised them, the young Ithacan "prayed and made sacrifice to Athena at the ship's stern" (*Odyssey* 15.222–223).

Thucydides included the more elaborate ritual that the Athenians performed in 415 BCE as they launched their fleet for a naval campaign against Sicily (6.32). A trumpet blast called for silence, and customary prayers were then recited. Mixing bowls, ashore and on each ship, were tended, and the fleet set sail as the last drops of wine were poured into the sea. While well-wishers ashore raised the paean, garlands were tossed from departing ships, and vows for successful journeys were fulfilled with repeated rituals and gifts.[113]

The prayers uttered at the outset of a voyage evolved into the stylized *propemptikon* in Greco-Roman literature, where authors expressed their wishes for safe voyages for friends traveling on water.[114] Best known is Horace's *propemptikon* for Vergil on his homeward return from Attica (*Odes* 1.3). Horace invoked the Cyprian goddess (Venus) and the Dioscuri to protect Vergil at sea, bading all the winds, except for the favorable westerly, Zephyr, to keep well away.[115]

At Syracuse (and perhaps elsewhere), as soon as sailors could no longer see the shield on Athena's temple (that is to say, when they lost sight of land), they would toss into the sea a terracotta cup (from a temple of Olympian Gaia at the island's very edge) filled with flowers, honeycomb, frankincense, and spices (Athenaeus 11.462c). Both by virtue of the material of the cup (earth) and the sanctuary from which the cup was taken (Gaia), the gift was a sacrifice of earth to the sea, linking sailors who traveled on water with the land that they had just left behind. It was an act of trust and subordination by which the sailors committed themselves to the mercy of the waves. The Syracusan sacrifice, initiated from the water on behalf of a particular crew, inversely evokes *Ploiaphesiae*, launched from the shore on behalf of the entire sailing community.

Sometimes rituals were omitted, and Propertius (2.25.25–26) admonished his readers that it was hardly appropriate to make vows for safe travel mid-storm. Such pledges might include libations, prayers, or the sacrifice of a bull (sacred to Poseidon).

Maritime Amulets

Ships and sailors were (and still are) protected by apotropaic devices. The bows of many ships were decorated with painted or ceramic eyes allowing ships to "see" where they were sailing as well as ward off the evil-eye of ill-wishers.[116] Among these are the famous Exekias cup showing Dionysus' ship rendered with an eye is rendered on the side of the ram at the water line. A black-figure jug shows a ship equipped with multiple sets of eyes: one eye with a small brow-line on the ram, and a large eye with five smaller eyes at the top of the rail.[117] Scenes 79–80 of Trajan's column show three small warships (*liburnae*) sailing from an Adriatic harbor. An altar adorns the bow of the topmost ship. The ram of the middle ship is decorated with a capricorn.[118] An eye is clearly etched at the water line below the gunwale of the lower ship. Anchors were also apotropaically inscribed. Furthermore, coins were placed within the mast-step of many ships— on the evidence of at least a dozen Roman shipwrecks—for good luck, a practice that continues into the modern era but derives from the tradition of the consecration of Greek temples.[119]

Sailors and travelers also wore personal amulets. Leukothea's purple scarf protected Odysseus as he made his way in open water to Scheria (Homer, *Odyssey* 5.346–350).[120] The superstitious Augustus, moreover, carried a sealskin amulet for protection against stormy weather (Suetonius, *Augustus* 90). There were even rituals to ward against pirates. The people of Syedra had been advised by the oracle of the Clarion Apollo:

> Erecting the likeness of bloody man-killing Ares in the middle of the city and keeping him in the iron bonds of Hermes, perform sacrifices next to it; on the other side Justice shall lord over him and judge him, but he should appear as if he were begging.[121]

Such a statue group, a bound Ares standing between Hermes and Justice, is confirmed on second- and third-century-CE coins from Syedra.[122] Other amulets against storms at sea include the heart, eye, scalp, or wing-tips of eagles, as well as the hearts of hoopoes. The eyes of sparrows were thought to assure skillful steering during a storm.[123]

Apotropaic symbols, including *phalloi*, were common on ships.¹²⁴ Terracotta *phalloi* have been recovered from Roman shipwrecks, suggesting that sailors viewed them as good-luck charms. A *phallus* together with a crescent moon (evoking the Dioscuri), was also embossed on a bronze sheath that would have adorned the figurehead of a late-second-century-CE ship found in the Rhine.¹²⁵ From an early-first-century-CE shipwreck of a Roman merchantman found near Marseilles, a togate Priapus, carved only on the front, was likely intended to decorate the niche of an onboard shrine. Priapus with his well-endowed *phallus* protected both ships and harbors, and statuettes of the godling were positioned on rocks as navigational aids.¹²⁶

It was, furthermore, believed that some stones possessed metonymic powers that safeguarded sailors from storms and other maritime dangers, as detailed by the so-called *Nautical Lapidary* of Byzantine date.¹²⁷ Protective powers seemingly derive from the superficial properties that link the stone with the sea, as in color (transparent, green, or blue) or maritime danger. Good voyages are guaranteed for travelers with dryops (with a white spot in the middle) or sea-blue beryl intaglios, especially those carved with an image of Poseidon in seahorse-drawn biaugae. Carbuncle and chalcedony protect the wearer if a ship goes down; blue-green diamonds, resembling hail, ward off typhoons and large waves. Also effective against storms are the "snake-bellied" stone from Egypt and coral, sewn into a sealskin, which should be tied to the mast-head to protect the entire ship.

Piscine Sacrifices

Although the Greeks preferred mammalian blood sacrifices (owing to their large supply of blood), fish were sacrificed to Poseidon (and other marine deities), but animal votives carried metonymic value:¹²⁸ fish move skillfully through water, horses are swift (and expensive). Gifts might be plunged into the sea or conveyed three times around the deck. The people of Halai (in Lokris) offered a θυνναῖον (*thynnaion*: tuna offering), the first-fruits of their tuna catch to Poseidon (Athenaeus 7.297e). Tuna, whose flesh is redder than most fish—because of the oxygen-binding molecules of myoglobin in their muscle tissue¹²⁹—might have been viewed as an acceptable gift.¹³⁰

Red mullet, moreover, was sacred to Hekate, the goddess of roads (crossroads), glances, and graveside dinners. Her associations included the moon and black magic. Her epithets, Trioditis (Τριοδῖτις: *Trioditis*: "Goddess of the Crossroads") and Triglenos (Τρίγληνος: *Triglenos*: "Three-Eyed"?), are evocative of the Greek word for red mullet (τρίγλη: *trigle*), and the fish were offered to the goddess on the thirtieth day of the month (ταῖς τριακάσι: *tais triakasi*: Athenaeus 7.325a). Thus, the fish had a triple etymological association with the goddess, a magical number that enhanced and guaranteed efficacy.[131]

The people of Phaselis in Lycia made annual offerings of smoked fish (or salted fish: *tarikhon thuousi*) to their local hero, a shepherd who dwelt far from the sea (Athenaeus 7.298c), a ritual act that speaks to the perishability of fresh fish and the commercial importance of salted fish. Artemis received grilled red mullet and hake from a fisherman seeking her favor and hoping that she would, in return, fill his nets with fish (Apollonides, *Greek Anthology* 6.105). Small fish, "off the fire," were offered to heroes at Cos.[132] At the *Ludi Piscatorii* in June, live fish were sacrificed by holocaust to Vulcan at Rome, possibly in his capacity as protector of the Tiber, in order to guarantee human life.[133] Eels were also sacrificed to "unknown gods" at Lake Copais.[134] In contrast, certain species of fish were inviolate, such as the "guide-fish" (*pompimos*), which accompanied ships into port and was thus considered sacred at Samothrace.[135]

Conclusion

Seeking protection on the water, travelers propitiated unseen forces with simple and complex vows, prayers, and rituals. Gods of the sea were numerous and multivalent. Poseidon was not just the divine embodiment of the savage ocean, but he also protected those who sailed. Apollo was the god of settlement whose adherents must travel over water. Aphrodite's aqueous origins rendered her sympathetic to the sea and those who traveled on it. Dionysus was the missionary god who journeyed widely to spread the gospel of the grape. Specialized cults also arose, including to the Dioscuri, whose epiphany comforted those distressed by storms at sea, and to the *Theoi Megaloi* whose center received sailor pilgrims, among other supplicants, seeking initiation into the mystery rites at Samothrace. The Gallizenae of Brittany may also have overseen a mystery cult that attracted pilgrims (Chapter 8), and sailors in the Euxine made pilgrimages to Achilles Island. These rites were extended to include powerful rulers who sought the authority of the gods and ensured nautical victory or safe travel over the water.

10

Conclusion

In this volume we have interrogated the effects of water on Greek and Roman thought. Water is a material foundation and an agent of change in the world. Water was conceived as a fundamental building block of the cosmos. Ocean was one of the earliest features to emerge from Near Eastern creation accounts as expressed also in the Greek tradition by Hesiod. Water, furthermore, is one of the four material *rhizomata* (roots) of the cosmos in the four-element theory, accounting for all the physical objects in the human world.

Water was, moreover, a convenient means of organizing the terrestrial world. The theory of a circumambient Ocean endured even as geographical knowledge improved. Seas and rivers were recognized as the borders of continents and territories. As a cartographic principle, water divided and arrayed peoples, just as it connected peoples through travel, trade, and politics (*UCWW*).

Water was discerned by Greek thinkers as balanced with earth. Land and water engaged in an eternal, complex, reciprocal exchange, with effects both in the short term (shaping coastlines, giving way to alluvial deposits) and long term (as proved by the inland deposition of fossilized marine fauna). Just as water affected the very land through which it flowed, watercourses also molded the creatures who dwelt near them, with the ability to inflict disease, foster health, or—in some places—even change the coat-color of livestock who imbibed there. Greco-Roman thinkers investigated and explained these phenomena (weather and disease) as owing to the effects of visible and invisible forces inherent within the elements.

The perceived nature of water, consequently, provides a context for the elucidation of natural phenomena and the spiritual realm. Water is not only an agent of change, it is by its very nature, polymorphic and metamorphic. The surface of water is malleable and fluid, changing hue with shifting light, changing texture with fluctuating winds and currents. The refraction of light, moreover, distorts the appearance of the submarine environment, producing images of animals (and objects) that seem to be collections of disjointed (or mis-joined) pieces. Visibility, furthermore, decreases with depth. It is thus difficult (if not impossible) to observe the aqueous sphere with any clarity, and it remained a font of mystery giving rise to fantastical and fearsome biological and mythical creatures who terrorized sailors. Marine fauna mirror the characteristics of their aqueous realm. Their composite nature seems as if inspired by the refraction of light, and many were thought to be amphibious, inhabiting both water and land (basking seals) or air (leaping dolphins). Owing to its mysterious murkiness, water was a bridge between the physical world and the realm of the imagination, linking humans and gods, living and deceased.

Water provides the substrate upon which life as we know it exists. In greater or lesser degrees, water is found mixed with air and earth, respectively as precipitation and as watercourses streaming through the landscape. Unseen water also exists in the ground and as humidity in the air. The Mediterranean Basin, together with its mountainous terrain, defines the physical world of the Greeks and Romans. Water was thus a central component in how ancient thinkers viewed and interpreted their world, providing a prism through which to observe and consider natural philosophy and watery phenomena.

The behavior of water, in fact, inspired the earliest conceptions of natural philosophy, of how the world is constructed, and how it works. Thales seems to have proposed water as the material substrate of the world as well as the catalyst for change. The Presocratic era, which Thales instigated, was an exciting period in Greek intellectual history. With each subsequent thinker, the questions became increasingly complex and subtle. Of what material(s) is the world composed? How do those materials interact? What causes generation and decay? How did animal life come about? Anaximander even proposed a watery origin of all terrestrial life in order to explain the embryonic sacs in which mammals gestate (*TEGP* 37–39). These early theories were forged from a combination of theory and empiricism. The inland deposition of marine fossils, for example, has far-reaching consequences for understanding erosion and natural cycles, as interrogated by Xenophanes and Herodotus.

Eventually the conversation shifted from the abstract to the practical, and the provocative focus on natural philosophy gave way to debates on the metaphysical nature of the soul. Plato's theories of *physis* (nature) and water were presented in support of his ethics. Aristotle synthesized and curated the thought of his intellectual predecessors, often dismissing early theories of the natural world as risible or immature. He thus did violence to the earliest sources that now survive exclusively as extractions and redactions in Aristotle, Seneca, Pliny, and a few other ancient thinkers. Aristotle essentially suppressed voices with whom he disagreed, and his own theories then largely dominated the intellectual landscape. Subsequent thinkers were compelled to answer to Platonic ethics or Aristotelian physics.

This is not to say that advances in natural philosophy were entirely stunted. Knowledge in many scientific subfields accrues with the escalation of contact between distant peoples and the exchange of ideas between them. War, settlement, and exploration are powerful catalysts in many spheres of knowledge. Thales and his successors were working when Greek *poleis* were actively exploring and settling the Mediterranean (seventh through fifth centuries BCE), and engaging with Near Eastern advances in mathematics, astronomy, and cartography. Alexander's expeditions (fourth century BCE) stretched the reach of Greek intellectual endeavor into the Punjab to the east and Alexandria to the south. In Alexander's wake, Alexandria became an international center of scholarly endeavor: manuscripts on all manner of topics were collected by the Museum, and scholars throughout the Greek-speaking world, from Sicily to Asia Minor, convened in Alexandria in order to avail themselves of the library and stimulating academic debate that naturally would have occurred there. Soon thereafter, Posidonius succeeded in marshaling Stoic theory and empirical evidence into a remarkably accurate theory of tidal behavior that aligned both with empirical evidence and Stoic physics. Despite the prosaic aim of the encyclopedia—a Roman innovation—both Seneca the Younger (*Natural Questions*) and Pliny the Elder (*Natural History*) employed new data and their own independent observations in presenting the history of knowledge. Although Pliny's work is largely a compilation, it is not entirely devoid of originality.

This conversation, furthermore, did not occur in isolation. Greek theories of cosmogony were sparked by contact with the Near East, where the world was envisioned as developing in accordance with the methodical extraction of physical components from the primordial morass. Near Eastern cosmogony was, in turn, a product of observing the mutual effects of water and earth in real time.

In other areas of thought, moreover, the Greek conception of nature was grounded within a larger Mediterranean convention. For example, hybridism— the modality by which sea creatures were understood—is a universal means of explaining and controlling liminal spheres, the intersection of human and divine, the intersection of land and water. In early Egyptian thought, the gods are theriomorphic hybrids that bridge the human and natural (divine) world (Anubis the jackal, Bastet the cat). The Assyrian Lamassu (human, bird, bull/ lion) came to symbolize the power of the king (who was viewed as a god on earth). The elephant-headed, humanoid Ganesha is the beloved divine patron of art and science in the Hindu tradition. Although Greco-Roman gods were fully human with metamorphic capabilities (thus they were hybrids of a different sort and linked to the natural world more subtly), composite creatures of classical antiquity are entrenched in the larger Mediterranean heritage of liminalism, crossing the boundaries between human and divine, water and earth. Hybridism provides a space to imagine combinations that do not exist in the human realm but might dwell elsewhere, on the edges of human habitation, unknown spheres that demand explanation and control. The man-bull Minotaur was a monster of faraway Crete, the horse-men centaurs dwelled in distant Thessaly. Enigmatic creatures that dwell in inscrutable waters are interpreted as being both familiar (akin to land animals) but reflecting their ever-changing fluid environment. Even their names are often composite (e.g., dogfish). Animals on the edges do not behave like animals in the center, nor do the laws of physics, and we have seen that waters on the edges often have paradoxical natures, as reported in our ancient sources. This is a powerful comment on the Greco-Roman self-image and identity, and the superiority of the center over the edges. Those in the center provide the metric for social norms and beliefs. For this reason, theories of heliocentrism were considered blasphemous, since heliocentrism removes humankind from that center.[1]

The Greeks and Romans inhabited an intellectual world that was earnestly anthropocene. Humanity prevailed at the core of the physical world, and nature existed for the purpose of serving and advancing human endeavor. In "Whether Fire or Water is more Useful" (*Moralia* 955d–958e), Plutarch had debated the relative merits of fire and water, concluding that fire was more useful because it is the catalyst of *techne* (technology). Fire (which sustains *techne*) enables humans to manipulate—if not control—nature according to a litany of skills and specialized knowledge that privileges the human race to overcome capricious natural forces: carpentry and architecture, astronomy and the agricultural calendar, mathematics and literacy, animal husbandry and yoking, navigation, medicine, prophecy (divination), and metallurgy (Aeschylus, *Prometheus Bound* 442–50). For the ancients, control over nature comes first with understanding and discernment, skills valued above all other human traits by the philosopher, Heraclitus (*TEGP* 8, 10, 18) and the Titan Prometheus who gifted humanity with *techne*-inducing fire (*Prometheus Bound* 450–71, 476–506). Fire makes possible human control over nature, which in turn allows for inquiry into the nature of the physical world.[2] But this *techne* is applied to the interrogation and manipulation of the watery realm, which fosters growth, organizes the *oikoumene*, provides a conduit for the exchange of *techne*, and links the realms of the living and the dead, the spheres of human and divine, and the worlds of earth and water.

Appendix

Major scientific, technical, historical, and medical writers and thinkers of Ancient Greece and Rome[1]

Thinker	Provenance	Language	Approximate dates of activity	Major scientific works or area(s) of scientific contributions
Claudius Aelianus (Aelian)	Praeneste	Greek	195–235 CE	Historical Miscellany On the Nature of Animals (NA)
Anaxagoras	Clazomene	Greek	480–428 BCE	Cosmology, Physics
Anaximander	Miletus	Greek	580–545 BCE	Cosmology, Physics
Anaximenes	Miletus	Greek	555–535 BCE	Cosmology, Physics
Archimedes	Syracuse and Alexandria	Greek	287–212 BCE	Mathematics, Engineering
Aristotle	Stagira and Athens	Greek	384–322 BCE	*Metaphysics* *Meteorology* *On Generation and Corruption* (GC) *On the Enquiry into Animals* (HA) *On the Generation of Animals* (GA) *Cosmos* *On the Parts of Animals* (PA) *Physics* and others
Flavius Arrianus (Arrian)	Nicomedia	Greek	120–170 CE	*Periplus of the Black Sea* *Indica*

Thinker	Provenance	Language	Approximate dates of activity	Major scientific works or area(s) of scientific contributions
Artemidorus	Ephesus	Greek	fl. ca. 100 CE	Geography
Atheneaus	Attaleia (or Tarsus)	Greek	Second century CE	*Deipnosophistae* ("Dinner Sophists")
Athenodorus	Tarsus	Greek	60–20 BCE	Wrote on tides
Basil	Caesarea	Greek	330–379 CE	Neoplatonism, Christian monk
Gaius Julius Caesar (Caesar)	Rome	Latin	100–44 BCE	*Commentaries on the Gallic Wars* (BG) *Commentaries on the Civil Wars* (BC)
Marcus Porcius Cato (Cato the Elder)	Rome	Latin	Lived 234–149 BCE	*On Agriculture*
Aulus Cornelius Celsus (Celsus)	Rome	Latin	15–35 CE	*De Materia Medica* (*On Medical Matters*) (*On Medical Matters*)
Marcus Tullius Cicero (Cicero)	Rome	Latin	Lived 80–43 BCE	*Letters* *On Divination* *On Fate* *On the Nature of the Gods* *Tusculan Disputations*
Lucius Junius Moderatus Columella (Columella)	Rome	Latin	Lived 4–70 CE	*On Agriculture*
Democritus	Abdera	Greek	440–380 BCE	Cosmology, Mathematics, Physics
Diodorus Siculus	Sicily	Greek	80–20 BCE	*Bibliotecha* (*Universal History*)
Empedocles	Acragas	Greek	460–430 BCE	*On Nature* (fragmentary) *Purifications* (fragmentary)
Epicurus	Samos and Athens	Greek	Lived 310–270 BCE	*Letter to Herodotus* (Physics) *Letter to Pythocles* (Astronomy, Meteorology) *On Nature* (fragmentary)

Thinker	Provenance	Language	Approximate dates of activity	Major scientific works or area(s) of scientific contributions
Eratosthenes	Cyrene and Alexandria	Greek	245–195 BCE	*Geographica* *Measurement of the Earth*
Euclid	Alexandria	Greek	300–260 BCE	*Elements*
Sextus Julius Frontinus (Frontinus)	Rome	Latin	Lived ca. 40–103/4 CE	*On the Aqueducts*
Galen	Pergamum and Rome	Greek	129–215 CE	Physician, Anatomy, Surgery
Heraclitus	Ephesus	Greek	510–490 BCE	Cosmology, Physics
Herodotus	Halicarnassus	Greek	circa 484–430/420 BCE	*Histories of the Persian Wars*
Herophilus	Chalcedon	Greek	lived circa 330–260 BCE	Physician, Anatomy
Hesiod	Ascra	Greek	ca. 700 BCE	*Theogony* *Works and Days*
Hipparchus	Nicea	Greek	fl. ca.140–120 BCE	Astronomy, astrology, geography, mathematics
Hippocrates and his school		Greek		*Affections* *Airs, Waters, Places (AWP)* *Aphorisms* *Coan Prenotions* *Epidemics* *Internal Affections* *On Regimen in Acute Diseases* *Prorrhetic*
Homer	Unknown (Chios?)	Greek	750–700 BCE	*Iliad* *Odyssey*

Thinker	Provenance	Language	Approximate dates of activity	Major scientific works or area(s) of scientific contributions
Josephus	Jerusalem	Greek	37–100 CE	*Antiquities of the Jews*, *Jewish War*
Juvenal	Rome	Latin	Second century CE	Satirist
Lucian	Samosata	Greek	120–180 CE	Satirist and rhetorician
Titus Lucretius Carus (Lucretius)	Rome	Latin	lived circa 99–55 BCE	*De Rerum Natura* (*On the Nature of Things*) Epicurean Physics
Manilius	Rome	Latin	10–30 CE	Astrology
Pomponius Mela	Baetica and Rome	Latin	fl. 30–60 CE	Geography
Oppian	Anazarbos in Cilicia	Greek	176–180 CE	*On Fishing* (*Fishing*), *On Hunting*
Publius Ovidius Naso (Ovid)	Rome	Latin	43 BCE–17/18 CE	*Fasti*, *Heroides*, *Metamorphoses*, *Tristia ex Ponto*
Parmenides	Elea	Greek	490–450 BCE	Cosmology
Philo	Byzantium	Greek	fl. 240–200 BCE	Engineering and Mechanics *Mechanical Collection* (*Lever, Harbor Construction, Pneumatics, Artillery Construction, Siege Preparations, Siege Craft*)
Plato	Athens	Greek	Lived 428/427–348/347 BCE	*Timaeus* Astronomy, Cosmology, Mathematics
Gaius Plinius Secundus (Pliny the Elder)	Novum Comum	Latin	Lived 23–79 CE	*Natural History*

Thinker	Provenance	Language	Approximate dates of activity	Major scientific works or area(s) of scientific contributions
Plutarch	Chaeronea	Greek	Lived 46–120 CE	*On the Cleverness of Animals* (*Cleverness*) *Table Talk* *Biographies*
Polybius	Megalopolis	Greek	180–120 BCE	*Histories*
Posidonius	Apamea	Greek	Lived 135–50 BCE	Cosmology, Geography, Hydrology, Mathematics, Meteorology
Protagoras	Abdera	Greek	487–412 BCE	
Claudius Ptolemaius (Ptolemy)	Alexandria	Greek	127–after 146 CE	Astronomy, Mathematics
Pythagoras	Samos, Croton	Greek	570–495 BCE	Cosmology, Mathematics, Physics
Pytheas	Massilia	Greek	fl. ca. 340–290 BCE	
Lucius Annaeus Seneca (Seneca the Younger)	Cordoba, Rome	Latin	Lived 4 BCE/1 CE–65 CE	*Natural Questions*
Seleukos	Seleukia (Tigris River)	Greek	165–135 BCE	Astronomy, mathematics
Socrates	Athens	Greek	469–399	Ethics
Strabo	Amasia	Greek	Lived 64 BCE–24 CE	Geography
Strato	Lampsakos	Greek	fl. 295–268 BCE	Natural philosophy, mechanics
Thales	Miletus	Greek	fl. 600–545	Cosmology, Physics
Theophrastus	Eresus and Athens	Greek	fl. 340–387/6 BCE	*On the Causes of Plants* (*CP*) *On the Enquiry* (*Historia*) *into Plants* (*HP*)
Marcus Terentius Varro (Varro)	Reate	Latin	Lived 116–27 BCE	*On Farming*

Thinker	Provenance	Language	Approximate dates of activity	Major scientific works or area(s) of scientific contributions
Vegetius Renatus (Vegetius)	Unknown	Latin	Circa 450 CE	On Military Matters
Publius Virgilius Maro (Vergil)	Mantua and Rome	Latin	Lived 70–19 BCE	Aeneid Eclogues Georgics
Marcus Vitruvius Pollio (Vitruvius)	Rome	Latin	Lived 85–20 BCE	On Architecture
Xenophon	Athens	Greek	400–355 BCE	Anabasis Economics Hellenika Symposium
Xenophanes	Colophon	Greek	Lived 570–478 BCE	Physics
Zeno	Elea	Greek	Lived 490–430 BCE	Physics
Zeno	Citium	Greek	333–262 BCE	Founder of the Stoic school of philosophy

Notes

Introduction

1. Boyce 2002: 154–84; Weintraub 2018: 77–88. Cf. Kargel 2004; Harland 2005; NOAA, "Are there oceans on other planets?": https://oceanservice.noaa.gov/facts/et-oceans.html (retrieved March 7, 2019).
2. Dyches and Chou, "The Solar System and Beyond is Awash with Water": https://www.nasa.gov/jpl/the-solar-system-and-beyond-is-awash-in-water (retrieved March 7, 2019).
3. Water-rich exoplanets are, in fact, common: about 35 percent of the known exoplanets (larger than the earth) should be water-rich: Zeng 2018.
4. Homer, *Odyssey* 4.410–455; Ovid, *Art of Love* 1.761, *Fasti* 1.369.
5. For example, Vergil, *Aeneid* 1.81–123; Lucan 1.498–503; Seneca the Elder, *Suasoria* 1.15, citing Albinovanus Pedo.
6. Dowden 1989: 708–9; Oleson 2008b: 129.
7. Sharples and Gutas, 1998: 190–219.
8. Diogenes Laërtius, *Life of Theophrastus* 47; Sharples 1992: 347–85.
9. Beaulieu 2016: 25: *pontus* indicates a path that is difficult to cross, as in Sanskrit pántāḥ (a road marked by obstacles). Indeed poets refer to the "paths of the sea": see *UCWW*: Chapter 5.
10. E.g., the Aegean (πέλαγος Αἰγαῖον: Aeschylus, *Agamemnon* 659); the Sardinian (τὸ Σαρδόνιον καλεόμενον πέλαγος: Herodotus 1.166); the Adriatic and Tyrrhenian Seas (ἐκ τοῦ Ἀδρίου καὶ ἐξ ἑτέρου πελάγους ὃ καλεῖται Τυρσηνόν: Pausanias 5.25.3); the Euxine (ἐν δ᾽ Εὐξείνῳ πελάγει: Pindar, *Nemean* 4.49).
11. ἡ θάλασσα: the Sea (e.g., *Iliad* 2.294, *Odyssey* 5.413); ἡ μεγάλη θάλασσα: the Great Sea (Arrian, *Indika* 2.7); ἡ ἡμέτερα θάλασσα: Our Sea (Plato, *Phaedo* 113a; Aelian, *NA* 12.25; Strabo 1.2.29); this sea: ἥδε ἡ θάλασσα (Herodotus 1.1, 185, 4.39); τῆς νῦν ἑλληνικῆς θαλάσσης: the Hellenic Sea (Herodotus 5.54, Thucydides 1.5; Plutarch, *Cimon* 13); ἡ θάλαττα ἡ καθ᾽ ἡμάς: the sea around us (Polybius 1.3.9); ἡ παρ᾽ ἡμῖν θάλασσα: the Sea near us (Plato, *Phaedo* 113a; Strabo 2.5.18); the deep sea of salt water (πέλαγός τε θαλάσσης: Apollonius, *Argonautica* 4.608).
12. Roller 2010b: 178–80, 193–4.
13. Arrian, *Indika* 32.9–13; Pearson 1960: 83.
14. Nearchus f1b=Strabo 15.2.12; cf., Roller 2018: 848; Chapter 7.
15. Arrian, *Indika* 31.6–8; Roller 2018: 877. The Nereid seems to be a doublet for Circe.

16 Juba II (ruled 29–25 BCE) used Onesicritus as a source for his own *Libyka*: Roller 2003: 194, 228, 240n93.
17 Two chapters (Chapters 2 and 3) are expanded from Irby 2016b. Cf., Irby forthcoming.

Chapter 1

1 No culture exists in isolation, and the Greek intellectual journey was very much shaped by contact with the Mesopotamians, Egyptians, and others. Helen was reputed to have imported Egyptian tranquilizers: Homer, *Odyssey* 4.219–233. Helen's drug was perhaps opium, which becomes laudanum when mixed with alcohol: Hughes 2005: 233. Thales of Miletus was said to have learned geometry from Egyptians (*TEGP* 1). Seneca intriguingly ascribed a four-element theory to the Egyptians (*NQ* 3.14.2).
2 Smith 1994: 85; McCall in Irby, McCall, and Radini 2016: 297.
3 Kragerud 1972: 41 takes Tiamat and Abzu as representing not salt and fresh waters but instead deep (lower) and upper waters.
4 Crawford 1998: 8–10.
5 See Crawford et al.1997 for an archaeological analysis of the site.
6 Burkert 1992: 92–3; cf., Jacobsen 1968: 105.
7 Burkert 1992: 93; Penglase 1994: 4. Cf., Introduction.
8 *Iliad* 14.201, 302: "whence the gods have risen"; *Iliad* 14.246: "the seed of all."
9 *Enuma Elish* 2.19–28 (cf., 3.24–29); trans. King 1902.
10 The parallels between the *Theogony* and Hurrian/Hittite, Mesopotamian, and Jewish cosmological myths have been discussed at length elsewhere: E.g., Naddaf 1986: 339–64; Solmsen 1989: 413–22; Mondi 1990: 141–98. Known to the Greeks by the fifth century (Hecataeus f373 = Diodorus Siculus 40.3.8), at least in part, the Jewish creation tradition envisions a tripartite cosmogonic framework consisting of sky, earth, and water, and YHWH's spirit hovers over the surface of the waters: Gen. 1:2.
11 *Theogony* 233–264. Hesiod listed 51 Nereids by name.
12 Herington 1963: 190; cf., Aeschylus *Eumenides* 904–906.
13 Lloyd 1970: 5. Far superior to the awkward Babylonian, Greek, and pre-Caesarian Roman civil calendars, this system was implemented in Rome under Julius Caesar in 45 BCE (Pliny 18.212). Losing time at the nearly imperceptible rate of one day every century and a half, Caesar's calendar was not substantially revised until 1582.
14 Strabo 14.1.6. Thales' student Anaximander of Miletus may have founded a colony, perhaps on the Black Sea: Hahn 2001: 202–3.
15 Kirk, Raven, and Schofield 1983: 105.
16 Rihll 1999: 8–9.

17 *TEGP* 15; Graham 2010: 39.
18 Lloyd 1979: 11; e.g., Anaximenes, *TEGP* 36. Even in the "scientific" Aristotelian corpus, matters of theology received considerable attention: Barnes 1995: 67, 106. "Science" in the ancient world never lost its sense of wonder nor its connection with ethics and the divine, and "science" was perpetually negotiating the pervasive tension between tradition and innovation.
19 Seneca described Thales' theory of earthquakes as "inadequate" (*inepta*): *NQ* 3.14.1 = Thales, *TEGP* 19.
20 Other interpretations are possible: Graham 2010: 90.
21 Cf., *TEGP* 50; Homer, *Iliad* 7.99.
22 Graham 2010: 327.
23 Cf., *TEGP* 26 (*rhizomata*), 114, 115.
24 Homer, *Iliad* 3.278, 15.187–188; Hesiod, *Theogony* 678–683, 736–745.
25 Cf., Homer, *Iliad* 18.483; *Odyssey* 1.52–54, 5.293–296; Irby-Massie 2008: 144–5.
26 Kingsley 1995: 348.
27 *HH* 2 *Demeter* 47–50. Persephone also fasted, thus refusing Hades' hospitality. During her abduction, she abstained from food, with the exception of the six pomegranate seeds that she consumed either absentmindedly or by force (371–4, 411–13). She was consequently obligated to him.
28 Palmer 2016: 30–54. This interpretation is fraught and much debated: Gregory 2016: 18. See Aristotle, *GC* 1.1.315a3–19; cf., Plato, *Sophist* 242e3–243a2.
29 Plato's Demiurge derives from the governing principles of earlier systems: *Logos* in Heraclitus (*TEGP* 8), *Nous* (Mind) in Anaxagoras (*TEGP* 30-34), *Anake* (Necessity) in Empedocles (*TEGP* 25).
30 Plato, *Phaedo* 109b–113d; Chapter 2.
31 On Aristotle's theory, the moon separates the supra-lunar from the sub-lunar regions, but he failed to account for the connection between the two regions. The divine, living, intelligent supra-lunar region is filled with an incorruptible fifth element (quintessence/aither, whose particles, arbitrarily, take the shape of a dodecahedron). Aither's natural motion is perfectly circular, and it altogether lacks terrestrial properties (hot/cold; wet/dry). This region is divided into concentric spherical shells that support the planets as they revolve in their eternal, invariable circles.
32 Aristotle, *GC* 2.4.331a7–b27; *Metaphysics* 2.2.994a20–b10.
33 Elements also have weight (heavy/light), a determinant of natural place on Aristotle's theory, settling into concentric spheres arranged according to weight and size: from smaller and heavier at the center to lighter and larger (earth, water, air, fire, aither/ quintessence: *Cosmos* 2.13.293a–34). Cf., Ovid, *Metamorphoses* 15.237–251 for an eloquent expression of the four-element theory where each is relegated to its proper place according to Aristotelian physics. Ovid's engagement with the Empedoclean

four-element theory is also treated in Ham 2013. See further Manilius 1.149–166, a passage that many interpret as an expression of Stoic physics. Manilius, like Ovid, however, seems to eschew the Stoic conception of fire as a catalyst that shapes the other three secondary elements into the physical world: Volk 2009: 30–1.
34 See Atomists, *TEGP* 10–12.
35 Epicurus, *To Herodotus* 43–44, 46–47, 61–62. See Lucretius 2.162, 222.
36 For swerve, Lucretius 2.216–250. For swerve and free will, Epicurus, *On Nature* 34.21–22, 26–30; Lucretius 2.251–293. For a hostile analysis of Epicurus' swerve: Cicero, *Fate* 46–48.
37 Epicurus, *To Menoecus* 123–124, *To Herodotus* 76–77; Lucretius 5.1161–1225.
38 Alexander of Aphrodisias, *On the Soul*, 26, 16; *On Mixtures* 224, 15. Sambursky 1959: 1–20; Sandbach 1975: 82–5.
39 See especially Aëtius 1.7.33. Sambursky 1959: 1–7.
40 Simplicius, *Categories* 165, 32–166, 29; cf., Sambursky 1959: 18; Sandbach 1975: 69–94.
41 The renowned Stoic polymath, Posidonius, also wrote five volumes exploring the theoretical and physical bases of prophecy: God causes omens, but humans must learn how to interpret those omens correctly (Cicero, *Divination* 1.64, 125–126, 2.35).
42 Simplicius 30a; Cf., Sanders 2011: 270.
43 For Aristotle and others, the ability to reason, think, and solve problems separates humans from other animals who lack those skills: Aristotle, *On the Soul* 2.3.414a32–b1; Bos 2010: 821–41.
44 Aeschylus, *Prometheus Bound* 450–471, 476–506. The debate over the play's authorship is beyond our scope: see further Irby-Massie 2008: 135–6.
45 Control of fire, together with agricultural advances—crop and animal domestication—leads to permanent settlements, food surpluses, and civilization: Humphrey, Oleson, and Sherwood 1998: 97–100, 133.
46 Margulis 1998: 121; Ball 1999: 228–34.

Chapter 2

1 Aeschylus, *Suppliant Maidens* 408; scholiast ad loc. The Aristotelian author of *Problems* also queried why divers' eardrums burst in the deep (32.2), and why some divers deliberately punctured their eardrums (32.5). Professional divers in premodern societies continue to puncture their eardrums to avoid the lengthy process of equalizing the pressure within the ears: Frost 1968: 182. Cf., *Problems* 32.3 (sponges prevent water from entering the ears with excessive force) and 32.11 (divers put oil into their ears before a dive). Men would also dive with a mouthful of olive

oil (Oppian, *Fishing* 5.638, 646), perhaps to protect their Eustachian tubes from exorbitant exposure to seawater (Frost 1968).
2 *Problems* 32.5. The average diver can spend two to four minutes underwater without mechanical aids (Frost 1968: 182).
3 Strabo 11.7.4. Polyclitus flourished ca. 360–300 BCE.
4 Herodotus 4.42; Kahanov 1999; Roller 2006: 23–6. Cf., *UCWW*: Chapter 5.
5 *Codex Palatinus graecus* 398; cf., Herodotus 4.196; Pliny 2.169; Arrian, *Indika* 43.10–12.
6 Roller 2006: 105.
7 Roller 2006: 100n72.
8 Strabo 4.2.1; Roller (2018: 188) observes that Korbilon, a major center for the Narbonians, may no longer have been in existence during the governorship of Julius Caesar who did not mention the site.
9 Roseman 1994: 127–31; Romm 1992: 22–3; Roller 2006: 85–6. The Vikings referred to these waters as a "sea of worms" (*madkasjo*), where, according to the Saga of Erik the Red, Bjarni the Icelander's ship was destroyed by marine worms (Kunz 1997: 18).
10 Polybius 34.3; Walbank 1972: 52, 126–7; Roller 2006: 101.
11 See Walbank 1972: 126: "he wants to ensure that Odysseus is not regarded as having anticipated himself, Polybius, in exploring the outer ocean."
12 Pliny is here conflating sources and data: Roller 2006: 101.
13 E.g., Strabo 2.2.3; Polybius 34.1.7–14.
14 Romm 1992; Roller 2006.
15 Strabo 3.5.5; Velleius Paterculus, *Historia Romana* 1.2.3. Margarida 2009: 113–30.
16 *Iliad* 7.421–422; *Odyssey* 19.433–434.
17 Hesiod, *Theogony* 240–2; Chapter 1.
18 Mourelatos 2008: 134–68.
19 Graham 2010: 136.
20 Only one redaction suggests that one "cannot step twice into the same river" (*TEGP* 63).
21 Kahn 1979: 139–44.
22 Aristotle, *Meteorology* 3.1.371a15–17; Pliny 2.133. See Kahn 1979: 141–2, for an analysis of evidence from Hesiod, Herodotus, Aristophanes, Xenophon, and Aristotle.
23 Heraclitus, *TEGP* 51 = DK22B31b.
24 Graham 2010: 189.
25 Heraclitus, *TEGP* 79. In Empedocles, fishes are "water-nourished" (*TEGP* 41).
26 Aristotle, *Meteorology* 2.1.353b = Atomists, *TEGP* 88.
27 The Epicureans, who adapted Democritus' atomic theory, believed in the senescence of the earth, exhaustion of the soil, and the eventual failure of agriculture: Lucretius 2.1144–1177.

28 E.g., Homer, *Odyssey* 19.109–114; cf., Plato, *Republic* 2.363c.
29 Wilson 2013: 179. Aristotle seems to criticize Plato for placing a *katabasis* at the center of the discussion on Ocean. We also note that Odysseus reached the underworld by first sailing to Ocean: *Odyssey* 11.13–14. Cf., Beaulieu 2016: 10.
30 Cp., Hesiod, *Theogony* 736–741. Hesiod likely had in mind something akin to Anaximander's *apeiron* or indeterminate sources of moisture.
31 Wilson 2013: 182; West 1966: 361.
32 Wilson 2013: 190.
33 Herodotus 4.8. For Hipparchus, Strabo 1.1.8-9 = f 214Kidd; Kidd 1972: 3.279.
34 Aristotle, *Metaphysics* 2.2.354b; Eratosthenes f39Roller; Posidonius f214Kidd; Strabo 1.1.8–9.
35 Eratosthenes f39; Roller 2010b: 156.
36 Aëtius 3.16.5; cf., Aristotle, *Meteorology* 2.2.354b15–17.
37 See Pliny 2.222.
38 Aristotle, *Meteorology* 2.2.358a16–26; Taub 2003: 102.
39 Aristotle, *Meteorology* 2.3.359a17–b3; Diodorus Siculus 19.98.
40 Strabo 16.2.42; cf., Pliny 2.226 on the bitumen-producing Lake Asphalitis in Judea. Lake Sirbonis is in eastern Egypt (Strabo 16.2.32); cf., Roller 2018: ad loc.; Kidd 1972: 2b.951–3.
41 https://www.nasa.gov/topics/earth/earthmonth/earthmonth_2013_04.html.
42 *Problems* 23.27, 30.
43 Salt does lower the freezing point of liquid, but relative temperature depends on many properties, including region and season.
44 *Odyssey* 12.1–2, 235–243; cf., *Iliad* 18.399; *Odyssey* 20.65.
45 Euthymenes: Seneca, *NQ* 4.2.22; Sataspes: Herodotus 4.43.6.
46 Caesar, *BG* 4.29; Mohler 1944–5: 189–91; Kidd 1972: 2b.774–5.
47 Manilius suggested further that the stars also affected tidal amplitude (2.89–92).
48 Polybius 10.14.2; Livy 26.45.8; Lovejoy 1972: 110–11.
49 Caesar, *BG* 3.12-13; Strabo 4.4.1.
50 For Hanno's attempted circumnavigation, cf., Aristotelian *On Marvelous Things Heard* (#37).
51 Kidd 1972: 2b.790.
52 Roller 2006: 76–7. The Jersey Channel Islands have tides up to 12.5 meters (41 feet), and tidal amplitude at the Bay of Fundy in Canada, renowned for its high tides, can reach up to 17 meters (56 feet). See Waddelove and Waddelove 1990: 253–66.
53 Aëtius 3.17.3; Roseman 1994: 60–2, 80–2.
54 Eckenrode 1975: 279–81, 286.
55 Athenodorus of Tarsus also composed a work on tides, in which he compared tidal behavior to breathing and explained tidal peaks as enhanced by sub-oceanic springs (f6a).

56 Plutarch, *Platonic Questions* 8.1. In his old age Plato had considered the merits of heliocentrism. Seemingly, Metrodorus and Crates also promoted helio-centrism (pseudo-Plutarch, *Placita* 2.15), and perhaps Anaximander, though the evidence is contradictory. Pliny seems to have hinted at heliocentrism (2.116).
57 Strabo 1.1.9; Dicks 1960: 115; Kidd 1972: 2b.760-3; Roller 2005: 115-17.
58 Kidd 1972: 1.xiii.
59 Posidonius visited southern Gaul, Sicily, Dalmatia, Greece, North Africa, and Italy on his grand tour(s) of the Mediterranean, (probably) in the 90s BCE. Finally, he settled at Rhodes, a prosperous crossroads that enthusiastically fostered scientific inquiry. There he received citizenship and pursued a career as a writer, teacher, and public servant, and he founded an important center of Stoic learning. He also served in the Prytany, an executive governing board, and was selected as ambassador to Rome in 87-86 BCE, that eventful year when Marius and Cinna had captured the city, Sulla rose to prominence, and civil war loomed (Plutarch, *Marius* 43). At Rome, Posidonius, who already enjoyed international renown, attracted an illustrious circle of admirers including Pompey, the powerful general and statesman, and the acclaimed orator, Cicero, an enthusiastic student of Stoicism. During their travels, both Romans trekked to Rhodes to attend Posidonius lectures: Plutarch, *Pompey* 42.5; *Cicero* 4.4.
60 Kidd 1972: 2b.776.
61 Kidd 1972: 2b.809-10.
62 Kidd 1972: 1.281, 2b.775-6.
63 Goodyear 1981: ad loc.
64 Eckenrode 1975: 280-2.
65 Vinson 1994: 31; McGrail 2001: ix.
66 Aristotle, *Meteorology* 1.13.351a9-15; cf., Pliny 2.224.
67 Oleson 2008b: 130.
68 Avienus, *Ora Maritima* ("The Sea Coasts") 122-9. The Gulf is generally quite shallow, with a maximum depth of 90 meters (295 feet) and an average depth of 50 meters (164 feet).
69 Strabo 17.3.20.
70 This submarine ridge divides the eastern and western Mediterranean Seas.
71 Herodotus 2.12 speculated on erosion and deposition after seeing sea shells in Egypt's uplands.
72 Archimedes, *On Floating Bodies* 1 proposition 2; cf., Pliny 2.163 with empirical proofs.
73 f16Roller; Roller 2010b: 132.
74 Beaujeu 2003: ad loc.
75 *Problems* 23.2; cf., 23.28 where "successive impetus is given to the adjoining water."
76 Vergil, *Aeneid* 3.414; Strabo 1.3.10; Mela 2.115; Pliny 3.87; Seneca, *Epistle* 79.

77 Procopius, *History of the Wars* 8.6.20; Justin Martyr, *Cohortatio ad graecos* ("Exhortation to the Greeks") 34bMigne; Procopius 8.6.20; Wilson 2013: 179.
78 See further El-Geziry and Bryden 2014: 39–46 for a model of both surface and deepwater currents in the Mediterranean.
79 Strabo 6.2.3; cp. Pliny 3.87.
80 Homer, *Odyssey* 12.104–5; Vergil, *Aeneid* 3.420–3; Strabo 1.2.16, 36.
81 Strabo 6.2.3; Seneca, *NQ* 3.29.7; Pliny 3.87.
82 Polybius 34.3.10 (without explanation); Eratosthenes f16Roller; Strabo 1.3.11.
83 Strabo 6.3.2; Seneca, *Epistle* 79.1–2; Silius Italicus, *Punica* 14.254–257.
84 Appian, *Civil War* 5.90. Petronius employed the whirlpool to wreck Encolpius' ship (*Satyricon* 114).
85 Triantafillidis and Koutsoumba 2017: 169–70.

Chapter 3

1 For Grand Canyon geology: Harris et al. 1990: 7–10; Price 1999; Beus and Morales 2003.
2 Strabo 9.4.4; Pliny 4.27. The earthquake of 426 BCE drastically rerouted the river's course: Strabo 1.3.20.
3 Willcock 1976: 28.
4 Herodotus 5.52; Evans 2005: 105–19.
5 Graham 2010: 132.
6 For historical changes in sea levels and their causes, see Rona 1995. See also Broodbank 2013 for a magisterial survey of geological, topographical, and ecological changes.
7 Strabo 3.5.5. The mythological account of the creation of the Pillars is repeated as late as Mela (1.27), in whom, according to oral tradition, "Hercules split the mountains, which had once been joined in a continuous ridge over Spain and Morocco, and then Ocean, previously shut out by the mole of those mountains, was let into those places that it now inundates."
8 Eratosthenes f15Roller; Roller 2010b: 129–30.
9 For the geoarchaeology of the Nile Valley: Hassan 1997.
10 Plato, *Timaeus* 25d; cf., Roller 2006: 20, 50.
11 Aristotle, *Meteorology* 2.1.354a22; cf., Vidal-Naquet 2005: 48.
12 Posidonius f49d; Strabo 2.3.6; cf., Pliny 2.205 who expressed skepticism: "if we believe Plato's story."
13 We observe this process today in the equally dramatic loss of wetlands in southern Louisiana—where swamps have fluctuated for thousands of years—because of coastal excavation, canal dredging, and the resultant processes of erosion at a rate of

16.57 square miles (about 43 square km) a year for the past quarter century: for the changes in the coastlines, see the USGS report that documents the loss of land from 1932 to 2010: https://pubs.usgs.gov/sim/3164/.
14 For the causes of meandering rivers: Hooke 2003.
15 An "oceanic island" with poor soil, Cyprus was never rich in vegetation: Knapp 2010.
16 Hughes 2014: 3. The Libyan government is currently mining underground aquifers to reintroduce cultivation—but this is, at best, a short-term solution. The so-called Great Man-Made River draws water from the Nubian Sandstone Fossil Aquifer by means of 2,820 km (1,750 miles) of pipes and aqueducts together with 1,300 wells, in order to supply 6,500,000 meters^3 of fresh water per day to the cities of Tripoli, Benghazi, Sirte, and elsewhere: Keys 2011.
17 Hughes 2014: 81–2.
18 Plato, *Phaedo* 112a–113d; Homer, *Iliad* 8.14. Cf., Chapter 2.
19 Wilson 2013: 267. Pliny, however, recognized seasonal fluctuation in groundwater: 31.54.
20 Seneca, *NQ* 3.11; cf., Theophrastus f216Fortenenbaugh. Earthquakes often affect the flow of subterranean watercourses: Wang and Magna 2015.
21 In March 2017, the New Zealand Parliament passed into law the status of personhood for the Te Awa Tupua River, "teardrop of Ranginui (Sky Father)," which the Maori tribes of Whanganui have been seeking to protect from industrial degradation since the 1870s: https://www.loc.gov/law/foreign-news/article/new-zealand-bill-establishing-river-as-having-own-legal-personality-passed/. This legislation recognizes the deep spiritual connection between the people and their ancestral river.
22 Diels 1879: 226–8.
23 Posidonius is not known to have visited the Rhineland or Britain, thus his accounts, if any, would have been second-hand.
24 Campbell 2012: 291–8.
25 Hecataeus, *FGrHist* 264.7; cf., Herodotus 4.36. Cf., Bridgman 2005.
26 Apollonius, *Argonautica* 4.257–260, 282–293; Strabo 1.2.39.
27 Green 2007: 303.
28 Eratosthenes f16Roller; Strabo 1.3.15, 4.6.9.
29 The Danube does support at least 30 major tributaries along its 1,780-mile course (2,860-km) and scores of lesser tributaries (nearly 200), branching into the Danube Delta in three distributaries before discharging into the Black Sea.
30 Cf. Lordkipanidze 2000: 13–36, without comment on the river's reputed paradoxa.
31 Aelius Aristides, *To Rome* 82; Liddle 2003: 99.
32 Strabo 11.3.4; Pliny 6.13.
33 Homer, *Iliad* 2.754; Aristotelian *Problems* 23.6; Polybius 4.42; Arrian, *Periplus* 8; cf., Ovid, *Pontus* 4.10.63–64. Pliny (6.12) and Strabo (11.2.16–17, 11.3.4) are curiously silent about the unusual properties of the water.

34 Liddle 2003: 99; King 2004: 17.
35 India was possibly known to Scylax of Caryanda and perhaps even Hecataeus: Roller 2018: 847.
36 For crocodiles in the Indus: Herodotus 4.44. The hippopotamus was lacking from the Indian rivers (Strabo 15.1.13), although Alexander's helmsman, Onesicritus, claimed to have seen "this horse" there.
37 See Strabo 15.1.22–24; Pearson 1960: 104. Aelian, *NA* 8.21.
38 Beagon 1996: 290; Murphy 2004: 142–4; Williams 2008: 231.
39 Roller 2003: 192–6.
40 Coleman-Norton (1950: 240n15) speculates that Cicero's *Catadupa* may have been the second cataract, whose descent was greater than that of the first cataracts, and whose roar could be heard a mile and a half away. Coleman-Norton reports the drop at the first cataract as 80 feet, but gives no data for the second cataract. From autopsy, Winston Churchill reported the 9-mile-long second cataract had a total descent of about 30 feet (1902: Chapter 6).
41 Wainwright 1953.
42 Arrian, *Anabasis* 6.1.1; Strabo 15.1.25; Burstein 1976.
43 The source of the Blue Nile, from which flows most of the water that feeds the Blue and White Nile Rivers and their fertile alluvium, was finally confirmed in 1875 by Henry Morton Stanley when he circumnavigated Lake Victoria and reported the great outflow at Ripon Falls on the Lake's northern shore, thus validating the theory of John Hanning Speke, the first European to reach Lake Victoria: Speke 1863; Maitland 1971; Jeal 2011.
44 Eratosthenes f99Roller; Posidonius f222; Strabo 17.1.5.
45 Anaxagoras, *TEGP* 56 = Seneca, *NQ* 4a 2.17. Aeschylus, *Suppliant Maidens* 559; Sophocles fragment 882Radt; Euripides, *Helen* 3, *Archelaus* fragment 228.3-5Nauck.
46 See Graham 2003: 291–310.
47 Williams 2008: 218. Seneca, who had visited Egypt before 31 CE, composed *On the site and rites of the Egyptians* (Fr. VII [12] Haase = T19 Vottero). Nero would later send an expedition to search for the river's source: Pliny 6.181, 12.19; Dio 63.8.1–2; Hine 2006: 63n88.
48 Now submerged, Philae was near Aswan, but Seneca has confused the site with Meroë: Hine 2010: 199n27; Oltramare 1961: ad loc.
49 See Williams 2008 for flattery as a trope within Seneca's broader discussion of natural philosophy.
50 Aristotle, *Meteorology* 1.13.349b27–350a2; cf., Eratosthenes f96Roller.
51 Eratosthenes f 140Roller; Roller 2010b: 215. For the Alpheus, Smith 2004.
52 Pliny 2.225; cf., Ovid, *Metamorphoses* 5.599–641; Pausanias 7.2.
53 Moschus fragment 3 (Stobaeus 4.20.55); Ovid, *Metamorphoses* 5.572–641; *Greek Anthology* 9.362, 683; Lucian, *Dialogues of the Sea Gods* 3: Poseidon and Alpheus.

54 Eratosthenes f87Roller; Seneca, *NQ* 3.26.4.
55 Roller 2010b: 190.
56 Eratosthenes f110Roller; Pliny 6.58; Roller 2010b: 19, 206.
57 Hesiod, *Theogony* 2; Pindar, fragment 95; Horace, *Odes* 1.12.5; Ovid, *Tristia* 4.49-50; Statius, *Silvae* 1.2.4; Pausanias 9.31.4.
58 Pliny 2.228-235; Vitruvius 8.3; Mela 1.39.
59 Seneca, *NQ* 3.26.5; for hot springs, Chapter 1.6.
60 In vogue was the so-called third style of Pompeian wall painting, which spotlighted delicately wrought, chimeric architectural features, as well as fantastical hybrid animals in Egyptianizing landscapes. In this context Petronius composed his picaresque *Satyricon*, a novel of ostentatious and grotesque extravagance, and—a little later—Apuleius penned his *Metamorphoses*, a tale of a man turned into a donkey turned back into a man after conversion to the cult of Isis. Dwarfs were popular both in the arena and as mascots at court. In Pliny (7.75), we learn that Augustus' granddaughter, Julia, kept as a "pet" a dwarf whom she called Conopas.
61 Ovid, *Metamorphoses* 15.307-355. For the spring at Ammon, cf., Mela 1.39; Pliny 5.36.
62 Neleus was a son of Poseidon who, together with his twin brother, Pelias, was abandoned by their mother, Tyro. Neleus was banished to Messenia in the Peloponnese where he ruled a maritime kingdom and eventually became king of Pylos: Apollodorus, *Library* 1.9.8-9. Cerona is otherwise unattested.
63 Pliny 31.13-16. Pliny aimed to persuade the reader of his authority with generous citations from well-known, respected experts, including the Greek authorities, Theophrastus and Ctesias, and his Roman predecessor, Varro.

Chapter 4

1 Hughes 2014: 9.
2 McCormick et al. 2012; Manning 2013; Leveau et al. 2016. Debate over climactic conditions in Roman North Africa continues: Drake et al., 2011; Kouki 2013.
3 Wilson 2013: 262-3.
4 Cf., Franconi 2017 for a collection of essays that interrogate climactic effects on riparian environments.
5 Taub 2003: 78.
6 Peripatetic, *On Weather Signs* 19, 40; Pliny 18.85, 87.
7 For a lively, accessible introduction to becoming "weather-wise": Sloane 1980.
8 See Lehoux 2007: 3-10.
9 Taub 2003: 123, 143, 163.
10 Cf., Taub 2003: 138-9; Chapter 1.

11 Hankinson 1995: 117. See Chapter 1.
12 Aristotle, *Physics* 2.7.198b16–21, 198b34–199a8.
13 Aristotle, *HA* 8.19.601b9–34; cf., Taub 2003: 60.
14 The reader is referred to Daiber 1992 for a translation and scholarly commentary (translations used herein are Daiber's). Kidd 1992 questions whether the treatise is authentically Theophrastean.
15 Taub 2003: 98–102, 117, 124, 188.
16 Kidd 1992: 300.
17 Cloud collision was a common explanation for thunder in early rational explanations: e.g., Democritus *TEGP* 82. Cf., Aristophanes, *Clouds* 383 where Socrates explains the source of thunder: "the Clouds, when full of moisture, dash against each other and clap by reason of their density."
18 Sider 2002: 108.
19 Taub 2003: 37, 132.
20 Epicurus, *Letter to Pythocles* (= Diogenes Laërtius 10.98–99).
21 Although fundamentally a moralizing natural philosopher, Seneca uniquely seems to have expressed genuine interest in meteorology for its own sake: Hine 2010: 6.
22 Pliny 2.82–153 and 18.340–365, the latter section appended to his book on farming.
23 The authority of Homer and Hesiod remained unimpugnable: Taub 2003: 93–6; cf., Hine 2010: 6 for Seneca's doxography.
24 Theophrastus, *On Winds* 48, trans. Wood 1894.
25 Pliny 18.210–2170; *On Weather Signs* 4.
26 Taub 2003: 177–8.
27 Taub 2003: 37; Lehoux 2007: 59.
28 Sloane 1980: 4–5.
29 *On Weather Signs* 10. This weather wisdom is repeated by Matthew's Jesus:

> When it is late, say, "The weather will be fair; for the sky is fiery red."
> And when it is early, "Today there will be stormy weather, for the sky is fiery and overcast."
>
> Mt. 16:2–3

and by Shakespeare,

> Once more the ruby-colour'd portal open'd,
> Which to his speech did honey passage yield;
> Like a red morn, that ever yet betoken'd
> Wreck to the seaman, tempest to the field,
> Sorrow to shepherds, woe unto the birds,
> Gusts and foul flaws to herdmen and to herds.
>
> *Venus and Adonis*, stanza 75

30 Aratus, *Phaenomena* 758–1,141 (799–814: moon, 954–955: oxen, 963–966: crows). See Kidd 1984; Kidd 1997: ad loc.
31 *Georgics* 1.351–463. Vergil was praised and largely imitated by Vegetius 4.41.
32 Aristotle, *Meteorology* 2.9.369b1–370a11; Seneca, *NQ* 2.8, 2.27.
33 Aristotle, *Cosmos* 4.395a16–21; cf., Pliny 2.142.
34 Epicurus, *Letter to Pythocles*, Diogenes Laërtius 10.102; see Kidd 1992: 302.
35 Aristotle, *Meteorology* 3.1.370b5–16.
36 Kidd 1992: 300; f135; cf., Seneca, *NQ* 2.54.
37 Wilson 2013: 152.
38 Theophrastus, *Metarsiology* [9] 2–11; cf., Posidonius f11Kidd.
39 Anaximenes, *TEGP* 26; Graham 2010: 321.
40 Aristotle, *Meteorology* 1.12.347b14–349a11; cf., Wilson 2013: 152–5.
41 Wilson 2013: 154.
42 Posidonius f136 = Seneca, *NQ* 4b.3; cf., Kidd 1992: p. 298.
43 Kidd 1972: 2.513–14.
44 ἐννοσίγαιος: Homer, *Odyssey* 5.423, 6.326, 9.518, 11.102, Hesiod, *Theogony* 442, 818, 930; Agathias Scholasticus of Asian Myrina, *Greek Anthology* 4.4.62; ἐνοσίχθων: *Odyssey* 1.74, 5.282, 5.339, 5.366, 7.56, 9.283, 11.252, 12.107, 13.146, 13.162; σεισίχθονος: Bacchylides 18.22.
45 E.g., Homer, *Odyssey* 5.365–368, 7.273–275.
46 Thales, *TEGP* 20; Seneca, *NQ* 3.14, 6.6. The sources are vague, and Thales' view on the cause of earthquakes is far from certain: Taub 2003: 73.
47 Cf., Taub 2003: 90.
48 Aristotle, *Meteorology* 2.8.366a5–23; cf., Theophrastus, *Metarsiology* [15] 26–27; Taub 2003: 165.
49 Seneca, *NQ* 6.18.6–7; Taub 2003: 148–52.
50 In Seneca, earthquakes are not predictable: *NQ* 6.6.1.10. Pliny merely planted a seed of doubt: "if these things are true...:" *quae si vera sunt* (2.192).
51 Smid 1970.
52 On December 26, 2004 a tsunami devastated the lands along the Indian Ocean, claiming the lives of up to 230,000 people in fourteen countries. The Sumatra-Andeman Earthquake that generated the tsunami registered between 9.1 and 9.3, lasting 8.3 to 10 minutes. With an epicenter between Simeulue and Sumatra, the after-effects were felt across the Pacific to Alaska and South Africa (8,500 km [5,300 miles] away), which experienced a tidal surge of 1.5 meters (5 feet) about 16 hours after the quake. The greatest wave amplitude was recorded for the west coast of Aceh in Indonesia at 15–30 meters (49–98 feet), where three waves reached the shore within 20 minutes of the quake: Borrero 2005.
53 Smid (1970: 102) observes that the sources omit mention of foul weather and other meteorological phenomena in conjunction with these unusually large waves.

54 Smid 1970: 102.
55 Aristotle, *Meteorology* 2.8.368a33–b11; cf., *Meteorology* 1.6.343b2; Taub 2003: 92, 169.
56 On December 26, 2004, three elephants on the Patanangala beach at Yala National Park in Sri Lanka were observed fleeing, and other animals in the tsunami zone sought shelter to the best of their abilities. Bats flew away before the wave hit, and dogs refused to go outside. Ravi Corea, president of the Sri Lanka Wildlife Conservation Society, reported that no animal carcasses were found within the park, where at least sixty human visitors lost their lives to the tsunami: Mott 2005.
57 Seneca, *NQ* 3.27–29; Inwood 2002; cf., Ovid, *Metamorphoses* 1.262–92. The motif is nearly universal, and flood accounts feature prominently in the *Atrahasis*, *Gilgamesh* (tablet 11.1–4), and the Hebrew Bible (Genesis 6.9–9.17).
58 Vegetius 4.41, following Vergil, *Georgics* 1.393–463.
59 Appian, *Civil War* 5.100; Dio 48.48; Powell 2002: 115.
60 Dio 50.31.2; Plutarch, *Antony* 65.1, 66.1–2; Orosius 6.19.10; Powell 2015: 89, 92. Vergil, *Aeneid* 8.710: cf., *UCWW*: Chapter 10.
61 *AE* 1973 #475; Šašel and Šašel 1986: 2. #468.
62 Aldrette 2007.
63 Ammerman 1990, 1998. Under Domitian, the ground level at the Campus Martius was raised in response to flood threats (Belati 1999: 16), perhaps causing the inaccuracy of Augustus' monumental sundial after only 50 years (Pliny 36.72–3).
64 Wilson 2013: 270–1.
65 Furius Crassipes was Cicero's son-in-law, the second husband of Tullia, from whom she was divorced before 51 BCE. They had been betrothed in 56 BCE (*To Quintus* 10 [2.6]). Cicero seems to have enjoyed visiting Crassipes' suburban villa: *To Quintus* #8 (2.4); Treggiari 2007: 76–7.
66 Responses to natural disasters were mostly inconsistent and inadequate: Hughes 2014: 213–14. Apamea in Mysia received earthquake relief from both Alexander the Great and Mithridates VI (Strabo 12.8.18). After the earthquake that devastated Rhodes (and destroyed the colossal statue that guarded the city's harbor), the first recorded, collective royal effort provided economic relief by exempting Rhodian merchants from customs dues: Duncan-Jones 1990: 37. Herod the Great was detained from participating in the naval battle at Actium because of earthquake relief in Judea in 31 BCE: Josephus, *Jewish Wars* 1.19.3–4. In 17 CE, when twelve populous cities in Asia were devastated by a nocturnal earthquake, Tiberius extended a grant of 10,000 sesterces and tax relief (Tacitus, *Annals* 2.47). Titus appointed a senatorial commission to organize relief efforts for the victims of Vesuvius' 79 CE eruption, including funds for rebuilding and housing displaced residents (Suetonius, *Titus* 8.3).
67 Pliny 31.53; Hughes 2014: 197.

Chapter 5

1 Nutton 2013: 26.
2 Many public latrines were equipped with channels for storing sponges, as well as basins and trenches for washing them: Jansen et al., 2011: 8, 20, 23–4, 27. People may have also employed stones, shells, or foliage to clean themselves: Lambton 1995: 7.
3 Celsus 6.6; Pliny 21.138. For oil lamps and eye disorders: Donahue 2016: 613. Celsus also provided a detailed description of cataract surgery: 6.14.
4 Livy 37.23.2; cf., Polybius 1.19.1.
5 Nutton 2013: 25.
6 Cf., Retief and Cilliers 2000.
7 E.g., Harper and Armelagos 2009; Wilbur et al., 2009: 1,990–7; Nikita et al., 2016.
8 Cf., Retief and Cilliers 2000: 267.
9 Grmek 1989: 16. See Hippocratic *Prorrhetic II*. 22 for the mortality rate of weaned children.
10 Hippocratic *On Regimen in Acute Diseases* 18–19; Celsus 4.18.1. Biraben 1998: 325–6.
11 Hippocratic *Epidemics* 5.10; Pliny 20.122; Plutarch, *On Borrowing* 831b; Galen, *Method of Medicine* 7.513 Kühn. Among the prescriptions for "cholera" in Pliny are mint (20.146, 150), coriander with honey and raisins (20.218), blackberries with hypocisthis and honey (24.120), grapes kept in rainwater (23.12), pearl barley (20.67), and quinces (23.101).
12 Grmek 1989: 89.
13 Grmek 1989: 354–5; cf., Hippocratic *Epidemics* 7.24, 7.55, *Epidemics* 3, *Coan Prenotions*.
14 Oerlemans and Tacoma 2014: 229–34.
15 Grmek 1989: 346–50.
16 Grmek 1989: 437n21; cf., Ongaro 1967.
17 Plutarch, *Alexander* 75–76; Oldach et al. 1998; Pope et al. 2012: 49.
18 See Celsus 4.15.3: "nothing is more harmful to the liver than cold things."
19 Buchan 1937: 161–2.
20 During the campaigns against the Numantines in 140 BCE, unfamiliar food, water, and climate caused soldiers to fall ill with dysentery, some of whom died (Appian, *Spanish Wars* 6.78).
21 Cunha and Cunha 2008: 194–9.
22 Mueller et al. 2007: 278–83; Collins 2012: 107–21.
23 *AWP* 7; Hippocratic *Aphorisms* 3. 21, 22. Grmek 1989: 281. See Nikita et al. 2016: 472. Cf., Angel 1966: 760–3; Grmek 1989: 265–83. Craik 2017; Oerlemans and Tacoma 2014: 219–24.

24 For the malarial swamps of Camerina in Sicily: Servius, *ad Aeneid* 3.701; *Greek Anthology* 9.685. See Silius Italicus 8.575 for the noxious Frengenae (Maccarese) swamp north of Ostia.
25 Sallares 2002: 103–14; Radcliffe 1921: 230, citing J.A. Thomson's lecture before the Royal Institution, January 6, 1921.
26 Juvenal, *Satire* 4.56–57: where the ill hope for a quartan fever from which they can recover. Cf., Martial 2.40.
27 Cicero, *Nature of the Gods* 3.63; Seneca, *Apocolocyntosis* 4–6.
28 Hippocratic *Epidemics I*, case 1; *Epidemics III*, case 3; Grmek 1989: 295–304.
29 Thucydides was likely describing *P. falciparum* malaria (6.64, 7.47.1–2): Grmek 1979: 141–63; Sallares 2002: 36; Hughes 2014: 162.
30 Retief and Cilliers 2006: 685–8. Cicero, *To Atticus* #128 (7.5), #154 (8.6); Suetonius, *Caesar* 1.
31 Cf., Velleius Paterculus 2.84.1; Plutarch, *Antony* 67.7, 68.4–5; Powell 2015: 85–6.
32 Galassi et al. 2016: 84–7.
33 Sallares 2002: 277.
34 See Hughes 2014: 200.
35 Tan Sy 2008.
36 Diogenes Laërtius 8.70. The drainage program is attested on fifth century coinage from Selinus. On one example, the horned river god holds a lustral branch in his left hand and phiale in his right, while he sacrifices at an altar. In front of the altar is a rooster, and behind the muscled, nude, river god, standing on a pedestal, is a small bull possibly representing "the previously untamed river." BM 1924, 1220.1; Russell 1955: Figure 15.
37 Sallares 2002: 89; Hughes 2014: 200.
38 Tertian and quartan fevers in malaria patients arise owing to the fact that the sporozoan parasites, as they absorb nutrients from their host, complete their growing cycle in erythrocytes every two or three days.
39 Hippocratic *Affections* 17–19; *AWP* 7; Pappas et al. 2008: 347–50.
40 "The clearest and most exact description of the two clinical types of tertian fevers are to be found in the works of Celsus:" Marchiafava and Bignami 1894: 231–2.
41 Celsus 3.3.1. See Sallares et al. 2004: 311–28; Oerlemans and Tacoma 2014: 213–41.
42 Quintus Serenus Sammonicus, *Liber Medicinalis* (*De Medicina Praecepta Saluberrima*) 51. This is the first attested use of ABRACADABRA.
43 Wootton 1910: 1.164–166.
44 Thucydides 3.87. See Kallet 2013.
45 Thucydides 2.58. The Peloponnese was not severely affected.
46 Thucydides 2.47–54. It is not our aim to identify the plague's modern equivalency, which some scholars have connected with smallpox (Littman and Littman 1969); or even an outbreak of ebola (Kazanjian 2015). Cf., Craik, 2001.

47 Scholion on Aristophanes, *Birds* 997 quoting Phrynicus Monotropos fragment 376k. Meton had constructed the wells in the Piraeus before 414 BCE. Cf., Longrigg 2000.
48 Rose and Masago 2007: 41–8; Pope et al. 2012: 56. Papagrigorakis et al. 2006. Both the modality and the results of the dental-pulp study are fraught: Shapiro et al. 2006.
49 Littman 1984; Vanotti 1989; Angeletti and Gazzaniga 2002.
50 Pliny 25.17; 8.152; Marcus Aurelius, *Meditations* 6.57.
51 Dioscorides 3.91; cf., Beck 2017: *ad loc.*
52 Pliny 25.20–21; Davies 1970a: 92. See Vitruvius 8.3.23 for a water supply at Susa that causes imbibers to lose their teeth.
53 Koloski-Ostrow 2015: 84–101.
54 Scobie 1986: 422n178.
55 Fagan 1999: 176–88; Fagan 2000; Knapp 2011: 45–6; Rogers 2018: 44.
56 Artemidorus of Daldis, *Interpretation of Dreams* 2.26.
57 Kaplan 2002; Rothschild, et al. 2000; Farhi and Dupin 2010.
58 Ammianus Marcellinus 18.4; Dio, 72.14.3–4; Eutropius 31.6.24. Gilliam 1961: 228–9; Littman and Littman 1973: 243. Measles has been ruled out as a cause of the Antonine epidemic: Furuse et al. 2010; Bruun 2007.
59 Grmek 1989: 89.
60 McNeill 1976: 109.
61 Cockburn and Cockburn 1983; Waldron 1987; Metcalfe 2007; Pope et al. 2012: 52–5.
62 In a controlled experiment with six groups of mariners, Scottish physician, James Lind, attached to the *HMS Salisbury*, supplemented the diets of sailors suffering from scurvy with a variety of foodstuffs (including hard cider, vitriol, vinegar, seawater, two oranges and one lemon, and an "electuary" of garlic, mustard seed, balsam of Peru, dried radish root, and gum myrrh, tamarinds, and cream of tartar). Lind concludes: "As I shall have occasion elsewhere to take notice of the effects of other medicines in this disease, I shall here only observe that the results of all my experiments was, that oranges and lemons were the most effectual remedies for this distemper at sea:" Lind, *A Treatise of the Scurvy: Containing an Inquiry into the Nature, Causes, and Cure, of That Disease Together with a Critical and Chronological View of What Has Been Published on the Subject* (1748): 196. Lind failed to recognize the connection. James Cook was independently concerned with the effects of scurvy, and he mandated that the shipboard diet include antiscorbutics, such as malt and cabbage, "scurvy grass," and wild celery, along with strict attention to hygiene and clean water. In 1795, physician Gilbert Blane convinced the British Royal Navy to issue some form of lemon juice to its sailors: Lloyd 1961: 129–30.
63 Grmek 1989: 36.
64 Aristophanes, *Women at the Thesmophoria* 882; Petronius, *Satyricon* 103; Plutarch, *Precepts of Statecraft* 798d; Plutarch, *On the Tranquility of the Mind* 466c (larger ships are more comfortable for those prone to seasickness); Lucian, *Dialogues of the*

Dead 6 (crossing the River Styx); Lucian, *On Friendship* 19 (rough weather); Athenaeus 2.37d (rough weather); Alciphron, *Letters of the Courtesans* 4.9. Marius' flagship had to put to shore when the general succumbed to seasickness (Plutarch, *Marius* 36).

65 Semonides, *Women* 54 (f7); Galen, *Natural Faculties* 3.12 Kühn.
66 Cf., Meeusen 2016: 412.
67 Reynolds and Ward-Perkins 1952: #171 (Africa proconsularis).
68 Galen, *Preserving Health* 3.4 Kühn.
69 Cf., 8.3.32; Dvorjetski 2007: 84–5 and note 6.
70 Leprosy is treated in the modern world with a long-term cocktail of antibiotics.
71 For example, Vitruvius 8.3.6 (gout); Celsus 4.12 (on paralysis); Pliny 31.3–8.
72 Cf., Strabo 5.3.2; Martial 1.12.
73 Thermal waters are heated either geothermally from the earth's mantle or by the decay of naturally radioactive elements.
74 Croon 1967: 230, 244; Jackson 1990: 5–9.
75 Malkin 2003: 228; Croon 1952; 1956: 210–17.
76 For Thermopylae: Herodotus 7.176; Strabo 9.4.13; cf., Dvorjetski 2007: 93–4.
77 Deyts 1985; Cunliffe 1988: 359–62; Allason-Jones 1989: 156–7. For physicians at other spas, Dvorjetski 2007: 109–10.
78 Davies 1970b: 103 and note 14; Dvorjetski 2007: 106. For Roman soldiers at Bath, *RIB* 139, 143–4, 146, 147, 152, 156–60.
79 Goldsworthy 2003: 99, 142–3.
80 *CIL* 3.12336; Dvorjetski 2007: 106–7.
81 *AE* 1968: 323; Nesselhauf and Petrikovits 1967: 268–79.
82 Hartmann 1973.

Chapter 6

1 Anaximander, *TEGP* 37–39. Cp. Anaxagoras, *TEGP* 37 where animals arise from warm, moist earth; cf., Chapter 1.
2 According to Pliny (9.143, 154, 156), creatures in hard, flinty shells, like oysters, lack a sense of feeling, but some species are susceptible to diseases.
3 In Aelian, *NA* 9.50, a marine mammal roars from the rocks. The Greek word (καστορίδες: *kastorides*) is generally taken to mean "seal," but the terminology is imprecise.
4 Aristotle, *PA* 4.5.679a10–15; Pliny 9.84; Aelian, *NA* 7.11; Plutarch, *Cleverness* 978e.
5 Aristotle, *HA* 9.37.621b30–622a6; Pliny 9.86. Cf., Cunningham 1999: 17–41.
6 The author of the British-produced *Aberdeen Bestiary*, one of the finest examples of the type, tells us that the *Bestiary's* purpose is "to improve the minds of ordinary

people, in such a way that the soul will at least perceive physically things which it has difficulty grasping mentally: that what they have difficulty comprehending with their ears, they will perceive with their eyes" (25v).

7 Lévi Strauss 1964: 89.
8 Aelian, NA 6.31; Plutarch, *Cleverness* 961e. Although crabs figure prominently in art (especially the coinage of Acragas: *UCWW*: Chapter 8) they are little represented in the literature. Their distinctive "backwards scuttling" movement is noted: *Greek Anthology* 6.196 and *Vassal Treaty of Esarhaddon* 9 ("May…your sons and daughters go backward like a crab").
9 *SVF* 2.821 (Chrysippus); Diogenes Laërtius, *Epicurus* 32; Gilhus 2006: 23.
10 Aristotle, *PA* 1.1–3.642b5v644a12; *GA* 2.1.732b15–28.
11 Aristotle, *HA* 2.15.505b25–32. Cf., Voultsiadou et al. 2017: 468–78.
12 Aristotle, *PA* 4.10.565b2–12. Cf., Popa 2016: 284.
13 Kitchell 2014: 53.
14 Aelian, NA 16.18; Kitchell 2014: 54.
15 The Knossos dolphin is possibly the striped dolphin: Kitchell 2014: 56. Llewellyn-Jones and Lewis (2018: 412) consider these depictions as lacking anatomical accuracy, with misaligned dorsal and pectoral fins, and tail fins consistently shown from an aerial view even when the dolphins swim in profile. Inaccuracies might be because of the fact that dolphins were not hunted, and it would be rare to see a live (or recently deceased) specimen on land. Depictions became more stylized and less anatomically accurate over time: Ridgway 1970. Some scholars have unconvincingly interpreted the Minoan color scheme as a clue to their species: Gill 1985: 65–7; Morgan 1988: 61.
16 Ridgway 1970.
17 Reliefs from Ostia show dolphins escorting ships into harbor. Pilot fish were also observed guiding whales through the depths "like loyal friends" and protecting the "monsters" from fishermen's hooks: Oppian, *Fishing* 5.67–102; Plutarch, *Cleverness* 980f; Aelian, NA 2.15, 15.23. Another type of "pilot-fish," tuna follow behind ships under sail for several miles, not dissuaded even when javelins are tossed (Pliny 9.51).
18 Lucan 5.552; Artemidorus 1.16, 110; 2.16.
19 *SEG* 27.468; Toynbee 2013: 207–8.
20 Aelian, NA 11.12, 12.45. Cf., Aristophanes, *Frogs* 1317; Euripides, *Elektra* 435–6.
21 *HH* 7 *Dionysus*; Oppian, *Fishing* 1.649–653; Apollodorus, *Library* 3.5.3; Nonnus, *Dionysiaca* 45.105–168; Lucian, *Dialogue of the Mariner* 8.
22 Spivey and Rasmussen 1988: 2–8. Repatriated to the Italian government from Toledo, Ohio in 2012.
23 Aelian, NA 11.12; Oppian, *Fishing* 5.416–47; Plutarch, *Cleverness* 984a–985c.
24 Pliny 9.29–32; Oppian, *Fishing* 5.416–47; Aelian, NA 2.8. In Brazil at Laguna (Santa Catarina) and Inibé/Tramandaí (Rio Grande do Sul), dolphins in alliance with fishermen wrangle fish, retreating while the nets are cast: Kitchell 2014:55.

25 Plutarch, *Dinner of the Seven Sages* 19; Kitchell 2014: 55.
26 Aelian, *NA* 6.15; cf., Pliny 9.27; Plutarch, *Cleverness* 984e–f.
27 Dolphins are known to come close to shore and interact with human swimmers, as in New Zealand in 1955/56 when the bottle-nose dolphin Opo, who had lost his mother, began swimming near the coast of Opononi and became a local celebrity: Kitchell 2014: 55.
28 Aelian, *NA* 5.6; cf., Oppian, *Fishing* 5.448–518.
29 Aristotle, *HA* 9.48.631a7–b4; Aelian, *NA* 12.6.
30 Oppian, *Fishing* 2.628–41; Antipater of Thessalonica, *Greek Anthology* 7.214–216; Kitchell 2014: 56; Llewellyn-Jones and Lewis 2018: 413.
31 Aristotle, *HA* 1.5.489a35–b7; 2.15.506b3–5; 3.21.521b21–26; 4.9.535b32–536a4; 6.12.566b2–26; 8.2.589a31–b11; *GA* 2.1.732b24–27; *PA* 4.13.697b1f. Dolphin gestation is closer to 11–12 months.
32 Viviparous and suckling: Aristotle, *GA* 1.9.718b28–31; Pliny 9.21; Aelian, *NA* 11.37. For the function of the blowhole: Aristotle, *HA* 1.4.489a35–b7; 3.20.521b21–4, 6.12.566b13–18, 8.2.589a32–b11; Pliny 9.19; Aelian, *NA* 2.52, 10.8; Oppian, *Fishing* 1.733–734. Aristotle also recognizes the differences between dolphins and porpoises (found in the Black Sea): *HA* 6.12.566b9–16; Pliny 9.34; see Frantzis et al. 2001.
33 Blowholes allow cetaceans to breath. Over time, the organ moved to the top of the head, which is more frequently exposed to air than the snout. The spray (or blow) that results when cetaceans expire results from the forceful expulsion of air into the lower-pressure, colder atmosphere, where water vapor condenses, giving the appearance of a watery spray.
34 Aristotle, *Respiration* 18.476b14–30, *HA* 8.2.589a31–b20, *PA* 4.13.697a16–22.
35 Aelian, *NA* 11.22; Plutarch, *Cleverness* 28 979d.
36 Homer, *Iliad* 21.22–24 compares Achilles to a dolphin chasing fish to shore.
37 Aristotle, *HA* 9.48.631a22; Aelian, *NA* 12.12.
38 Pliny 9.20; Oppian, *Fishing* 2.533–52.
39 Xenophon, *Anabasis* 5.4.27; Oppian, *Fishing* 5.519–588; Strabo 12.3.19.
40 Monachus: Johnson and Lavigne 1999: 20–9, 30–2; Kitchell 2014: 166.
41 Aristotle, *PA* 4.13.691a8–10, 697b1–4; *HA* 6.12.566b27–567a14; Kitchell 2014: 167; Llewellyn-Jones and Lewis 2018: 22, 408.
42 Aristotle, *HA* 4.12.566b 27–567a15; Plutarch, *Cleverness* 34 982c.
43 Oppian, *Hunting* 5.376–391; Llewellyn-Jones and Lewis 2018: 407. Seals occasionally turn up as bycatch.
44 Pliny 8.111, 32.112, 120, 130; Aelian, *NA* 3.19.
45 Niarchos Collection, Athens. Llewellyn-Jones and Lewis 2018: 407; *LIMC* Ketos 26; See Johnson and Lavigne 1999: Figures 6–7; Kitchell 2014: 167.
46 Toynbee 2013: 205–6.
47 Calpurnius Siculus, *Eclogue* 7.65–66; *Acts of Paul and Thecla* 1.88; Kitchell 2014: 167.

48 Homer, *Odyssey* 15.480-1; Schneider 2001: 45-57.
49 Llewellyn-Jones and Lewis 2018: 417.
50 Jon. 1:11-2:1-6; Lucian, *True Story* 1.30-2.2. After a year and eight months, Lucian's 1,500-stade long (ca. 220 km!) whale is killed from the inside out as the hostages dig and cut their way, setting fire to the tail which, after seven days finally begins to erode. On the twelfth day the passengers are able to prop open the dying whale's mouth and escape. See also Boardman 1987.
51 Papadopoulos and Ruscillo 2002: 216-21.
52 Aristotle, *HA* 3.12.519a23-25; Llewellyn-Jones and Lewis 2018: 419-20.
53 Strabo 15.2.12; Arrian, *Indika* 30; Pliny 2.5.
54 Britain (Juvenal 10.14); the Atlantic (Ausonius, *Mosella* 144-9). Mosella's whale was gentle compared to the typical Atlantic whales that could displace large quantities of water and whose crests could block out the mountains.
55 Pliny 9.12-15; see Oppian, *Fishing* 5.109-35 for another graphic description of a whale hunt.
56 In Petronius, the witch Quartilla drove off a eunuch with her whalebone rod (*ballaenaceam virgam*) whose magical properties, if any, are unclear: *Satyricon* 21.2. See also Papadopoulos and Ruscillo 2002.
57 *Porphyrios* ("Purple") refers either to the whale's color or is meant to elicit royalty. Procopius' whale, whose career spanned fifty years, may actually have been several animals, who swam consecutively in that territory: Llewellyn-Jones and Lewis 2018: 420.
58 Oppian, *Fishing* 1.373-382; Aelian, *NA* 1.55; Llewellyn-Jones and Lewis 2018: 673.
59 Reese 2001: 287-8 and Figures 226-229.
60 Aristotle, *HA* 5.5.540b15-28, 6.10-11.565a12-566b2; Pliny 9.78.
61 Oppian, *Fishing* 1.734-741; Aelian, *NA* 2.55; Athenaeus 7.294a; Bodson 1983: 394-5.
62 Reese 1984: 190-1.
63 Ischia: Museo Archeologico, Pithekoussai 618813; Boardman 1998: Figure 161; Llewellyn-Jones and Lewis 2018: 674. Under excavation since 2012, a mosaic floor at the synagogue at Huqoq, three miles west of the Sea of Galilee, shows several biblical scenes, including the story of Jonah who is swallowed by a fish. On the mosaic, the prophet's fish is swallowed by a larger one, in turn swallowed by yet a larger fish: Romney 2018.
64 Plutarch, *Phocion* 28.1-3. See also Aeschines 3.130; Parker 2005: 109. The shark attack was interpreted as presaging that the *polis* would lose the part of the city closest to the Aegean.
65 Betancourt 2007: 86-8.
66 Forty-five examples of octopods are cited in the Beazley Archive, including on Athena's shield on a Panathenaic amphora series (Taranto 4319, 4320).
67 Attic red-figure cup, Berlin Pergamonmuseum F2264, from Vulci, attr. Oltos, c. 490 BCE; Detienne and Vernant 1969: 291-317; Llewellyn-Jones and Lewis 2018: 676.

68 Toynbee 2013: 213–14.
69 Aristotle, *HA* 9.37.621b30–622a6; *PA* 4.5.679a8–32; Oppian, *Fishing* 3.156–165; Plutarch, *Cleverness* 978a–b, e–f.
70 Aristotle, *HA* 5.6.541b1–11, 5.18.549b32–550a9.
71 The female octopus is able to retain the sperm in a separate sac until her eggs are ready. She carefully lays the eggs and then discharges the sperm, and she will incubate the nest for two to five months, depending on the species, defending the eggs against predators, and keeping the nest clean and oxygenated by wafting water over the eggs. The female dies of exhaustion and malnutrition shortly after her fry hatch. Octopods can lay up to 200,000 eggs in a clutch, and the survival rate is about one percent. Life expectancy varies by species, ranging from six months to five years.
72 Aristotle, *PA* 4.9.685a14–b16. The legendary strength of the octopod's grip provided for striking similes: Odysseus clung to the rocks like an octopus to its lair (*Odyssey* 5.432), and the nymph Salmacis held the youth Hermaphroditus as an octopus would engulf its prey (Ovid, *Metamorphoses* 4.366–367).
73 Aristotle, *HA* 4.1.524a14–20; Pliny 9.85.
74 Llewellyn-Jones and Lewis 2018: 678–9.
75 Wood and Anderson 2004: 95, 99; Llewellyn-Jones and Lewis 2018: 679. For tree-climbing octopods see Athenaeus 7.317c; Aelian, *NA* 9.45; Oppian, *Fishing* 4.264–307.
76 Pliny 9.91; Würzburg, inventory #527; Llewellyn-Jones and Lewis 2018: 675–9.

Chapter 7

1 Llewellyn-Jones and Lewis 2018 pp. 675, 679.
2 Cf., Oppian, *Fishing* 5.46–49 where sea monsters keep to the ocean's floor because of their bulk and weight.
3 Boardman 1987: 78 and plates 23.8–9; Boston Museum of Fine Arts 6.67 (Archaic, Proto-Attic vase).
4 Ostia Antica: Room Four, Baths of Neptune, second century CE; Van Duzer 2013: 9.
5 See Boardman 1987: 74–6. The artistic record is rich: a south Italian vase from Ruvo shows a *ketos* with flippers (Ruvo J 1500; Boardman plate 22.3). Fins emanate into streamers resembling wings on a gem (New York, Metropolitan Museum 41.160). On the Mildenhall Dish, a *ketos* has an upturned snout (BM 1946, 1007.1). Thetis, furthermore, cuddles a *ketos* (with a scrunched snout and feathery dewlap) on her lap on the Portland Vase (BM 1945,0927.1).
6 Nigidius Figulus, *Sphaera Graecanica* 122–25; Hölscher 1965; Dwyer 1973.
7 Emphasizing the restoration of peace on land and sea, the capricorn was widely employed on Augustan coinage, public monuments, and as a *signum* for military

units: e.g., *Legio II Augusta*, whose *dies natalis* fell on Augustus' birthday: *CIL* VII 103, *addit.* 306 (*RIB* 327); cf., Zanker 1990: 82–5; cf., Terio 2006: 97–110. For the Capricorn and Augustus' apotheosis: Germanicus, *Aratea* 554–60; Irby 2019b; cf., *UCWW*: Chapter 10.

8 Biological seahorses are mentioned in Aelian, *NA* 14.20 as a toxin. Burnt seahorses mixed with marjoram and liquid pitch or lard were recommended as a remedy for *alopekia* (Dioscorides 2.3; Pliny 32.67; Galen, *On Simples* 11.41 [12.362K]). In Dioscorides they are small. In various preparations, seahorses were an antidote to the poison of the sea hare and cures for the removal of leprous sores, pains in the side, and incontinence (Pliny 32.58, 84, 93, 109; cp., Aelian, *NA* 2.45). In Oppian, seahorses frequent the shoals (*Fishing* 1.93–101). According to Aelian (*NA* 9.60), pipefishes are slender, womb-less creatures whose fetuses are ejected. But he seems unaware of the link between pipefish and seahorses. Aristotle may have watched pipefishes. He listed them among his "elongated" fishes and noted that they produce an egg pouch (*HA* 2.15.506b9, 6.9.543b12–14). Cf., Pliny 9.156, where pipefish have so many fry that the belly bursts open when spawning but then grows back together after birth. The mating and birthing habits of shy, biological seahorses were likely not observed first-hand, and they are not described in the ancient sources.

9 Cf., Simon 2014: 39.

10 BM 1946, 1007.1; black-figure cup, sixth century BCE, BM 428 (Beazley 302378).

11 Athenian black-figure vase (sixth century BCE), BM 1836, 0224.66.

12 Apulian red-figure plate: State Hermitage Museum, Saint Petersburg GR-10334, ca. 320–310 BCE.

13 Job 3:8, 40:15–41:26; Amos 9:3; Ps. 74:13–23, 104:26; Isa. 27:1; cf., Wakeman 1969. The motif is replicated in the *Enuma Elish*, where Marduk defeats the sea deity, Tiamat, from whose body Marduk creates the world (Chapter 1). In the Canaanite tradition Baal defeats the sea god Yam ("sea" in Hebrew).

14 Beaulieu 2016.

15 E.g., Chalcidian black-figure hydria ca. 540–530 BCE, Staatliche Antikensammlungen, Inv. 596. For Medieval sea monsters: Van Duzer 2013.

16 Apollodorus, *Library* 2.2; Pausanias 37.4 (Corinth); six heads are shown on bronze fibulae ca. 700 BCE; nine heads: Alcaeus fragment 443; fifty heads: Simonides fragment 569.

17 Euripides, *Herakles* 1274–1275; Diodorus Siculus 4.11.5.

18 See Dio 9.28 for the "hydra-like" resolve of the Roman army that rebounds in the face of defeat.

19 John Tzetzes, *On Lycophron* 45; Servius, *ad Aen.* 3.420; Hyginus, *Fabulae* 199. Cf., Boardman 1987: 76. The wronged-nymph version may be Ovid's invention (*Metamorphoses* 14.1–74), synthesizing the sea monster of Homer with an eponymous nymph created by Hellenistic poets: Athenaeus 7.296e. The Hellenistic

Scylla was inspired by the daughter of the legendary diver, Skyllias of Skione, who accomplished a 9-mile underwater swim to sabotage the Persian fleet (Herodotus 8.7). The Skyllias legend grew, and he soon acquired a daughter, Hydne, who was also a talented diver: Lowe 2011.

20 Sharks have 3–50 rows of teeth, depending on the species. Aristotle ascribed a triple row of teeth to "martichoras," striped, blue-eyed man-eaters that resemble lions in size and paws, but men in face and ears, with a scorpion-like tail, complete with stinger and spines (*HA* 2.3.501a27–28). But the literary sources are silent on the arrangement of shark teeth. For shark-tooth amulets: Reese 1984: 190–1.

21 Boeotian red-figure bell-crater with three dogs, 450–425 BCE. Louvre, CA 1341; terracotta plaque from the Cyclades with two dogs projecting from her belly, 450 BCE. BM 1867, 0508.673.

22 Ogden 2013: 123–9. In Ovid, Perseus' slaughter of the Aithiopian *ketos* is also an etiology for the origin of coral, petrified seagrass: *Metamorphoses* 4.604–803.

23 Corinthian, black-figure amphora, 575–550 BCE from Cerverteri (Italy), Antikensammlung Museum, Berlin.

24 Red-figure *loutrophoros*, ca. 350 BCE from southern Italy, Getty Museum (84. AE.996).

25 Apollodorus, *Library* 2.5.9; Diodorus Siculus 4.42.6–7. Laomedon had promised Hesione's hand in marriage to Herakles in remuneration for slaying the *ketos*. But, the faithless king again refused payment. Herakles later returned to Troy, slaying Laomedon and enthroning Priam. The earliest representation shows the hero shooting arrows and the princess throwing rocks at a monster that does not resemble the typical serpentine *ketos* but instead may record the discovery of a large, fossilized, prehistoric mammal: Boston Museum of Fine Arts, 63.420; Mayor 2000. Cf., Ahlberg-Cornell 1984; Ogden 2013: 118–23. Boardman (1987: 77) connects Andromeda's *ketos* with Jonah's, who had set sail from Jaffa, where Andromeda was bound. Cp., Lucian, *True Story* 1.30–2.2.

26 Caeretan black-figure hydria, sixth century BCE, Stavros S. Niarchos Collection.

27 We omit discussion of the snakes sent from Tenedos to destroy Neptune's libidinous priest, Laocoön: Vergil, *Aeneid* 2.203–227. Although they must traverse an expanse of sea, their behavior is more snaky than fishy.

28 Hesiod, *Theogony* 178–200; Euripides, *Hippolytus* 415, 1210–1212.

29 Seneca, *Phaedra* 1035–1049. See Mader 2002; Kitchell 2014: 199.

30 Ovid, *Metamorphoses* 7.445; Pseudo-Apollodorus, *Bibliotheca* Epitome 1.2; Hyginus, *Fabulae* 38; Scholia. ad Euripides, *Hippolytus* 976.

31 Palace of King Sargon II, Khorsabad, Assyria. Louvre, AO 19889–19890. Van Duzer 2013: 13; Albenda 1983; Fontan 2001.

32 Vergil, *Aeneid* 1.144–145; Valerius Flaccus, *Argonautica* 1.679–680.

33 Heraclitus, *Allegories/Homeric Questions* 64–7; Beaulieu 2016: 38–9.

34 Cf., Vergil *Georgics* 4.432. For their stench, din, and malformation: Aristotle, *Movement of Animals* 18.714b.12–14; Aristophanes, *Wasps* 1035.
35 Trinquier 2009: 72–4; Llewellyn-Jones and Lewis 2018: 408; Beaulieu 2016: 27v30, 36–8. See also Lucian, *Dialogues of the Sea Gods* 4: *Proteus and Menelaus*.
36 Pedley 1970: 45v53.
37 Beaulieu 2016: 37.
38 Thetis knew that her son must die soon after the death of the Trojan hero, Hector (*Iliad* 18.95–96).
39 Berlin, Antikensammlung Museum F2279. See Hesiod, *Theogony* 240–4; Homer, *Iliad* 1.358, 18.36, 24.60; Pindar, *Pythian* 3.92; Apollodorus, *Library* 1.11.
40 Rudhardt 1971; Beaulieu 2016: 39–40.
41 In 44 BCE, Caesar's enemies may have deliberately chosen that day, the Ides of March (March 15), to execute Caesar, when so many of his supporters from the people would have been intoxicated on the banks of the Tiber.
42 Van Duzer 2013: 31.

Chapter 8

1 Hesiod, *Works* 109–201; Heraclitus, *TEGP* 53; Empedocles, *TEGP* 41. See Chapter 1.
2 Lucretius 2.1144–1174; see Slavena-Griffin 2016.
3 Heraclitus, *TEGP* 40; Empedocles, *TEGP* 51–52; Plato, *Timaeus* 28a6; Aristotle, *Metaphysics* 12.8.1074a15–16; for pneuma: Aëtius 46a.
4 See Parker 1983; Carbon and Peels-Matthey 2018.
5 In the Ancient Near East, the health of the land is inextricably interwoven with the moral compass of the people, a theme that permeates the biblical literature of the Babylonian captivity in the sixth century BCE: e.g., Leviticus 20:22; see McCall in Irby et al. 2016: 298–301.
6 Robertson 2010. For the purity regulations at Cyrene: Dobias-Lalou 2000: 297–309; Dobias-Lalou 2017: # 16700.
7 For the *Lex Sacra* of Selinunte: Faraone and Obbink 2013; Kotansky 2015.
8 E.g., *delabrum* at the Suburban Baths of Hercules: Facchinetti 2008: 47; Rogers 2018: 77.
9 Homer, *Odyssey* 3.445; cf., Burkert 1985: 75–7.
10 Vergil, *Aeneid* 6.212–235; Servius, *ad Aen.* 6.218.
11 Lindsay 2000: 115.
12 Festus-Paulus 3L s.v. *aqua et igni*.
13 Wachsmuth 1967: 260–2; Blakely 2017: 376–7.
14 Cf., Dionysius of Halicarnassus, *Roman Antiquities* 2.67; Livy 2.42.
15 Livy, *Periochae* 20; Valerius Maximus 8.1.5; Pliny 28.12; Augustine, *City of God* 10.16.

16 Ovid, *Fasti* 4.179-375. Whenever stresses affected the Roman people, the senate consulted the Sibylline Books, which generally advised importation of Greek cults: Orlin 1997: 76-115.
17 *October Horse*: Polybius 12.4; Dio 43.24.4; *Armilustrium*: Varro, *Latin Language* 5.15.3; Plutarch, *Romulus* 23.3. See Ovid, *Fasti* book 3; Scullard 1981: 193-6.
18 See further *UCWW*: Chapter 1; Plutarch, *Whether Fire or Water is more useful* 955d-958e.
19 Aeschylus, *Glaucus of the Sea* fragment 26Radt; Ovid, *Metamorphoses* 13.898-965; Beaulieu 2016: 41-2; Chapter 7.
20 Sein, Pointe du Raz, Finistère, off the coast of Brittany.
21 Silberman 1988: ad loc.
22 Cassandra: Aeschylus, *Agamemnon* 1198-1208; the Cumaean Sibyl: Ovid, *Metamorphoses* 14.132-133.
23 Attempts to link the Gallizenae with Druidism uniformly lack documentation and are unconvincing: e.g. MacKillop, *A* 2004), s.v. Gallizenae.
24 Vergil, *Georgics* 1.466-468; Ovid, *Metamorphoses* 15.785-786.
25 Petosiris, f7; Irby-Massie and Keyser 2002: 89.
26 Cicero, *Divination*; cf., Denyer 1985: 1-10.
27 Taub 2003: 140, 157.
28 Gehlken 2012.
29 Luck 2006: 312.
30 Lucian crafted a parody of Demeter's oracle at Patrai: whoever descends into the shallow well on the moon could then hear all that is said on the earth. Those gazing into the mirror could also see all the cities of the earth, "just as if standing over it" (*True Story* 1.26).
31 *Greek Magical Papyri* PGM.5. Serapis was a syncretized Greco-Egyptian deity introduced to Alexandria by Ptolemy I Soter as a cult of cultural unification.
32 *Greek Magical Papyri* PGM 4.222-234; Betz 1992: 42; Taylor 2008: 106.
33 Taylor 2008: 107, Figure 56.
34 Varro, *Farming* 3.17.4.
35 Pliny 32.17; Aelian, *NA* 8.5; Chapter 6.
36 Strabo 14.2.23. Cf., Herodotus 5.119; Pliny 32.16.
37 ὑπονήχηται: the ancient description cannot be identified with a modern species.
38 Hecataeus, *FGrHist* 1.27; Xenophon, *Anabasis* 6.2.2; Ovid, *Metamorphoses* 7.406-419, Mela 1.92; Woodford 1971: 211-25; Stafford 2012; Chapter 7.
39 Rudhardt 1971; Beaulieu 2016: 15.
40 Gallini 1963: 61-90.
41 Homer, *Odyssey* 10.507-515; Beaulieu 2016: 53-7.
42 Mela 3.31-33, Pliny 4.89.
43 Beaulieu 2016: 152.

44 Pindar, *Pythian* 10.27–30; Beaulieu 2016: 46–7.
45 Blakely 2012; Plato, *Phaedo* 109b–113d; cf., Chapter 2.
46 Homer, *Iliad* 12.385–386, 16.742–743; *Odyssey* 12.413–41; Beaulieu 2016: 149–50.
47 Vermeule 1979.
48 Herodotus 1.24; Aelian, *NA* 12.45; Pausanias 3.25.7; *Greek Anthology* 9.88, 308.
49 Kingsley 1995: 352–3. Cf., Chapter 1 for Nestis, Empedocles' goddess of water.
50 Richardson 1974: 18–19, 181–2 n99. At Philikos, *Hymn to Demeter* 37–40, the grieving Demeter makes a "spring with her tears."
51 Strabo 6.2.9; See further Kingsley 1995: 348–58.

Chapter 9

1 Étienne and Braun 1986.
2 Herodotus 7.189–191; cf., Blakely 2017: 367 and n29.
3 Aeschylus, fragment 1–2 Nauck; Apollodorus, *Library* 3.28; Nonnus, *Dionysiaca* 10.45–107; Ovid, *Metamorphoses* 4.481–542; Beaulieu 2016: 161–6.
4 Burkert 1985: 136; Simon 2014: 37; Mylanopoulos 2003, 2006.
5 Homer, *Iliad* 20.404; Herodotus 1.148; Diodorus Siculus 15.49.
6 Homer, *Iliad* 15.186–193; Hesiod, *Theogony* 456.
7 For the equine Arion: Homer, *Iliad* 23.346; Apollodorus, *Library* 3.6.8; Statius, *Thebaid* 6.301; Pausanias 8.25.7–9.
8 Simon 2014: 39.
9 *HH* 22 *Poseidon*; Pausanias 7.21.9; Burkert 1985: 138.
10 Theophrastus in Athenaeus 6.261d; Plutarch, *Dinner of the Seven Wise Men* 163b; Eustacius 1293.8.
11 Nilsson 1956: 237.
12 Florus 2.18.3; Dio 48.48.5; Appian, *Civil War* 5.100.416; Powell 2002: 123.
13 In light of recent finds of human skeletons in sacrificial ritual contexts in Crete and at Mt. Lykaion, the prevalence of human sacrifice remains a matter of debate in the historical era. Although rare, the evidence for human sacrifice in European antiquity is secure (e.g., Livy 22.55–57; Hughes 1991), triggered by particular stresses within the community. In the Celtic world, archaeological evidence suggests sporadic ritual killing between the invasions of Caesar (55 and 54 BCE) and Claudius (43 CE), a trope foregrounded in classical authors to emphasize the savagery of the "other:" cf., Aldhouse-Green 2001. In 2016, archaeologists uncovered human remains near a sacrificial altar of Zeus at Mt. Lykaion: Romano and Voyatzis 2018. See also Burkert 1985: 64–6.
14 Ptolemaeus Hephaistos or Khennos, *New History* 4: 194–5 Westermann; Sextus Empiricus, *Against the Mathematicians* 1.264; Servius, *ad Aen.* 2.44; Photius, *Bibliotheca* 150a16.

15 In the Homeric version, the deceased blind seer, Tiresias, advised Odysseus on assuaging Poseidon's wrath. After returning to Ithaca, the hero should walk inland with an oar until he met someone who mistook the nautical implement for a winnowing fan (in other words, until he met folk who had no knowledge of the sailor's arts). Odysseus was then supposed to plant the oar in the ground and next sacrifice a ram, bull, and boar to the god. The hero could return home, where he was to sacrifice hecatombs to each of the gods (*Odyssey* 11.119–134). Odysseus would thus expand Poseidon's cult to inland worshippers. Noting the resemblance between Odysseus and the *Mahabharata's* traveling Sanskrit hero, Arjuna, Allen (1995) sees the second (public) sacrifice of hecatombs to all the gods (presumably including Poseidon) as analogous to a horse sacrificed in penance to the trident-wielding Shiva by Arjuna. In the Sanskrit tradition, the elaborate, year-plus-long horse sacrifice is "the highest of the royal rituals and establishes the cosmic supremacy of a king." The ritual occurs in two parts: first a fine stallion is selected and then allowed to roam through India with warriors pursuing him to protect the stallion from harm. If the stallion survived, he was sacrificed in the capital together with other victims. Allen suggests that the two versions derive from a common source in which the "unmotivated" equine metamorphosis in the Greek version could be linked with the sacrifice of the wandering horse in the more fully developed Sanskrit version.

16 Burkert 1983: 159. Horses were also, at least symbolically, sacrificed at the Roman festival of the October Horse (Timaeus, *FGrHist* 566f36 = Polybius 12.4b). Timaeus argued that, by stabbing a horse, the descendants of the Trojan fugitives thus avenged the fall of their mother city, which was destroyed by a horse (the wooden horse on wheels: *Odyssey* 8.493, 512; Vergil, *Aeneid* 2.57–267). But, according to Polybius (12.4b), many "barbarian" peoples, entirely unconnected with the Trojan War, sacrificed horses before engaging in war (cf., Livy, *Epitome* 49). Burkert (1983: 159–60) accepts Timaeus' evidence. Although the cultic elements eventually evolved into a mechanical trick, regardless of the real connection, there was an implicit link between the city's fall and the sacrifice of a horse: "the fact that Troy's fall, at the fateful feast when the Trojans accepted the wooden horse, was linked to the sacrifice of a horse by means of a spear attests to a deeper understanding": Pascal 1981. The death of Aeneas' helmsman, furthermore, is yet another metonymic human sacrifice of thanksgiving to the sea after a long, perilous voyage: Vergil, *Aeneid* 5.814–815; Nicoll 1988.

17 Hipponax 65; Tzetzes, Commentary on *Iliad*.

18 Oppian, *Fishing* 5.680. Poseidon is also invoked as Ἀσφάλιός (*Asphalios*) at Aristophanes, *Acharnians* 682; Apollonius of Tyana 4.9; Pausanias 7.21.7 (Calydon); Heliodorus 6.7.

19 *AE* 1951, #71.

20 *CIL* 8.26491.

21 *AE* 1946, #71 = *AE* 1987, #1069.
22 *CIL* 10.3813.
23 *CIL* 6.536. A Neptune shrine was also incorporated into the Circus' *spina* where dangerous (and sometimes deadly) chariot races occurred.
24 *RIB* 91; Cleere 1974.
25 Hadrianic (*RIB* 1319, 1694, 1990, *CSIR* 1.6.89) and Antonine (*RIB* 2105, 2149). Near the sea: Lympne (*RIB* 66); Maryport (*RIB* 839); Newcastle (*RIB* 1319). Cf., *UCWW*: Chapter 10.
26 Dinsmoor 1971: 2–4.
27 Herodotus 6.87; *SEG* 33.147; Lupu 2005: no. 1.
28 See Herodotus 8.53; Paga 2016: 184.
29 Herodotus 8.1210. Two other triremes were captured: one dedicated to Poseidon at the Isthmus of Corinth, and the other at Salamis to its tutelary hero, Ajax.
30 *PECS*, s.v. Sounion; Paga 2016: 185.
31 Apollonius, *Argonautica* 4.1695–1705; cf., Albis 1995: 104–9.
32 *Syll.* 3 29; Umholz 2002: 268–9 and n33; Walsh 1986: 321.
33 Tarn 1910: 209–22; cf., Constantakopoulou 2017: 94.
34 Pritchett 1979: 268–9; Westcoat 2005: 170. Otto Benndorf, who excavated at the site, speculated that the Samothracian Nike was dedicated by the Macedonian dynast, Demetrios I Poliorcetes (337–283 BCE), to commemorate his victory at Salamis on Cyprus: cf., Lawrence 1926: 213–18.
35 The practice endured. Into the seventeenth century, many coastal English churches received ship models votives from returning sailors and voyagers by sea. These thanksgivings were often hung from the ceilings (as, for example, at All Hallows by the Tower in London and Notre-Dame-de-Bon-Secours Chapel in Montreal, Quebec). Graffiti prayers and ships were also etched into the walls of churches serving maritime communities: see Champion 2015.
36 Blakely 2017: 366.
37 Hasaki and Nakas 2017: 67–8.
38 Diodorus Siculus 4.53.3; Eumelus 22; Apollodorus, *Library* 1.9.27; Dio Chrysostom 37.15.
39 Aratus, *Phaenomena* 342–66; Manilius 5.32–45.
40 Suetonius, *Nero* 40.3; Dio 63.19.2; Bradley 1975: 305–7.
41 Cagnat 1901–1927: 4.1539.
42 See Carratelli 1965: 281–4.
43 Hesiod, *Theogony* 178–200; Euripides, *Hippolytus* 415.
44 Piraeus: Pausanias 1.1.3; Knidos: *IosPE* I^2,168: first century CE; Delos: Dürrbach 1923–1937: 2132: after 166 BCE Demetriou 2010: 67–89, 74; Constantakopoulou 2017: 79–81. Merchants also worshipped Aphrodite Euploia, e.g., at Halicarnassus: Laumonier 1958: 625–6.

45 Solon f19 West.
46 Welter 1938: 489–90, 497, Figure 11.
47 Torelli 1977: 435; Demetriou 2010: 79.
48 See also Barbantani 2005: 141–2. Demetriou (2010: 81–5) sees Aphrodite's roles as maritime deity and goddess of sex as intertwined and her power over sex as a metaphor for her navigational abilities. Men notoriously squandered their wealth on *hetairai* when they were in port, especially at Corinth and other centers of temple prostitution (Herodotus 2.135; Strabo 12.2.36; Athenaeus 12.596b–c).
49 *Greek Anthology* 10.21; Callimachus f5.
50 *SEG* 23.170, Roman imperial period.
51 Euripides, *Hippolytus* 522; cf., Graf 1985: 261; Demetriou 2010, 74–5.
52 *Greek Anthology* 9.601; *SEG* 28.838; Plutarch, *Greek Questions* 303c–d.
53 Athenaeus 15.675f–676c. See also Totelin 2014: 17–29.
54 Athenaeus 5.207e; Barbantani 2005: 144n30.
55 Kallixeinos of Rhodes *FGrHist* 627; Roller 2010a: 66.
56 Casson 1995: #125.
57 Lucian, *Navigium* 5–6; Casson 1995: #147–50, 156.
58 Inv. No. Б. 3269. Saint-Petersburg, The State Hermitage Museum; cf., Ovid, *Metamorphoses* 10.708; Cyrino 2010: 123 and Fig. 6.3.
59 Bricault 2000: 136–49. For Isis as Pharia: Tibullus 1.3.32; Ovid, *Art of Love* 3.635; Martial 10.48; Statius, *Silvae* 3.2.102.
60 *Corpus Inscriptionum Graecarum* 4683b; Bricault 2000: 139.
61 Bricault 2000: 139.
62 Barbantani 2005: 135–65.
63 Robert 1966: 192–208: *Ox.* 2465; Barbantani 2005.
64 Poseidippos 119 = Athenaeus 7.318d; Hauben 1970: 42–6.
65 Callimachus *Epigram* 15 Pfeiffer = 14 Gow-Page.
66 See Barbantani 2005: 148.
67 Barbantani 2005: 146.
68 Barbantani 2005: 147 and n42.
69 Philo, *On the Embassy to Caligula* 150–1; Weinstock 1971: 289; Rocca 2015: 51. Augustus was also aligned with Zeus Soter on the Athenian acropolis: *IG* II–III^2, 3173; Thompson 1966: 171–87.
70 Blakely 2017: 364. For the long tradition of the sacred nature of signal fire: Tuck 2008: 325–41.
71 *Inscriptiones Graecae ad res Romanas pertinentes* 4.1542; Blakely 2017: 363–4.
72 Weber 1907: 190 and n679.
73 Ovid, *Heroides* 16.329; Plutarch, *Theseus* 29.
74 Ovid, *Metamorphoses* 6.414; Pausanias 8.46.6; Hyginus, *Fabulae* 173.
75 Varro, *Latin Language* 5.58; Eusebius, *Preparation for the Gospel* 1.10.
76 Sixth to fourth centuries BCE. Brixhe 2006: 1–20; Blakely 2012: 60.

77 Kühr 2006: 91–106.
78 Clinton 2001: 29; Dimitrova 2008: #72, 87, 100, 101, 104.
79 Dimitrova 2008: #66.
80 Mowery 2011: 113.
81 Dimitrova 2008: #58, 127.
82 Dimitrova 2008: #46, 53.
83 Dimitrova 2008: #49.
84 Lewis 1958: #229g, 158; Mowery 2011: 117.
85 Lehmann and McCredie 1998: 32–3.
86 Lehmann and McCredie 1998: 36–7; Dimitrova 2008: 89. For a reconstruction of the ritual space: Cole 1984: 26; Lehmann and McCredie 1998: 96–8; Blakely 2012.
87 Dimitrova 2008: #29.
88 Eleusinian Mysteries: Gagné 2009; Filonik 2013.
89 In a rare first-person account, Lucretius 6.1044–46 described "Samothracian iron" as "dancing" and iron filings in copper vessels as "raging" when exposed to a magnet stone; cf., Pliny 33.23 (plated with gold). Magnetization was reputedly discovered at Samothrace: Zenobius 4.22; Cole 1984: 30. See also Plato, *Ion* 533d–e. See also Sedley 1998: 52–4 for a discussion of the phenomenon as a product of magnetic attraction (not repulsion as Lucretius implies): effected by the motion of a lodestone underneath a bronze bowl, causing the iron shavings to stand or fall. Sedley continues, "But Lucretius' mistake makes slightly better sense if he had also witnessed genuine magnetic repulsion, of which the 'dancing' iron objects may therefore represent an authentic case."
90 Blakely 2012: 62.
91 Cole 1984: 30. For the apparently magical properties of the lodestone: Plato, *Ion* 533d–e.
92 Zenobius 4.80; Blakely 2012: 63.
93 For the magnetic properties of the local geology at Samothrace: Size et al. 2014.
94 Needham 1962: 4.279–93.
95 Carlson 1975.
96 Alexander Neckham, ca. 1180; Paine 2013: 382.
97 Witt 1997: 181; Beresford 2013: 40–1. During the Roman Republic, *vota publica* "for the health of the state" were sworn annually by the new consuls entering office at the beginning of the year (Livy 21.63). This eventually was augmented with a vow for the emperor's safety (Dio 51.19), sworn both at Rome (Tacitus, *Annals* 16.22) and in the provinces (Pliny the Younger, *Epistle* 10.35, 36).
98 See Plutarch, *How to Tell a Flatterer from a Friend* 50. Beresford also suggests that the Romans, seeking favorable sailing conditions, propitiated the Dioscuri annually at Ostia on January 27. Our single textual source, Ammianus Marcellinus (330–400 CE), however, does not indicate an annual festival, but only that, during rough winter conditions, the seas grew calm allowing grain ships to enter Ostia after the urban

prefect made sacrifices in the temple of the Castori (19.10.4). Mention of such a festival is lacking from Ovid's *Fasti* (secure for January). See also Dumézil 1970: 412–14. Nonetheless, an annual festival to Neptune (Neptunalia) is attested for late July or early August: *CIL* 6.2305 (*ILS* 8745); *CIL* 6.2306; cf., Scullard 1981: 168; Beard et al. 1998: 325.

99 Vegetius 4.39; Witt 1997: 165–84.
100 The Mediterranean waters were never entirely "closed to sailing": see *UCWW*: Chapter 5.
101 John Lydus, *On the Mind* 4.45; Witt 1997: 178.
102 *CIL* 6.32503 (*ILS* 8745).
103 *Navigium / Vene/ris*: CAIG-69-02, 366.
104 Witt 1997: 183.
105 *HH* 7 *Dionysus* 51–3; Exekias black-figure cup 214449. München, Staatliche Antikensammlung; Burkert 1985: 166 and n38.
106 Beazley 4319, London, BM B79; cf., Boardman 1958: 4–12.
107 Casson 1994: 155.
108 Museo Torlonia #430; Testaguzza 1970: 171.
109 Wachsmuth 1967: 145–7.
110 Maiuri 1958.
111 Casson 1995: 182n70. Ca. 120 BCE, a 115-foot-long (35m) cargo ship, loaded with amphorae of wine, went down in the strait between Corsica and Sardinia. The wreck was discovered in 1939, and the remains are on display at the Nino Lamboglia Museum in La Maddelena, in northern Sardinia.
112 Marangou 2006; Blakely 2017: 372–3.
113 Burkert 1985: 266–7.
114 Theocritus, *Idyll* 7; Propertius 1.8; Ovid, *Art of Love* 1.177–228; Ovid, *Amores* 2.11; Statius, *Silvae* 3.2.
115 Cf., Cairns 2006: 404–43.
116 Carlson 2009.
117 BM: B508, late sixth century BCE.
118 Irby 2019a; Zanker 1990: 82–5. Cf., Chapter 7.
119 Burkert 1992: 54; Carlson 2007.
120 Wachsmuth 1967: 276.
121 Merkelbach and Stauber 1996: #15.3–7.
122 Graf 2007.
123 *Cyranides* 1.85.22–24, 3.81.8–12; Wachsmuth 1967: 198; Blakely 2017: 376.
124 Wachsmuth 1967: 82–97, 243–6.
125 Neilson 2002.
126 *Greek Anthology* 10.1–2, 4–9, 14–16; *IG* 12.3.421c.
127 Yébenes 2010: 457–86.

128 Blakely 2017: 374.
129 Levine 2011.
130 Durand 1989: 127–8.
131 For triple repetition and threes in magical contexts: Faraone and Obbink 1997: 41–2, 177, 183n12, 191–2, 194, 223.
132 Paton and Hicks 1891: 75, #36 b, lines 4 and 24.
133 Festus 274L; see Varro, *Latin Language* 6.3.
134 According to Agatharchides in Athenaeus 7.297d; Marzano 2013: 201.
135 Pseudo-Erinna f2; Athenaeus 7.283a; Aelian, *NA* 15.23; cf., Blakely 2017: 368.

Chapter 10

1 Rihll 1999: 73; Irby 2019a: 81 and n5.
2 The control of fire together with agricultural advances—crop and animal domestication—leads to permanent settlements, food surpluses, and civilization: Humphrey, Oleson, and Sherwood 1998: 97–100, 133.

Appendix

1 For further information about individual authors, the interested reader is invited to consult *Encyclopedia of Ancient Natural Scientists: The Greek Tradition and Its Many Heirs* edited by P.T. Keyser and G.L. Irby-Massie, Routledge (2008); *Oxford Classical Dictionary*; *Internet Encyclopedia of Philosophy*: https://iep.utm.edu/; or *Stanford Encyclopedia of Philosophy*: https://plato.stanford.edu/

Bibliography

Ahlberg-Cornell, G (1984) *Herakles and the sea-monster in Attic black-figure vase-painting*, Skrifter utgivna av Svenska Institutet i Athen. 4°, XXXIII, Stockholm.
Albenda, P (1983) "A Mediterranean seascape from Khorsabad," *Assur* 3, pp 1–17.
Albis, RV (1995) "Jason's prayers to Apollo in 'Aetia 1' and the '*Argonautica*,'" *Phoenix* 49 (2), pp 104–9.
Alcock, JP (2001) *Food in Roman Britain*, Stroud.
Aldhouse Green, MJ (2001) *Dying for the Gods: Human Sacrifice in Iron Age & Roman Europe*, Stroud.
Aldrete, GS (2007) *Floods of the Tiber in Ancient Rome*, Baltimore.
Allason-Jones, L (1989) *Women in Roman Britain*, London.
Allen, NJ (1995) "Why did Odysseus Become a Horse?" *Journal of the Anthropological Society of Oxford* 26 (2), pp 143–54.
Ammerman, AJ (1990) "On the Origins of the Forum Romanum," *AJA* 94 (4), pp 627–45.
Ammerman, AJ (1998) "Environmental archaeology in the Velabrum, Rome: interim report," *JRA* 11, pp 213–23.
Angel, JL (1966) "Porotic Hyperostosis, Anemias, Malarias, and Marshes in the Prehistoric Eastern Mediterranean," *Science* 153 (3,737), pp 760–3.
Angeletti, LR and V Gazzaniga (2002) "La peste di Siracusa (396 a.C.) in Diodoro Siculo (XIV,70)," *Annali di igiene: medicina preventiva e di comunità* 14, pp 7–13.
Ball, P (1999) *Life's Matrix: A Biography of Water*, New York.
Barbantani, S (2005) "Goddess of Love and Mistress of the Sea: Notes on a Hellenistic Hymn to Arsinoe-Aphrodite ('P. Lit. Goodsp. 2,' I–IV)," *Ancient Society* 35, pp 135–65.
Barnes, J (1995) "Metaphysics," in Jonathan Barnes (ed.) *The Cambridge Companion to Aristotle*, Cambridge, pp 66–108.
Beagon, M (1996) "Nature and views of her landscapes in Pliny the Elder," in G Shipley and J Salmon (eds.) *Human Landscapes in Classical Antiquity: Environment and Culture*, London and New York, pp 284–329.
Beard, M, J North and S Price (1998) *Religions of Rome*, Cambridge.
Beaujeu, J (2003) *Pline L'Ancien: Histoire Naturelle: Livre II*, Paris.
Beaulieu, M-C (2016) *The Sea in the Greek Imagination*, Philadelphia.
Beck, L (2017) *Dioscorides: De Materia Medica*, Third edition, Hildesheim.
Belati, M (1999) "Le inondazioni del Tevere nell'antichità," *Forma Urbis* 4.7/8, pp 12–17.
Beresford, J (2013) *The Ancient Sailing Season*, Leiden.

Betancourt, P (2007) *Introduction to Aegean Art*, Philadelphia.
Betz, HD (1992) *The Greek Magical Papyri in Translation, Including the Demotic Spells*, Second edition, Chicago.
Beus, SS and M Morales (2003) *Grand Canyon Geology*, Second edition, New York°Biraben, J-N (1998) "Diseases in Europe: Equilibrium and Breakdown of the Pathocenosis," in MD Grmek (ed.), *Western Medical Thought from Antiquity to the Middle Ages*, Cambridge, MA, pp 319-53.
Blakely, S (2012) "Toward an Archaeology of Secrecy: Power, Paradox, and the Great Gods of Samothrace," *Archeological Papers of the American Anthropological Association* 21, pp 49-71.
Blakely, S (2017) "Maritime Risk and Ritual Responses: Sailing with the gods in the Ancient Mediterranean," in de Souza and Arnaud, pp 362-79.
Boardman, J (1958) "A Greek Vase from Egypt," *JHS* 78, pp 4-12.
Boardman, J (1975) *Athenian Red Figured Vases: The Archaic Period*, London.
Boardman, J (1987) "Very like a whale – Classical sea monsters," in AE Farkas (ed.), *Monsters and demons in the ancient and medieval worlds: papers presented in honor of Edith Porada*, Mainz, pp 73-84.
Boardman, J (1998) *Early Greek Vase Painting, 11th-6th Centuries BC: A Handbook*, London.
Bodson, L (1983) "Aristotle's Statement on the Reproduction of Sharks," *Journal of the History of Biology* 16 (3), pp 391-407.
Bos, A (2010) "Aristotle on the Differences between Plants, Animals, and Human Beings and on the Elements as Instruments of the Soul (*De Anima* 2.4.415b18)," *The Review of Metaphysics* 63, pp 821-41.
Boyce, JM (2002) *The Smithsonian Book of Mars*, Washington DC.
Bradley, KR (1975) "Two notes concerning Nero: I: Two Neronian inscriptions from Cyrene; II: A note on Q. Veranius, cos. A.D. 49," *Greek, Roman and Byzantine Studies* 16, pp 305-7.
Bricault, L (2000) "Un phare, une flotte, Isis, Faustine et l'annone," *Chronique d'Égypte* 75, pp 136-49.
Bridgman, TP (2005) *Hyperboreans: Myth and History in Celtic-Hellenic Contacts*, London.
Brixhe, C (2006) "Zone et Samothrace: Lueurs sur la langue thrace et nouveau chapitre de la gramaire comparee"? *Comptes Rendus de l'Academie des Inscriptions et Belles-Lettres* 150, pp 1-20.
Broodbank, C (2013) *The Making of the Middle Sea: A History of the Mediterranean from the Beginning to the Emergence of the Classical World*, Oxford.
Bruun, C (2007) "The Antonine Plague and the 'Third-Century Crisis,'" in O Hekster, G de Kleijn, D Slootjes (ed.), *Crises and the Roman Empire: Proceedings of the Seventh Workshop of the International Network Impact of Empire*, Nijmegen, June 20-24, 2006, Leiden, pp 201-18.
Buchan, J (1937) *Augustus*, London.

Burkert, W (1983) *Homo Necans: The Anthropology of Ancient Greek Sacrificial Ritual and Myth*, trans. Peter Bing, Berkeley.
Burkert, W (1985) *Greek Religion*, Cambridge, MA.
Burkert, W (1992) *The Orientalizing Revolution: Near Eastern Influences on Greek Culture in the Early Archaic Age*, Cambridge, MA.
Burstein, SM (1976) "Alexander, Callisthenes and the Sources of the Nile," *Greek, Roman, and Byzantine Studies* 17, pp 135–46.
Cagnat, René (ed.) (1901–1927) *Inscriptiones graecae ad res romanas pertinentes*, Paris.
Cairns, F (2006) *Sextus Propertius: The Augustan Elegist*, Cambridge.
Campbell, B (2012) *Rivers and the Power of Ancient Rome. Studies in the History of Greece and Rome*, Chapel Hill.
Carbon, JM and S Peels-Matthey (eds.) (2018) *Purity and Purification in the Ancient Greek World. Texts, Rituals, and Norms*, Kernos, supplément 32.
Carlson, DN (2007) "Mast-step coins among the Romans," *The International Journal of Nautical Archaeology and Underwater Exploration* 36, pp 317–24.
Carlson, DN (2009) "Seeing the Sea: Ships' Eyes in Classical Greece," *Hesperia* 78, pp 347–65.
Carlson, JB (1975) "Lodestone Compass: Chinese or Olmec Primacy?: Multidisciplinary analysis of an Olmec hematite artifact from San Lorenzo, Veracruz, Mexico," *Science* 189, pp 753–60.
Carratelli, GP (1965) "Theoi apobaterioi," in A Frova (ed.), *Studi in onore di L. Banti*, Roma, pp 281–4.
Casson, L (1994) *Travel in the ancient World*, Baltimore.
Casson, L (1995) *Ships and Seamanship in the ancient World*, Baltimore.
Champion, M (2015) "Medieval Ship Graffiti in English Churches: Interpretation and function," *The Mariner's Mirror* 101, pp 343–50.
Churchill, W (1902) *The River War: An Historical Account of the Reconquest of the Soudan*, London abridged.
Cleere, H (1974) "The Roman Iron Industry of the Weald and its Connexions with the Classis Britannica," *Archaeology Journal* 131, pp 172–99.
Clinton, K (2001) "Initiates in the Samothracian Mysteries, September 4, 100 B.C.," *Chiron* 31, pp 27–36.
Cockburn, A and E Cockburn (1983) *Mummies, Disease and ancient Cultures*, Cambridge.
Cole, SG (1984) *Theoi Megaloi: The Cult of the Great Gods at Samothrace*, Leiden.
Coleman-Norton, PR (1950) "Cicero and the Music of the Spheres," *CJ* 45, pp 237–41.
Collins, WE (2012) "Plasmodium knowlesi: A malaria parasite of monkeys and humans," *Annual Review of Entomology* 57, pp 107–21.
Constantakopoulou, C (2017) *Aegean Interactions: Delos and its Networks in the Third Century*, Oxford.
Cordovana, OD and GF Chiai (eds.) (2017) *Pollution and the Environment in Ancient Life and Thought*, Stuttgart.

Corvisier, JN (2008) *Les Grecs et La Mer*, Paris.
Craik, EM (2001) "Thucydides on the Plague: Physiology of Flux and Fixation," *CQ* 51 (1), pp 102–8.
Craik, EM (2017) "Malaria and the Environment of Greece," in Cordovana and Chiai, pp 153–62.
Crawford, HEW (1998) *Dilmun and Its Gulf Neighbours*, Cambridge.
Crawford, HEW, R Killick, and J Moon (eds.) (1997) *The Dilmun Temple at Saar: Bahrain and Its Archaeological Inheritance*, London.
Croon, JH (1952) *The Herdsmen of the Dead: Studies on Some Cults, Myths and Legends of the Ancient Greek Colonization-area Area*, Utrecht.
Croon, JH (1956) "Artemis Thermia and Apollo Thermios (With an Excursus on the Oetean Heracles-Cult)," *Mnemosyne* 9 (1), pp 193–220.
Croon, JH (1967) "Hot Springs and Healing Gods," *Mnemosyne* 20 (3), pp 225–46.
Cunha, CB and BA Cunha (2008) "Brief history of the clinical diagnosis of malaria: from Hippocrates to Osler," *The Journal of Vector Borne Diseases* 45, pp 194–9.
Cunliffe, B (1988) *The Temple of Sulis Minerva at Bath: The Finds from the Sacred Spring, II:*, Oxford.
Cunningham, A (1999) "Aristotle's Animal Books: Ethology, Biology, Anatomy, or Philosophy?" *Philosophical Topics* 27 (1), pp 17–41.
Cyrino, MS (2010) *Aphrodite*, Abingdon.
Daiber, H (1992) "The *Meteorology* of Theophrastus in Syriac and Arabic Translation," in Fortenbaugh and Gutas, pp 166–293.
Darwin, GH (1898) *The Tides and Kindred Phenomena in the Solar System: The Substance of Lectures delivered in 1897 at the Lowell Institute, Boston, MA*, Boston.
Davies, RW (1970a) "The Roman Military Medical Service," *Saalburg-Jahrbuch* 27, pp 84–104.
Davies, RW (1970b) "Some Roman Medicine," *Medical History* 14, pp 101–6.
De Souza, P and P Arnaud (eds.) (2017), *The Sea in History: The Ancient World*, Woodbridge.
Demetriou, D (2010) "Τῆς πάσης ναυτιλίης φύλαξ: Aphrodite and the Sea," *Kernos* 23, pp 67–89.
Denyer, N (1985) "The Case against Divination: An Examination of Cicero's *De Divinatione*," *Proceedings of the Cambridge Philological Society* 31, pp 1–10.
Detienne, M and J-P Vernant (1969), "La métis du renard et du poulpe," *Revue des études grecques* 82, pp 291–317.
Deyts, S (1985) *Le Sanctuaire des Sources de la Seine*, Dijon.
Dicks, DR (1960) *The Geographical Fragments of Hipparchus*, London.
Diels, H (ed.) (1879) *Doxographi Graeci*, Berlin.
Dimitrova, NM (2008) *Theoroi and Initiates in Samothrace: The Epigraphical Evidence*, Athens.
Dinsmoor, WB Jr (1971) *Sounion*, Athens.
Dobias-Lalou, C (2017) *Inscriptions of Greek Cyrenaica*, Bologna.

Dobias-Lalou, C (2000) *Le dialecte des inscriptions grecques de Cyrène*, (Karthago 25), Paris.

Donahue, JF (2016) "Culinary and Medicinal Uses of Wine and Olive Oil," in Irby 2016a, pp 605-17.

Dowden, K (1989) "Pseudo-Callisthenes, The Alexander Romance, translated with introduction and notes," in BP Reardon (ed.), *Collected Ancient Greek Novels*, Berkeley, pp 650-735.

Drake, N, R Blench, S Armitage, C Bristow, K White, and O Bar-Yosef (2011) "Ancient watercourses and biogeography of the Sahara explain the peopling of the desert," *Proceedings of the National Academy of Sciences of the United States of America* 108, pp 458-62.

Dumézil, G (1970) *Archaic Roman Religion, With an Appendix on the Religion of the Etruscans*, trans. Philip Krapp, Chicago.

Duncan-Jones, R (1990) *Structure and Scale in the Roman Economy*, Cambridge.

Durand, J-L (1989) "Greek Animals: Towards a Typology of Edible Animals," in M Detienne and J-P Vernat (eds.), *The Cuisine of Sacrifice Among the Greeks*, Chicago, pp 87-118.

Dürrbach, F (ed.) (1923-37), *Inscriptions de Délos*, Paris.

Dvorjetski, E (2007) *Leisure, Pleasure and Healing: Spa Culture and Medicine in Ancient Eastern Mediterranean*, Leiden.

Dwyer, E (1973) "Augustus and the Capricorn," *MDAI* 80, pp. 59-67.

Eckenrode, TR (1975) "The Romans and Their Views on the Tides," *Rivista di cultura classica e medioevale* 17, pp 269-92.

El-Geziry, TM and IG Bryden (2014) "The circulation pattern in the Mediterranean Sea: Issues for modeller consideration," *Journal of Operational Oceanography* 3, pp 39-46.

Étienne, R and J-P Braun (1986) *Ténos, I: Le Sanctuaire de Poséidon et d'Amphitrite*, Paris.

Evans, R (2005) "The Cruel Sea?: Ocean as Boundary Marker and Transgressor in Pliny's Roman Geography," *Antichthon* 39, pp 105-19.

Facchinetti, G (2008) "Offrire nelle acque: Bacini e altre strutture artificiali," in H di Giuseppe and M Serlorenzi (eds.), *I riti del costruire nelle acque violate*, Rome, pp 43-67.

Fagan, GG (1999) *Bathing in Public in the Roman World*, Ann Arbor.

Fagan, GG (2000) "Hygienic Conditions in Roman Public Baths," in GCM Jansen (ed.), *Cura Auqarum in Sicilia. BABesch* Supplement 6, Leiden, pp 281-7.

Faraone, CA and D Obbink (1997) *Magika Hiera: Ancient Greek Magic and Religion*, Oxford.

Faraone, CA and D Obbink (eds.) (2013) *The Getty Hexameters: Poetry, Magic, and Mystery in Ancient Selinous*, Oxford.

Farhi, D and N Dupin (2010) "Origins of Syphilis and Management in the Immunocompetent Patient: Facts and Controversies," *Clinics in Dermatology* 28, pp 533-8.

Filonik, J (2013) "Athenian Impiety Trials: A Reappraisal," *Dike* 16, pp 11–96.
Fontan, E (2001) "La Frise du Transport du Bois, Décor du Palais de Sargon II à Khorsabad," in C Doumet-Serhal (ed.), *Cedrus Libani, Archaeology and History in Lebanon* 14, pp 58–63.
Fortenbaugh, WW and D Gutas (eds.) (1992) *Theophrastus: His Psychological, Doxographical, and Scientific Writings*, New Brunswick, NJ.
Franconi, T (ed.) (2017) *Fluvial Landscapes in the Roman World, JRA* Supplement 104.
Frantzis, A, J Gordon, G Hassidis, and A Komnenou (2001) "The enigma of harbour porpoise presence in the Mediterranean Sea," *Marine Mammal Science* 17, pp 937–44.
Frost, FJ (1968) "Scyllias: Diving in Antiquity," *Greece and Rome* 2nd ser. 15, pp 180–5.
Furuse, Y, A Suzuki, and H Oshitani (2010) "Origin of measles virus: Divergence from rinderpest virus between the 11th and 12th centuries," *Virology Journal* 7, pp 52.
Gagné, R (2009) "Mystery Inquisitors: Performance, Authority, and Sacrilege at Eleusis," *Classical Antiquity* 28 (2), pp 211–47.
Galassi, FM, R Bianucci, G Gorini, GM Paganotti, ME Habicht, and FJ Rühli (2016), "The Sudden Death of Alaric I (c. 370–410 AD), the Vanquisher of Rome: A Tale of Malaria and Lacking Immunity," *European Journal of Internal Medicine* 31, pp 84–7.
Gallini, C (1963) "Katapontismos," *Studi e materiali di storia delle religioni* 34, pp 61–90.
Gehlken, E (2012) *Weather Omens of Enūma Anu Enlil: Thunderstorms, Wind, and Rain (Tablets 44–9)*, Leiden.
Gilhus, IS (2006) *Animals, Gods and Humans: Changing Attitudes to Animals in Greek, Roman and Early Christian Ideas*, London.
Gill, MAV (1985) "Some observations on representations of marine animals in Minoan art, and their identification," *Bulletin de correspondance hellénique* 11, pp 63–81.
Gilliam, JF (1961) "The Plague under Marcus Aurelius," *AJP* 82, pp 225–51.
Giraudi, C, M Magny, G Zanchetta, and RN Drysdale (2011) "The Holocene climatic evolution of Mediterranean Italy: A review of the continental geological data," *The Holocene* 21, pp 105–15.
Goldsworthy, A (2003) *The Complete Roman Army*, London.
Goodyear, FDR (1981) *The Annals of Tacitus volume II (Annals 1.55-81 and Annals 2)*, Cambridge.
Graf, F (1985) *Nordionische Kulte. Religionsgeschichte und epigraphische Untersuchungen zu den Kulten von Chios, Erythrai, Klazomenai und Phokaia*, Rome.
Graf, F (2007) "The Oracle and the Image. Returning to Some Oracles from Clarus," *Zeitschrift für Papyrologie und Epigraphik* 160, pp 113–19.
Graham, DW (2003) "Philosophy on the Nile: Herodotus and Ionian Research," *Apeiron* 36, pp 291–310.
Graham, DW (2010) *The Texts of Early Greek Philosophy: The Complete Fragments and Selected Testimonies of the Major Presocratics*, 2 vols, Cambridge.
Green, P (2007) *The Argonautika: Apollonios Rhodios*, Berkeley.
Gregory, AD (2016) "The Creation and Destruction of the World," in Irby 2016a, pp 13–28.

Grmek, MD (1979) "Les ruses de guerre biologiques dans l'antiquité," *Revue des Études Grecques* 92, pp 141–63.

Grmek, MD (1989) *Diseases in the Ancient Greek World*, trans. M Muellner and L Muellner, Baltimore.

Hahn, R (2001) *Anaximander and the Architects: The Contributions of Egyptian and Greek Architectural Technologies to the Origins of Greek Philosophy*, Albany.

Ham, CT (2013) *Empedoclean Elegy: Love, Strife and the Four Elements in Ovid's Amores, Ars Amatoria and Fasti*, Ph.D. Dissertation, University of Pennsylvania.

Hankinson, RJ (1995) "Science," in J Barnes, J (ed.) *The Cambridge Companion to Aristotle*, Cambridge, pp 140–67.

Harland, D (2005) *Water and the Search for Life on Mars*, Heidelberg and London.

Harper, K (2017) *The Fate of Rome: Climate, Diseases, and the End of an Empire*, Princeton.

Harper, K and G Armelagos (2009) "Genomics, the Origins of Agriculture, and Our Changing Microbe-Scape: Time to Revisit Some Old Tales and Tell Some New Ones," *American Journal of Physical Anthropology* 57, pp 135–52.

Harris, AG, E Tuttle, and SD Tuttle (1990) *Geology of National Parks*, 4th ed, Iowa.

Harris, WV (ed.) (2013) *The Ancient Mediterranean Environment between Science and History*, Leiden.

Hartmann, M (1973) "Neue Grabungen in Baden-Aquae Helveticae," *Jahresbericht: Gesellschaft Pro Vindonissa*, pp 45–51.

Hasaki, E and Y Nakas (2017) "Ship Iconography on the Penteskouphia Pinakes from Archaic Corinth (Greece). Pottery Industry and Maritime Trade," in J Gawronski, A van Holk, and J Schokkenbroek (eds.) *Ships and Maritime Landscapes: Proceedings of the 13th International Symposium on Boat and Ship Archaeology, Amsterdam 2012* (2011–2012), Eelde, pp 66–72.

Hassan, F (1997) "The Dynamics of a Riverine Civilization: A Geoarchaeological Perspective on the Nile Valley, Egypt," *World Archaeology* 29 (1), pp 51–74.

Hauben, H (1970) *Callicrates of Samos: A Contribution to the Study of the Ptolemaic Admiralty*, Leuven.

Herington, CJ (1963) "A Study in the '*Prometheia*,' Part I: The Elements in the Trilogy," *Phoenix* 17 (3), pp 180–97.

Hine, HM (2006) "Rome, the Cosmos, and the Emperor in Seneca's 'Natural Questions,'" *JRS* 96, pp 42–72.

Hine, HM (2010) *Lucius Annaeus Seneca: Natural Questions*, Chicago.

Hohlfelder, RL (ed.) (2008) *The Maritime World of Ancient Rome: Proceedings of 'The Maritime World of Ancient Rome' Conference held at the American Academy in Rome 27–29 March 2003*, Ann Arbor.

Hölscher, T (1965) "Ein römischer Stirnziegel mit Victoria und Capricorn," *Jahrbuch des Römisch-Germanischen Zentralmuseums* 12, pp 59–73.

Hooke, J (2003) "River Meander Behaviour and Instability: A Framework for Analysis," *Transactions of the Institute of British Geographers* 28 (2), pp 238–53.

Hope, VM and E Marshall (eds.) (2000) *Death and Disease in the Ancient City*, London.
Horden, P and N Purcell (2000) *The Corrupting Sea: A Study of Mediterranean History*, Malden, MA.
Hughes, B (2005) *Helen of Troy: The Story Behind the Most Beautiful Woman in the World*, New York.
Hughes, D (1991) *Human Sacrifice in Ancient Greece*, London and New York.
Hughes, JD (2014) *Environmental Problems of the Greeks and Romans: Ecology in the Ancient Mediterranean*, Baltimore.
Humphrey, JW, JP Oleson, and AN Sherwood (1998) *Greek and Roman Technology: A Sourcebook*, London.
Inwood, B (2002) "God and human Knowledge in Seneca's *Natural Questions*," in D Frede and A Laks (eds.), *Traditions of Theology: Studies in Hellenistic Theology, its Background and Aftermath*, Leiden, pp 119–57.
Irby, GL (2012) "Mapping the World: Greek Initiatives from Homer to Eratosthenes," in RJA Talbert (ed.), *Ancient Perspectives: Maps and Their Place in Mesopotamia, Egypt, Greece, and Rome*, Chicago, pp 81–108.
Irby, GL ed (2016a) *A Companion to Science, Technology and Medicine in Ancient Greece and Rome*, Malden, MA.
Irby, GL (2016b) "Hydrology," in Irby 2016a, pp 181–96.
Irby, GL (2019a) "The Politics of Cartography: Foundlings, Founders, Swashbucklers, and epic Shields," in *New Directions in Ancient Geography: Proceedings of the Association of Ancient Historians* 14, edited by Duane W Roller, pp 80–102.
Irby, GL (2019b) "Tracing the *orbis terrarum* from Tingentera," in *New Directions in Ancient Geography: Proceedings of the Association of Ancient Historians* 14, edited by Duane W Roller (2019), pp 103–34.
Irby, GL (forthcoming) "Knowledges of the Sea in Classical Antiquity," in *A Cultural History of the Sea in Antiquity*, Bloomsbury Academic, edited by Marie-Claire Beaulieu.
Irby, GL, Robin McCall, and Anita Radini (2016) "'Ecology' in the ancient Mediterranean," in Irby 2016a, pp 296–312.
Irby-Massie, GL (2008) "*Prometheus Bound* and Contemporary Trends in Greek Natural Philosophy," *Greek Roman and Byzantine Studies* 48, pp 133–57.
Irby-Massie, GL and PT Keyser (2002) *Greek Science of the Hellenistic Era: A Sourcebook*, London.
Jackson, RPJ (1990) "Waters and Spas in the Classical World," *Medical History Supplement* 10, pp 5–13.
Jacobsen, T (1968) "The Battle between Marduk and Tiamat," *Journal of the American Oriental Society* 88 (1), pp 104–8.
Jansen, CMG, AO Koloski-Ostrow, and EM Moorman (2011) *Roman Toilets: Their Archaeology and Cultural History. BABesch* supplement 19, Leuven.
Jeal, T (2011) *Explorers of the Nile: The Triumph and Tragedy of a Great Victorian Adventure*, London.

Johnson, W M and D M Lavigne (1999) *Monk Seals in Antiquity. The Mediterranean Monk Seal (Monachus monachus) in Ancient History and Literature*, Leiden.

Kahanov, Y (1999) "Ma'agan-Michael ship (Israel)," in P Pomey and E Rieth (Eds.), *Construction navale maritime et fluviale Archaeonautica* 14, pp 155–60.

Kahn, C H (1979) *The Art and Thought of Heraclitus: An Edition of the Fragments with Translation and Commentary*, Cambridge.

Kallet, L (2013) "Thucydides, Apollo, the Plague, and the War," *AJP* 134 (3), pp 355–82.

Kaplan, G B (2002) "The (Columbian) Myth of Syphilis: A Textual Perspective," *Hispanófila* 134, pp 21–35.

Kargel, J S (2004) *Mars: A Warmer, Wetter Planet*, London.

Kazanjian, Powel (2015) "Ebola in Antiquity"? *Clinical Infectious Diseases* 61, pp 963–8.

Keys, D (2011) "Libya Tale of Two Fundamentally Different Cities," BBC Knowledge Asia Edition, 3.7.

Kidd, D A (1984) "Weather Lore in Aratus' Phaenomena," *Weather and Climate* 4 (1), pp 32–6.

Kidd, D A (1997) *Aratus: Phaenomena*, Cambridge.

Kidd, I G (1972) *Posidonius*, Cambridge.

Kidd, I G (1992) "Theophrastus' *Meteorology*, Aristotle and Posidonius," in Fortenbaugh and Gutas, pp 294–306.

King, C (2004) *The Black Sea: A History*, Oxford.

King, L W (1902) *The Seven Tablets of Creation*, New York.

Kingsley, P (1995) *Ancient Philosophy, Mystery, and Magic: Empedocles and Pythagorean Tradition*, Oxford.

Kirk, GS, JE Raven, and M Schofield (1983) *The Presocratic Philosophers: A Critical History with a Selection of Texts*, 2nd ed., Cambridge.

Kitchell Jr, K F (2014) *Animals in the Ancient World from A to Z*, London.

Knapp, A B (2010) "Cyprus's Earliest Prehistory: Seafarers, Foragers and Settlers," *Journal of World Prehistory* 23, pp 79–120.

Knapp, R C (2011) *Invisible Romans: Prostitutes, Outlaws, Slaves, Gladiators, Ordinary Men and Women ... The Romans that History Forgot*, Cambridge, MA.

Koloski-Ostrow, A O (2015) *The Archaeology of Sanitation in Roman Italy: Toilets, Sewers, and Water Systems*, Chapel Hill.

Kotansky, R D (2015) "The *Lex Sacra* from Selinous: Introduction, Translation, and Notes," in A Iannucci, F Muccioli, and M Zaccarini (Eds.), *La Città Inquieta. Selinunte tra Lex Sacra e Defixiones* (Diadema 2; Milano-Udine), pp 127–34.

Kouki, P (2013) "Problems of Relating Environmental History to Human Settlement in the Classical and Late Classical Periods: The Example of Southern Jordan," in Harris, pp 197–211.

Kragerud, A (1972) "The Concept of Creation in Enuma Elish," in C J Bleeker, S F G Brandon, and M Simon (Eds.), *Ex Orbe Religionum: Studia Geo Widengren*, Leiden, pp 39–49.

Kühr, A (2006) *Als Kadmos nach Boiotien kam: Polis und Ethnos im Spiegel thebanischer Gründungsmythen*, Stuttgart.
Lambton, L (1995) *Temples of Convenience and Chambers of Delight*, New York.
Laumonier, A (1958) *Les cultes indigènes en Carie*, Paris.
Lawrence, A (1926) "The Date of the Nike of Samothrace," *JHS* 46 (2), pp 213-18.
Lehmann, K and JR McCredie (1998) *Samothrace: A Guide to the Excavations and the Museum*. 6th ed., Thessaloniki.
Lehoux, Daryn (2007) *Astronomy, Weather, and Calendars in the Ancient World: Parapegmata and Related Texts in Classical and Near Eastern Societies*, Cambridge.
Lesky, A (1947) *Thalatta: Der Weg der Griechen zum Meer*, Wien.
Leveau, P, F Trément, K Walsh, and G Barker (eds.) (2016) *Environmental Reconstruction in Mediterranean Landscape Archaeology*, Oxford.
Levine, DB (2011) "Tuna in Ancient Greece and Modern Tuna Population Decline,": https://sites.uark.edu/dlevine/daniel-b-levine-tuna-lecture-copywrite-2011/.
Lévi Strauss, C (1964) *Totemism*, London.
Lewis, N (1958) *Samothrace: The Ancient Literary Sources*, New York.
Liddle, A (2003) *Arrian: Periplus Ponti Euxini*, London.
Lindsay, H (2000) "Death Pollution and Funerals in the City of Rome," in Hope and Marshall, pp 105-17.
Littman, RJ (1984) "The Plague at Syracuse: 396 B.C.," *Mnemosyne* 4th Ser. 37 (1), pp 110-16.
Littman, RJ and ML Littman (1969) "The Athenian Plague. Smallpox," *Transactions of the American Philological Association* 100, pp 261-75.
Littman, RJ and ML Littman (1973) "Galen and the Antonine Plague," *AJP* 94 (3), pp 243-55.
Llewellyn-Jones, L and S Lewis (2018) *The Culture of Animals in Antiquity: A Sourcebook with Commentaries*, London.
Lloyd, C (1961) "The Introduction of Lemon Juice as a Cure for Scurvy," *Bulletin of the History of Medicine* 35, pp 123-32.
Lloyd, GER (1970) *Early Greek Science: Thales to Aristotle*, New York.
Lloyd, GER (1979) *Magic, Reason, and Experience: Studies in the Origins and Development of Greek Science*, Cambridge.
Longfellow, Brenda (2012) "Roman Fountains in Greek Sanctuaries," *AJA* 116 (1), pp 133-55.
Longrigg, J (2000) "Death and Epidemic Diseases in Classical Athens," in Hope and Marshall, pp 44-50.
Lordkipanidze, O (2000) *Phasis: The River and City in Colchis*, Stuttgart.
Lovejoy, J (1972) "The tides of New Carthage," *CPh* 67, pp 110-11.
Lowe, D (2011) "Scylla, the Diver's Daughter: Aeschrion, Hedyle, and Ovid," *CPh* 106, pp 260-4.
Luck, G (2006) *Arcana Mundi: Magic and the Occult in the Greek and Roman Worlds: A Collection of Ancient Texts*. 2nd ed., Baltimore.

Lupu, E (2005) *Greek Sacred Law. A Collection of New Documents*, Leiden.
MacKillop, J (2004) *A Dictionary of Celtic Mythology*, Oxford.
Mader, GJ (2002) "'Ut pictura poesis:' Sea-bull and Senecan Baroque (*Phaedra* 1035-49)," *Classica et Mediaevalia* 53, pp 289-300.
Maitland, A (1971) *Speke: And the Discovery of the Source of the Nile*, London.
Maiuri, A (1958) "Navalia pompeiana," *Rendiconti della Accademia di Archaeologia di Napoli* 33, pp 7-34.
Malkin, I (2003) *Myth and Territory in the Spartan Mediterranean*, Cambridge.
Manning, SW (2013) "The Roman World and Climate: Context, Relevance of Climate Change, and Some Issues," in Harris, pp 103-70.
Marangou, C (2006) "Land and Sea Connections: The Kastro rock-cut Site (Lemnos Island, Aegean Sea, Greece)," in L Blue, F Hocker, and A Englert (eds.), *Connected by the Sea*, Oxford, pp 130-6.
Marchiafava, E and A Bignami (1894) *On summer-autumn malarial fevers*, trans. J Thompson, London.
Margarida, AA (2009) "Phoenician Colonization on the Atlantic Coast of the Iberian Peninsula," in M Dietler and C López-Ruiz (eds.), *Colonial Encounters in Ancient Iberia*, Chicago, pp 113-30.
Margulis, L (1998) *Symbiotic Planet: A New Look at Evolution*, London.
Marzano, A (2013) *Harvesting the Sea: The Exploitation of Marine Resources in the Roman Mediterranean*, Oxford.
Mayor, A (2000) "The 'Monster of Troy' Vase: The Earliest Artistic Record of a Vertebrate Fossil Discovery?" *Oxford Journal of Archaeology* 19, pp 57-63.
McCormick, M, U Büntgen, MA Cane, ER Cook, K Harper, P Huybers, T Litt, SW Manning, PA Mayewski, AFM More, K Nicolussi, and W Tegel (2012), "Climate Change during and after the Roman Empire: Reconstructing the Past from Scientific and Historical Evidence," *Journal of Interdisciplinary History* 43 (2), pp 169-220.
McGrail, S (2001) *Boats of the World: From the Stone Age to Medieval Times*, Oxford.
McNeill, WH (1976) *Plagues and Peoples*, New York.
Meeusen, M (2016) *Plutarch's Science of Natural Problems: A Study with Commentary on Quaestiones Naturales*, Leuven.
Merkelbach, R and J Stauber (1996) "Die Orakel des Apollon von Klaros," *Epigraphica Anatolica* 27, pp 1-54.
Metcalfe, NH (2007) "A description of the methods used to obtain information on ancient disease and medicine and of how the evidence has survived," *Postgraduate Medical Journal* 83, pp 655-8.
Mohler, SL (1944-1945) "Caesar and the Channel tides," *Classical World* 38, pp 189-91.
Mondi, R (1990) "Greek Mythic Thought in the Light of the Near East," in L Edmunds (ed.), *Approaches to Greek Myth*, Baltimore, pp 141-98.
Morgan, L (1988) *The Miniature Wall-Paintings of Thera: A study in Aegean Culture and Iconography*, Cambridge.

Mott, M (2005) "Did Animals Sense Tsunami was Coming?" *National Geographic* (January 4).

Mourelatos, APD (2008) "The Cloud-Astrophysics of Xenophanes and Ionian Material Monism," in P Curd and DW Graham (eds.), *The Oxford Handbook of Presocratic Philosophy*, Oxford, pp 134–68.

Mowery, RL (2011) "Theoroi and Initiates in Samothrace," *Biblica* 92 (1), pp 112–22.

Mueller, I, PA Zimmerman, and JC Reeder (2007) "Plasmodium malariae and Plasmodium ovale – the 'bashful' malaria parasites," *Trends in Parasitology* 23 (6), pp 278–83.

Murphy, T (2004) *Pliny the Elder's Natural History: The Empire in the Encyclopedia*, Oxford.

Mylanopoulos, I (2003) Πελοπόννησος οἰκητήριον Ποσειδῶνος. *Heiligtümer und Kulte des Poseidon auf der Peloponnes, Kernos supplement* 13, CIERGA, Liége.

Mylanopoulos, I (2006) "Von Helike nach Tainaron und von Kalaureia nach Samikon: Amphiktyonische Heiligtümer des Poseidon auf der Peloponnes," in K Freitag et al. (eds.), *Kult—Politik—Ethnos. überregionale Heiligtümer im Spannungsfeld von Kult und Politik*, Stuttgart, pp 121–55.

Naddaf, G (1986) "Hésiode, précurseur des cosmogonies grecques de type « évolutionniste »," *Revue de l'histoire des religions* 203 (4), pp 339–64.

Needham, J (1962) *Science and Civilisation in China*, Cambridge.

Neilson, HR (2002) "A terracotta phallus from Pisa Ship E: More Evidence for the Priapus deity as Protector of Greek and Roman navigators," *International Journal of Nautical Archaeology* 31 (2), pp 248–53.

Nesselhauf, H and H Petrikovits (1967) "Ein Weihaltar für Apollo aus Aachen-Burtscheid," *Bonner Jahrbücher* 167, pp 268–79.

Nicoll, W (1988) "The Sacrifice of Palinurus," *CQ* 38 (2), pp 459–72.

Nikita, E, A Lagia, and S Triantaphyllou (2016) "Epidemiology and Pathology," in Irby 2016a, pp 465–82.

Nilsson, MP (1956) *Geschichte der griechischen Religion*, Munich.

Nutton, V (2013) *Ancient Medicine*, 2nd ed., London.

Oerlemans, APA and LE Tacoma (2014) "Three Great Killers: Infectious Diseases and Patterns of Mortality in Imperial Rome," *Ancient Society* 44, pp 213–41.

Ogden, D (2013) *Drakōn: Dragon Myth and Serpent Cult in the Greek and Roman Worlds*, Oxford.

Oldach, DW, RE Richard, EN Borza, and RN Benitz (1998) "A mysterious Death," *New England Journal of Medicine* 338, pp 764–9.

Oleson, JP (2008b) "Testing the Waters: The Role of Sounding Weights in Ancient Mediterranean Navigation," in Hohlfelder, pp 119–76.

Oleson, JP (ed.) (2008a) *The Oxford Handbook of Engineering and Technology in the Classical World*, Oxford.

Oltramare, P (texte établi et traduit) (1961) *Sénèque: Questions Naturelles*, Paris.

Ongaro, G (1967) "Evoluzione storica del concetto di tifo," *Riforma medica* 6, pp 3–11.

Orlin, EM (1997) *Temples, Religion, and Politics in the Roman Republic*, Leiden.

Paga, J (2016) "Attic Sanctuaries," in MM Miles (Ed.), *A Companion to Greek Architecture*, Malden, MA, pp 178–93.
Paine, LP (2013) *The Sea and Civilization: A Maritime History of the World*, New York.
Palmer, J (2016) "Elemental Change in Empedocles," *Rhizomata* 4 (1), pp 30–54.
Papadopoulos, JK and D Ruscillo (2002) "A Ketos in Early Athens: An Archaeology of Whales and Sea Monsters in the Greek World," *AJA* 106 (2), pp 187–227.
Papagrigorakis, M, C Yapijakis, P Synodinos, and E Baziotopoulou-Valavani (2006) "DNA Examination of Ancient Dental Pulp Incriminates Typhoid Fever as a Probable Cause of the Plague of Athens," *International Journal of Infectious Diseases* 10 (3), pp 206–14.
Pappas, G, IJ Kiriaze, and ME Falagas (2008) "Insights into infectious disease in the era of Hippocrates," *International Journal of Infectious Diseases* 12 (4), pp 347–50.
Parker, R (1983) *Miasma: Pollution and Purification in Early Greek Religion*, Oxford.
Parker, R (2005) *Polytheism and Society at Athens*, Oxford.
Pascal, CB (1981) "October Horse," *Harvard Studies in Classical Philology* 85, pp 261–91.
Paton, WR and EL Hicks (1891) *The Inscriptions of Cos*, Oxford.
Pearson, L (1960) *The Lost Histories of Alexander the Great*, New York.
Pedley, JG (1970) "The Friedlaender hydria," *CPh* 74, pp 45–53.
Penglase, C (1994) *Greek Myths and Mesopotamia. Parallels and Influence in the Homeric Hymns and Hesiod*, London.
Popa, T (2016) "Zoology," in Irby 2016a, pp 281–95.
Pope, JM, MH Weir, and JB Rose (2012) "History of Water and Health," in AN Angelakis, LW Mays, and D Koutsoyiannis (Eds.), *Evolution of the Water Supply Through the Millennia*, London, pp 43–75.
Powell, A (2002) "'An Island Amid the Flame': The Strategy and Imagery of Sextus Pompeius," in A Powell and K Welch (eds.) *Sextus Pompeius*, London, pp 103–33.
Powell, L (2015) *Marcus Agrippa: Right-Hand Man of Caesar Augustus*, Barnsley.
Price, LG (1999) *An Introduction to Grand Canyon Geology*, Grand Canyon, Arizona.
Pritchett, WK (1979) *The Greek State at War: Part III: Religion*, Berkeley.
Radcliffe, W (1921) *Fishing from the Earliest Times*, London.
Reese, DS (1984) "Shark and Ray Remains in Aegean and Cypriote Archaeology," *Opuscula Atheniensia* 15, pp 188–92.
Reese, DS (2001) "Fish: Evidence from Specimens, Mosaics, Wall Paintings, and Roman Authors," in WF Jashemski and FG Meyer (eds.), *The Natural History of Pompeii*, Cambridge, pp 274–91.
Retief, FP and L Cilliers (2000) "Epidemics of the Roman Empire, 27 BC–AD 476," *South African Medical Journal* 90 (3), pp 267–72.
Retief, FP and L Cilliers (2006) "Periodic pyrexia and malaria in antiquity," *South African Medical Journal* 96 (8), pp 686–8.
Reynolds, JM and JB Ward-Perkins (1952) *The Inscriptions of Roman Tripolitania* (Rome). Electronically enhanced and reissued by G Bodard and C Roueché, 2009, http://inslib.kcl.ac.uk/irt2009/.

Richardson, NJ (1974) *The Homeric Hymn to Demeter*, Oxford.
Ridgway, BS (1970) "Dolphins and Dolphin-Riders," *Archaeology* 23, pp 86–95.
Rihll, TE (1999) *Greek Science*, Oxford.
Robert, L (1966) "Sur un décret d'Ilion et sur un papyrus concernant des cultes royaux," in AE Samuel (ed.), *Essays in Honor of C. Bradford Welles*, New Haven, pp 175–211.
Robertson, N (2010) *Religion and Reconciliation in Greek Cities. The Sacred Laws of Selinus and Cyrene*, Oxford.
Rocca, S (2015) *Herod's Judaea: A Mediterranean State in the Classical World*, Eugene, OR.
Rogers, DK (2018) *Water Culture in Roman Society*, Leiden.
Roller, DW (2003) *The World of Juba II and Kleopatra Selene: Royal Scholarship on Rome's African Frontier*, London.
Roller, DW (2005) "Seleukos of Seleukeia," *Antiquité classique* 74, pp 111–18.
Roller, DW (2006) *Through the Pillars of Herakles: Greco-Roman Exploration of the Atlantic*, London.
Roller, DW (2010a) *Cleopatra: A Biography*, Oxford.
Roller, DW (2010b) *Eratosthenes' Geography: Fragments collected and translated, with commentary and additional material*, Princeton.
Roller, DW (2014) *The Geography of Strabo: An English Translation, with Introduction and Notes*, Cambridge.
Roller, DW (2018) *A Historical and Topographical Guide to the Geography of Strabo*, Cambridge.
Romano, DG and ME Voyatzis (2018) "Excavating at the Birthplace of Zeus," *Archaeology* 71, pp 44–9.
Romm, JS (1992) *The Edges of the Earth in Ancient Thought: Geography, Exploration, and Fiction*, Princeton.
Romey, Kristen (2018) "Man-eating fish, Tower of Babel revealed on ancient mosaic: Archaeologists discover even more remarkable biblical scenes on the floor of a 1,600-year-old synagogue in Israel," *National Geographic*, November 15, 2018: https://www.nationalgeographic.com/culture/2018/11/jonah-tower-babel-huqoq-ancient-synagogue-mosaic/.
Rona, P (1995) "Tectonoeustasy and Phanerozoic Sea Levels. *Journal of Coastal Research*," pp 269–77.
Rose, JB and Y Masago (2007) "A toast to our health: our Journey toward safe water," *Water Science and Technology: Water Supply* 7 (1), pp 41–8.
Roseman, CH (1994) *Pytheas of Massalia: On the Ocean*, Chicago.
Rothschild, BM, FL Calderon, A Coppa, and C Rothschild (2000) "First European Exposure to Syphilis: The Dominican Republic at the Time of Columbian Contact," *Clinical Infectious Diseases* 31 (4), 936–41.
Rudhardt, J (1971) *Le thème de l'eau primordiale dans la mythologie grecque*, Berne.
Russell, PF (1955) *Man's Mastery of Malaria*, Oxford.
Sallares, R (2002) *Malaria and Rome: A History of Malaria in Ancient Italy*, Oxford.

Sallares, R, A Bouwman, and C Anderung (2004) "The Spread of Malaria to Southern Europe in Antiquity: New Approaches to Old Problems," *Medical History* 48 (3), pp 311–28.
Sambursky, S (1959) *Physics of the Stocis*, Princeton, NJ.
Sandbach, FH (1975) *The Stoics*, New York.
Sanders, KR (2011) "Strato on 'Microvoid,'" in M-L Desclos and WW Fortenbaugh (eds.), *Strato of Lampsacus: Text, Translation, and Discussion*, New Brunswick, NJ, pp 263–76.
Šašel, A and J Šašel (1986) *Inscriptiones Latinae quae in Iugoslavia inter annos MCMII et MCMXL repertae et editae sunt*, Ljubljana.
Schneider, H (2001) "Thekla und die Robben," *Vigiliae Christianae* 55 (1), pp 45–57.
Scobie, A (1986) "Slums, Sanitation, and Mortality in the Roman World," *Klio* 68, pp 399–433.
Scullard, HH (1981) *Festivals and Ceremonies of the Roman Republic*, Ithaca, NY.
Sedley, DN (1998) *Lucretius and the Transformation of Greek Wisdom*, Cambridge.
Shapiro, B, A Rambaut, and T Gilbert (2006) "No proof that typhoid caused the plague of Athens (a reply to Papagrigorakis et al.)," *International Journal of Infectious Diseases* 10 (4), pp 334–5.
Sharples, RW (1992) "On Fish," in Fortenbaugh and Gutas, pp 347–85.
Sharples, RW and D Gutas (1998) *Theophrastus of Eresus: Sources for His Life, Writings, Thought and Influence: Commentary Volume 3.1: Sources on Physics*, Leiden.
Sider, D (2002) "On *On Signs*," in W W Fortenbaugh and G Wörhle (eds.), *On the Opuscula of Theophrastus: Akten der 3. Tagung der Karl-und-Gertrud-Abel-Stiftung vom 19.-23. Juli 1999 in Trier*, Stuttgart, pp 99–111.
Silberman, A (texte établi, trad. et annoté) (1988) *Pomponius Mela: Chorographie*, Paris.
Simon, E (2014) "Poseidon in Ancient Greek Religion, Myth, and Art," in SD Pevnick (ed.), *Poseidon and the Sea: Myth, Cult, and Daily Life*, Tampa, FL, pp 37–49.
Size, WB, B Westcoat, and M Page (2014) "Impact of the Geology and Tectonic History on the Construction and Destruction of the Ancient Sanctuary of the Great Gods on the Island of Samothrace, Greece," 2014 Geological Society of America Annual Meeting in Vancouver, British Columbia (19–22 October 2014): https://gsa.confex.com/gsa/2014AM/webprogram/Paper243641.html.
Slavena-Griffin, S (2016) "Nature and the Divine," in Irby 2016a, pp 60–75.
Sloane, Eric (1980) *Eric Sloane's Weather Book*, Mineola, NY.
Smid, TC (1970) "*Tsunamis* in Greek Literature," *Greece & Rome* 17 (1), pp 100–4.
Smith, J (2004) "The River Alpheus in Greek, Christian and Byzantine Thought," *Byzantion* 74, pp 416–32.
Smith, MS (1994) *The Ugarit Baal Cycle. Volume 1. Introduction with Text, Translation and Commentary of KTU 1.1–1.2*, Leiden.
Solmsen, F (1989) "The two Near Eastern Sources of Hesiod," *Hermes* 117, pp 413–22.
Solomon, S (2010) *Water: The Epic Struggle for Wealth, Power, and Civilization*, New York.

Speke, JH (1863) *Journal of the Discovery of the Source of the Nile*, Edinburgh.
Spivey, N and T Rasmussen (1988) "Dionisio e I pirati nel Toledo Museum of Art," *Perspettiva* 44, pp 2–8.
Tan Sy, Sung H (2008) "Carlos Juan Finlay (1833–1915): of Mosquitoes and Yellow Fever," *Singapore Medical Journal* 49 (5), pp 370–1.
Tarn, WW (1910) "The Dedicated Ship of Antigonus Gonatas," *JHS* 30 (2), pp 209–22.
Taub, L (2003) *Ancient Meteorology*, London.
Taylor, RM (2008) *The Moral Mirror of Roman Art*, Cambridge.
Terio, S (2006) *Der Steinbock als Herrschaftszeichen des Augustus*, Münster.
Testaguzza, O (1970) *Portus: Illustrazione Dei Porti Di Claudio E Traiano E Della Cittaà di Porto a Fiumicino*, Roma.
Thompson, H (1966) "The Annex to the Stoa of Zeus in the Athenian Agora," *Hesperia* 35, pp 171–87.
Torelli, M (1977) "Il santuario greco di Gravisca," *La Parola del passato: rivista di studi antichi* 32, pp 398–458.
Totelin, L (2014) "Smell as Sign and Cure in Ancient Medicine," in M Bradley (ed.), *Smell and the Ancient Senses*, London, pp 17–29.
Toynbee, JMC (2013) *Animals in Roman Art and Life* 2nd ed., Barnsley.
Treggiari, S (2007) *Terentia, Tullia and Publilia: The Women of Cicero's family*, London.
Triantafillidis, I and D Koutsoumba (2017) "The harbour landscape of Aegina (Greece)" in J Gawronski, A Van Holk, and J Schokkenbroek (eds.), *Ships and Maritime Landscapes: Proceedings of the thirteenth international Symposium on Boat and Ship Archaeology, Amsterdam 2012* (2011–2012), Eelde, pp 165–70.
Trinquier, J (2009) "Protée en sa grotte ou le parti pris du phoque," in A Rolet (ed.), *Protée en trompe-l'oeil: genèse et survivances d'un mythe, d'Homère à Bouchardon*, Rennes, pp 63–103.
Tuck, SL (2008) "The Expansion of Triumphal Imagery beyond Rome: Imperial Monuments at the Harbors of Ostia and Lepcis Magna," in Hohlfelder, pp 325–41.
Umholtz, G (2002) "Architraval Arrogance? Dedicatory Inscriptions in Greek Architecture of the Classical Period," *Hesperia* 71 (3), pp 261–93.
Van Duzer, C (2013) *Sea Monsters on Medieval and Renaissance Maps*, London.
Vanotti, Gabriella (1989) "La Peste a Siracusa Nel Racconto Di Diodoro XIV, 70," *Contributi Dell'Istituto Di Storia Antica Dell'Università Del Sacro Cuore* XV, pp 35–42.
Vermeule, E (1979) *Aspects of Death in Early Greek Art and Poetry*, Berkeley.
Vidal-Naquet, P (2005) *L'Atlantide. Petite histoire d'un mythe platonicien*, Paris.
Vinson, S (1994) *Egyptian Boats and Ships*, Oxford.
Volk, K (2009) *Manilius and his Intellectual Background*, Oxford.
Voultsiadou, E, V Gerovasileiou, L Vandepitte, K Ganias, and C Arvanitidis (2017) "Aristotle's scientific contributions to the classification, nomenclature and distribution of marine organisms," *Mediterranean Marine Science* 18 (3), pp 468–78.

Wachsmuth, D (1967) *Pompimos Ho Daimōn: Untersuchung Zu Den Antiken Sakralhandlungen Bei Seereisen*, Berlin.
Waddelove, E and AC Waddelove (1990) "Archaeology and Research into Sea-level during the Roman Era: Towards a Methodology based on highest astronomical Tide," *Britannia* 21, pp 253–66.
Wainwright, GA (1953) "Herodotus II, 28 on the Sources of the Nile," *JHS* 73, pp 104–7.
Wakeman, MK (1969) "The Biblical Earth Monster in the Cosmogonic Combat Myth," *Journal of Biblical Literature* 88, pp 313–20.
Walbank, FW (1972) *Polybios*, Berkeley.
Waldron, T (1987) "The Relative Survival of the Human Skeleton: Implications for Palaeopathology," in A Boddington, AN Garland, and RC Janaway (Eds.), *Death, Decay and Reconstruction: Approaches to Archaeology and Forensic Science*, Manchester, pp 55–64.
Walsh, J (1986) "The Date of the Athenian Stoa at Delphi," *AJA* 90 (3), pp 319–36.
Wang, C-Y and M Manga (2015) "New streams and springs after the 2014 Mw6.0 South Napa earthquake," *Nature Communications* 6, pp 597.
Weber, W (1907) *Untersuchungen zur Geschichte des Kaisers Hadrianus*, Leipzig.
Weinstock, S (1971) *Divus Julius*, Oxford.
Weintraub, DA (2018) *Life on Mars: What to Know Before We Go*, Princeton.
Welter, FG (1938) "Aiginetica XII-XXIV," *Archäologischer Anzeiger*, pp 480–520.
West, ML (1966) *Hesiod: Theogony*, Oxford.
Westcoat, B (2005) "Buildings for Votive Ships on Delos and Samothrace," in JJ Coulton, M Yeroulanou and M Stamatopoulou (Eds.), *Architecture and Archaeology in the Cyclades: Papers in Honour of J.J. Coulton*. British Archaeological Reports International Series, 1,455, Oxford, pp 153–72.
Wilbur, A, A Bouwman, A Stone, C Roberts, L-A Pfister, J Buikstra, and T Brown (2009) "Deficiencies and Challenges in the Study of Ancient Tuberculosis DNA," *Journal of Archaeological Science* 36 (9), pp 990–7.
Willcock, MM (1976) *A Companion to the Iliad*, Chicago.
Williams, GD (2008) "Reading the Waters: Seneca on the Nile in *Natural Questions*, Book 4A," *CQ* 58 (1), pp 218–42.
Wilson, A (2013) "The Mediterranean Environment in Ancient History: Perspectives and Prospects," in Harris, pp 259–76.
Wilson, M (2013) *Structure and Method in Aristotle's Meteorologica*, Cambridge.
Witt, RE (1997) *Isis in the Ancient World*, Baltimore.
Wood, JB and RC Anderson, 2004) "Interspecific Evaluation of Octopus Escape Behaviour," *The Journal of Applied Animal Welfare Science* 7, pp 95–10.
Wood, JG, (trans.), (1894) *Theophrastus of Eresus on Winds and on Weather Signs*, London.
Woodford, S (1971) "Cults of Herakles in Attica," in DG Mitten, JG Pedley, and JA Scott (Eds.), *Studies Presented to George M A Hanfmann*, Mainz, pp 211–25.
Wootton, AC (1910) *Chronicles of Pharmacy*, 2 vols., London.

Yébenes, SP (2010) "Magic at Sea: Amulets for Navigation," in RL Gordon and FM Simón (eds.), *Magical Practice in the Latin West: Papers from the International Conference held at the University of Zaragoza, 30 Sept.-1 Oct. 2005*, Leiden, pp 457–86.

Zanker, P (1990) *The Power of Images in the Age of Augustus*, trans. Alan Shapiro, Ann Arbor.

Zeng, L, SB Jacobsen, DD Sasselov, and A Vanderburg (2018) "Survival function analysis of planet size distribution with Gaia Data Release 2 updates," *Monthly Notices of the Royal Astronomical Society* 479, pp 567–76

Index of Places Cited

Poleis and City States

Abdera 20
Acragas 21, 112, 213
Actium 106, 107, 115, 148, 175, 208
Aegina 62, 176
Aenus 134
Agia Irini 131
Agia Triadha 131
Aigion 64
Akrocorinth 178
Alauna (Maryport) 223
Alexandria 29, 45, 66, 79, 101, 107, 108,
    178, 179, 193, 220
Ammon 65, 80, 81, 205
Anamoreia 64
Antioch 111
Apamea 4, 9, 52, 208
Aquae Albulae 123
Aquae Granni 124
Aquae Helveticae 124
Aquae Sulis 123, 148
Argos 172
Aricia 165
Asine 64
Athens 27, 29, 62, 79, 85, 108, 117, 118, 140,
    141, 174, 181, 186, 214

Babylon 114
Baiae 122
Benghazi 203
Brundisium 79
Byzantium 138

Calydon 222
Camerina 210
Capua 174
Carteia 143
Carthage 37, 105, 107, 174
Cerverteri 218

Ceryneia 104
Chalkis 60, 64
Chemnis 82
Cicilia 179
Citium 27
Clazomenae 186
Clytor 80
Colchis 73
Colophon 20
Colossae 81
Constantinople 174
Corinth 174, 177, 180, 217, 223,
    224
Cyaneae 163
Cyme 63
Cytiliae 81
Cyzica 138
Cyzicus 43

Delphi 132, 173, 175
Dichaearchia (Puteoli) 143
Dodona 64, 79
Dougga 174
Doulichion 64
Dyrrhachion (Epidamanus) 129,
    182

Elea 20, 65
Eleusis 168, 183
Elis 64
Ephesus 20, 40, 75, 185
Epidauros 180
Eretria 141, 185
Euripus 59, 60
Eurymenae 81

Falerii 80
Fishbourne 132, 174

Gades (Cadiz, Gadir) 38, 45, 53, 79
Genethlium 61
Gonoëssa 64
Gordaia 78
Gravisca 176

Halai 189
Haliartos 78
Halicarnassus 219, 223
Helike 64, 104
Helos 64
Heraklea 166
Hermionis 180
Hestiaeotis 80
Hierapolis 129, 165
Hippo Diarrhytus 134
Hippo Diarrhytus 134, 164
Huqoq 215
Hyampolis 64
Hyperesia 64

Iasus 134
Iolkos 175
Isthmia 176

Jaffa 151, 218

Kastro (Lemnos) 187
Kenchreai 55, 185
Knidos 179, 223
Knossos 131, 213
Korbilon 37, 199

Labraunda 129, 165
Lympne 223
Lyncestis 80
Lyon (Lugdunum) 185

Magnesia on the Meander 62, 80, 106, 184
Massillia (Marseilles) 4, 36, 37, 49, 71, 189
Megalopolis 158
Memphis 66
Meroë 204
Messenia 205
Miletus 17, 18, 19, 68, 108, 196
Mitylene 173

Narbonensis (Nîmes) 132
Naukratis 177

Neapolis (Naples) 104
New Carthage 50
Nicea 45
Nonacris 80
Noviomagus Reginorum (Chichester) 174

Olympia 64, 78, 180
Opononi, New Zealand 214
Orchomenos 78
Ormenios 64
Orobiae 103
Oxyrhynchus 165

Paestum/Poseidonium 68, 172
Pagasai 193
Panarea (Euonymus) 104
Paphos 177
Patrai 163, 220
Pellene 64
Phaselis 190
Pheneus 78
Philae 76, 204
Pompeii 101, 132, 186
Poroselene 134
Potidea 118, 172
Priene 68
Ptolemais 104
Pylos 172, 187, 205

Ravenna 68
Rhion 133

Sabratha 122
Sarpana 73
Scaptopara 124
Selinunte 159, 219
Selinus 116, 210
Sirte 203
Smyrna 158
Sousis 6
Sparta 64, 101, 153, 175, 181
Stagira 18
Susa 80, 211
Sybaris 80
Syedra 188
Syene (Aswan) 54, 75, 204
Syracuse 65, 78, 112, 115, 118, 187

Tabariah 123
Taenaron 163, 166, 168
Taras 132
Tarentum 140
Tarsus 80, 200
Telephusa 173
Terracina 80
Thebes 182
Thermopylae 106, 123, 212
Thessalonica 140, 214
Thronium 64
Thurii 80
Tivoli 123
Tomi 182
Tourdetania 50, 138
Trachiniae 123
Tripoli 203
Troy 3, 49, 62, 70, 72, 80, 102, 105, 151, 153, 159, 172, 174, 181, 218, 219, 222
Tusculum 108
Tyre 38, 104

Verona 124
Vindonissa 124
Volterra 164
Vulcano (Hiera) 104
Vulci 132, 133, 154, 215

Territories, Provinces, and Countries

Achaea 103
Africa (Libya/Libyka) 2, 36, 38, 45, 47, 52, 55, 64, 66, 68, 75, 76, 81, 105, 106, 121, 122, 132, 134, 180, 196, 201, 203, 205, 212
Aithiopia 3, 61, 73, 76, 80, 99, 117, 218
Anatolia 65, 184
Arabia Felix 61, 99
Arcadia 80, 81, 173
Argolis 61, 180
Armenia 64, 65, 81, 107, 108
Asia Minor 81, 172, 193
Assyria 153, 193
Attica 68, 71, 103, 174, 187

Bahrain 14
Boeotia 33, 78, 79, 85, 89, 218
Brazil 213
Brittany 50, 190, 220

Campania 72
Cantabria 114
Cappadocia 80, 107
Caria 134, 165
Cilicia 64, 67, 183
Cyrene 55, 107, 159, 176, 219
Cyzica 138

Dacia 123
Dalmatia 201

Egypt/Egyptians 16, 17, 54, 66, 67, 68, 74–7, 81, 82, 101, 162, 163, 164, 165, 171, 173, 178, 179, 185, 186, 189, 193, 196, 200, 201, 204, 205, 220
Etruria 68, 72, 80, 176

Galatia 38
Gaul 54, 201
Germany (Germania) 54, 72

Hibernia 38

Iberia 45, 50, 143
Illyria 72, 79, 111, 129
India 4, 5, 6, 52, 67, 74, 76, 78, 81, 204, 222
Italy 20, 59, 65, 68, 69, 80, 81, 86, 115, 123, 159, 160, 164, 172, 201, 218

Judea 200, 208
Jutland 37

Kea 131

Lacedaemonia 64, 103
Lakonia 180
Larissa 36
Latium 81, 155
Lerna 54, 150
Levant 104
Lokris 64, 189

Lusitania 146
Lycia 61, 70, 112, 123, 163, 165, 190
Lydia 65

Macedonia 5, 18, 36, 55, 70, 77, 80, 81, 176, 182, 183, 223
Mauretania 75, 81, 159
Mesopotamia 13–15, 17, 33, 90, 163, 196
Moesia Superior 81, 107
Morocco 51, 202
Mysia 72, 208

Nabataea 61, 120
Numidia 174

Palestine 47
Peloponnese 78, 102, 103, 163, 205, 206

Pelousion 66, 101
Persia 45, 49, 80
Phoenicia/Phoenicians 36, 38, 66, 101, 174
Phrygia 65, 81, 160
Portugal 51
Punjab 5, 75, 193

Scythia 72, 81
Spain (Hispania) 38, 45, 50, 51, 52, 71, 114, 153, 202
Syria 49, 52, 104, 107, 111, 165, 173

Thessaly 80, 81, 194
Thrace 72, 81, 124, 132, 134, 182
Trogodytika 61, 81
Turkey 17, 80, 102

Mountains

Aetna 60
Atlas 38

Corycus 70

Epomeus 104

Haemus 70
Helicon 79, 161

Ida 184

Kasion 101

Lykaion 221

Mykale 172

Olympus 181

Pyrene 72

Taygetus 101

Vesuvius 107, 208

Mountain Ranges

Abnoua 72
Alps 80, 81, 108
Apennine 86
Atlas 38

Bybline (Papyrus) 75

Caucasus 73, 108, 154

Elaphantine 75

Himalayas 75

Mauretanian 75

Rhipaean (gusty) 72

Index of Places Cited 257

Capes and Peninsulas

Acherousia 175
Aremorica (Brittany) 50

Caldone 61
Crimea 102

Maceta 6

Pelorus 59

Sounion 174–5

Zephyrion 178

Islands

Achilles 190
Andros 80
Atlantis 52, 66, 102
Azores 135

Britain (Britannia) 36, 37, 49, 71, 73, 106, 138, 174, 203, 215

Canaries 135
Cassiterides (Tin) 37
Cerinthus 64
Chios 76, 78, 80, 129
Corfu 173
Corsica 226
Cos 136, 175, 190
Crete 70, 102, 131, 175, 194, 221
Cyclades 133, 218
Cyprus 27, 68, 166, 176, 177, 203, 223

Delos 81, 174, 175, 176, 179, 223
Dilmum 14
Diodorus 60

Echinades 64, 68
Elektra 184
Elephantine 54, 75
Euboea 59, 60, 103, 104

Hesperides 154

Isles of the Blessed 132, 154
Ithaca 171, 187, 222

Jersey Channel Islands 200

Lemnos 184, 187
Lesbos 61, 134, 183

Malta 65, 182

Naxos 133

Orkneys 37
Ortygia 78

Pharos 66, 68, 154
Phocaea 135
Pithekoussai 104, 122, 140, 215

Rhodes 3, 89, 134, 201, 208, 226

Salamis 174, 176, 223
Samos 178–9
Samothrace 14, 107, 162, 167, 174, 175, 182–5, 190, 225
Sardinia 55, 195, 226
Scheria 188
Sena 161, 182
Shetlands 37
Sicily 21, 22, 33, 38, 42, 53, 55, 59, 65, 104, 106, 116, 123, 168, 187, 193, 201, 210
Skyros 81
Sporades 81
Sri Lanka 208

Tenos 171
Thrinakia 105

Ultima Thule 4, 37

Harbors

Kantharos 140

Portus/Ostia 108, 147, 160, 178, 210, 213, 216, 225

"Old Men's Harbor" (Chios) 129

Piraeus 180, 211, 223

Spice Port 86

Springs and Natural Fountains

Arethusa 78, 79

Camenae 165

Diana 165

Egeria 165

Fountain of the Sun 79

Hippocrene 79
Hypereia 64

Kyane 168

Marsyas 81

Rivers

Achelous 67, 68
Acheron 43
Akesines 75
Alpheus 78, 122, 204
Amilo 159
Amisia (Ems) 54, 107, 127
Anias 78
Arnus 72

Boagrios (Ox-hide) 64

Cephisus 64
Cerona 80, 205
Cocytus 43
Colorado 63
Crathis 80

Eridanos 71
Euphrates 15, 64, 73, 111, 114

Hydapses 75

Ilissos 71
Indus (Sinthos) 6, 61, 72, 74, 75, 187, 204

Ister/Hister (Danube) 3, 72–3, 76, 81, 203

Jordan 123

Ladon 78
Liger (Loire) 37
Liris 72

Meander 67, 68, 80
Melas 78

Neleus 80
Nile 2, 3, 6, 35, 38, 54, 56, 64, 66, 67, 72, 74–7, 123, 133, 148, 164, 165, 202, 204
Numicus 155

Okeanos (circumambient Ocean) 5, 16, 39, 45
Orontes 52

Peneius 64, 73
Phasis 1, 41, 64, 72, 73–4, 77
Pyramus 67–8
Pyriphlegethon 43

Rhine 71, 72, 120, 189

Scamander 70, 102
Seine 123
Sequana 123
Simois 70, 102
Styx 80, 167, 212

Tagus 51
Tanais 2

Te Awa Tupua River, New Zealand 203
Thames 71
Tiber 108, 112, 120, 155, 160, 190, 219
Tigris 15, 51, 78
Timavus 80
Titaressus 64, 73

Volturnus 72

Xanthus 70

Lakes

Acherusian 43, 166
Alkyonian 54
Asphaltitis (Dead Sea) 47, 200

Copais 190

"Frenzied" (Insanus) 81

Lily 166

Sirbonis 47, 200
Stygian 43

Thopitis 78

Victoria 204
Vadimo 81–2

Gulfs/Bays

Ambrakia 106
Aornus 80
Argolic 64

Corinthian 55, 138

Bay of Fundy 200

Gades 79
Gomaros Bay 106

Malian 103

Persian/Arabian 14, 55, 61, 201

Syrtes 55, 105

Straits

Chalcidian 59

Euboean 61

Hellespont (Bosporus) 60, 65, 73, 106

Messina 38, 59, 60–1

Pillars of Herakles (Strait of Gibraltar) 1, 36, 37, 38, 41, 55, 65, 66, 143, 202

Symplegades 102

Seas and Oceans

Adriatic 72, 73, 77, 78, 107, 188, 195
Aegean 5, 55, 58, 64, 65, 78, 81, 102, 103, 133, 141, 176, 180, 182, 183, 195, 215
Atlantic (External Sea/Ocean) 4, 5, 36–8, 39, 47, 49, 50, 52, 55, 85, 106, 114, 127, 130, 135, 137, 143, 215

Baltic 37
Britanic 161

Caspian 36, 44, 78

Erythraean (Indian Ocean) 49–50, 51, 52, 56, 60, 61, 81, 86, 131, 137, 138, 139, 145, 207

Euxine (Pontus/Black) 2, 6, 16, 36, 54, 55, 58, 62, 65, 67, 72, 78, 81, 96, 102, 190, 195

Galilee 215

Hyrcanian 44

Maeotis (Sea of Azov) 55, 67

Pacific 207
Propontis (Marmora) 65

Sargasso 55
"Sea of Worms" 199

Tyrrhenian 55, 195

Index of Authors and Sources

Aberdeen Bestiary 212
Acts of Paul and Thecla 214
Aelian
  NA 35, 104, 127, 128, 129, 130, 131,
    133, 134, 136, 137, 138, 139, 141,
    142, 143, 146, 149, 150, 165, 166,
    173, 195, 204, 212, 213, 214, 215,
    216, 217, 220, 221, 227
Aelius Aristides
  To Rome 203
Aeschines 215
Aeschylus 5, 31, 76, 221
  Agamemnon 182, 195, 220
  Eumenides 196
  Glaucus of the Sea 153, 220
  Prometheus Bound 75, 184, 194, 198
  Suppliant Maidens 198, 204
Aëtius 51, 198, 200, 219
Agatharchides 76, 134
Albinovanus Pedo 107, 127, 195
Alcaeus 17, 217
Alciphron
  Fishermen 62
  Letters of the Courtesans 212
Alcman 17
Alexander Romance 3-4
Ammianus Marcellinus 211, 225
Anaxagoras 32, 69, 71, 76, 92-5, 98, 100,
  197, 204, 212
Anaximander 2, 17, 19, 20, 39, 41, 73, 93,
  98, 101, 192, 196, 200, 201, 212
Anaximenes 17, 19-20, 21, 91, 93-4, 97-8,
  100, 197, 207
Antonius Musa 114
Appian
  Civil War 202, 208, 221
  Mithridatic War 172
  Spanish Wars 209
Apollonius of Rhodes
  Argonautica 3, 72, 73, 76, 102, 136, 153,
    175, 181, 182, 184, 195, 203, 223
Apollonius of Tyana 222

Apollodorus
  Library 154, 205, 213, 217, 219, 221, 223
Apuleius
  Apology 163, 164
  Metamorphoses 185, 205
Aratus, *Phaenomena* 91, 207, 223
Archimedes 56, 57
  On Floating Bodies 55, 60, 201
Aristophanes, 199
  Acharnians 222
  Birds 211
  Clouds 206
  Frogs 167, 213
  Wasps 219
  Women at the Thesmophoria 211
Aristotle 5, 9, 18, 21, 24-5, 28, 29, 30, 36,
  44-5, 59, 87, 88, 128, 130-1, 144,
  157, 192-3
  Cosmos 56-7, 197, 207
  GA 214
  GC 69, 197
  HA 73, 129, 131, 134-6, 137, 139,
    140-2, 206, 212, 213, 214, 215, 216,
    217, 218
  Metaphysics 197, 219
  Meteorology 4, 41, 44-5, 46-7, 54-5, 59,
    66, 67, 71, 72, 75, 77, 78, 79, 91-101,
    103, 199, 200, 201, 202, 204, 207, 208
  Movement of Animals 219
  Metaphysics 200
  On the Soul 198
  PA 35, 212, 213, 214, 216
  Physics 206
  Respiration 214
Aristotelian
  Problems 36, 47, 48, 57-8, 61, 198, 199,
    200, 201, 203
Arrian 5
  Anabasis 49, 74, 176, 204
  Indika 5-6, 138, 187, 195, 199, 215
  Periplus of the Euxine Sea 4, 6, 73, 107,
    171, 203

Artemidorus of Daldis
  Interpretation of Dreams 211, 213
Athenaeus 61, 122, 129, 139, 175, 177, 178,
  187, 189, 190, 212, 215, 216, 217,
  221, 224, 227
Athenodorus of Tarsas 200
Atrahasis 15, 157, 208
Augustine of Hippo
  City of God 164–5, 219
Aulus Hirtius 180
Ausonius
  Mosella 215
Avienus
  Ora Maritima 55, 127, 201

Bacchylides 207

Callimachus 224
Callisthenes 6
Calpurnius Siculus
  Eclogue 214
Celsus 114, 116, 119, 120, 122, 123, 209,
  210, 212
Chrysippus 130, 213
Cicero, Marcus Tullius 28, 201
  Divination 50, 101, 162, 198, 220
  Fate 198
  Nature of the Gods 210
  Republic 75, 204
  To Atticus 108, 115, 210
  To Quintus 108–9, 208
Columella 86, 89, 115, 124

Democritus 26, 41, 67, 100, 199, 206
Dio Cassius 108, 109, 115, 186, 204,
  208, 211, 217, 220, 221, 223,
  225
Dio Chrysostom 223
Diodorus Siculus 47, 75, 76, 103–4, 112,
  118, 168, 179, 182–3, 196, 200, 217,
  218, 221, 223
Diogenes Laërtius 210, 224, 225, 228,
  231
  Life of Empedocles 210
  Life of Epicurus 207, 213
  Life of Theophrastus 195
Dionysius of Halicarnassus
  Roman Antiquities 219
Dioscorides 9, 120, 211, 217

Egyptian Book of the Dead 166
Empedocles of Acragas 4, 9, 21–2, 23,
  24, 27, 37, 41, 46, 93, 94, 95, 116,
  141, 146, 157, 171, 197, 199,
  219, 221
Enuma Anu Enlil 162–3
Enuma Elish 13, 196, 217
Ephorus of Cyme 63, 184
Epicurus and Epicureanism 9, 26–7, 88,
  130, 198, 199, 206, 207, 213
Eratosthenes 5, 36, 45, 55, 59–60, 64, 65,
  67, 68, 72, 74, 75–6, 78, 200, 202,
  203, 204, 205
Eumelus 223
Euripides 76
  Archelaus 204
  Elektra 213
  Helen 204
  Herakles 217
  Hippolytus 152, 218, 223, 224
Eusebius
  Preparation for the Gospel 224
Eutropius 211
Exekias 188, 226

Festus Paulus 219, 227
Florus 221

Galen
  Method of Medicine 209
  Natural Faculties 212
  Nature of Man 112
  On Simples 217
  Preserving Health 122, 212
Geminus of Rhodes 89
Germanicus
  Aratea 217
Greek Anthology 130, 171, 176, 179, 204,
  210, 213, 221, 224, 226
  Agathias Scholasticus of Asian Myrina
    207
  Alpheios of Mitylene 173
  Antipater of Thessalonica 140, 146, 214
  Antiphanes of Megalopolis 158
  Apollonides of Smyrna 158, 190
  Meleager 134
  Mnasalkes 176
  Theodoridas 146
Greek Magical Papyri 164, 220

Hanno of Carthage 36, 45, 50, 121, 200
Hebrew Bible
  Amos 217
  Genesis 157, 196, 208
  Isaiah 217
  Job 217
  Jonah 155, 215, 236
  Leviticus 219
  Psalms 217
Hecataeus 20, 72, 73, 75, 196, 203, 204, 220
Heliodorus 222
Heraclitus of Ephesus 9, 20, 23, 27, 28, 40–1, 45, 157, 194, 197, 199, 219
Heraclitus
  Allegories/Homeric Questions 218
Herodian of Antioch 111
Herodotus 15, 36, 45, 54, 62, 66, 72, 73, 74, 75, 76, 78, 102, 103, 106, 123, 174, 183, 192, 195, 199, 200, 201, 202, 203, 204, 212, 218, 220, 221, 223, 224
Hesiod of Askra 20, 89, 133–4, 168, 191, 206
  Theogony 13, 14, 15–16, 33, 39, 69, 70, 72, 73, 149, 150, 153, 154, 167, 196, 197, 199, 200, 205, 207, 218, 219, 221, 223
  Works and Days 85, 87, 91, 219
Hieronymus of Kardia 47
Hipparchus of Nicea 45, 63, 200
Hippocratic Corpus 25, 115
  Affections 210, 113, 116
  Aphorisms 209
  AWP 209, 210
  Coan Prenotions 209
  Epidemics 113, 209, 210
  Internal Affections 114
  On Regimen in Acute Diseases 209
  Prorrhetic 209
Hippon of Croton 69
Hipponax of Ephesus 222
Homer 3, 4, 5, 8, 16, 43, 87, 206
  Iliad 14, 35, 39, 49, 64, 69, 70, 72, 102, 148, 167, 197, 203, 214, 219, 221
  Odyssey 49, 59, 60–1, 66, 102, 105, 136, 139, 150–1, 161, 167–8, 173, 174, 187, 188, 195, 196, 200, 202, 207, 215, 217, 219, 220
Homeric Hymns
  Apollo 175
  Demeter 197

Dionysus 213, 226
Dioscuri 181–2
Poseidon 173, 221
Horace
  Odes 187, 205

Isidore of Seville 5

Josephus
  Jewish Wars 208
Julius Caesar
  Gallic Wars 49, 106, 200
Justin Martyr 202
Juvenal 210, 215

Kallikrates of Samos 178
Kallixeinos of Rhodes 224
Krates of Mallos 45–6

Libanius 174
Livy
  ab Urbe condita 68, 106, 108, 118, 160, 181, 183, 200, 209, 219, 221, 222, 225
Lucan
  Pharsalia 74, 105, 195, 213
Lucian 3
  Dialogue of the Mariner 213
  Dialogues of the Sea Gods 133, 142, 204, 211–12, 219
  How to write History 179
  Navigium 224
  On Friendship 212
  True Story 137, 173, 215, 218, 220
Lucretius 26–7, 88, 112, 118, 121, 198, 199, 219, 225
Lydus, John
  On Signs 163, 226

Mahabharata 222
Manilius 78, 87, 148, 162, 175, 198, 200, 223
Marcus Aurelius
  Meditation 28, 120, 211
Martial 210, 212, 224
Meton of Athens 211
Metrodorus of Chios 201
Mildenhall Dish 148, 216
Moschus 204

Index of Authors and Sources

Nearchus 4, 5–6, 67, 138, 187, 195
New Testament
   Acts 182
   Matthew 206
Nigidius Figulus 216
Nonnus
   Dionysica 153, 213, 221

Oinopides of Chios 76
Oppian
   Fishing 35, 62, 130, 132, 134, 136, 137, 138, 173–4, 199, 213, 214, 215, 216, 217, 222
Orosius 208
Ovid 5
   Amores 226
   Art of Love 195, 224, 226
   Fasti 155, 159, 160, 161, 165, 181, 220, 226
   Heroides 224
   Metamorphoses 3, 74, 80, 147, 150–1, 152, 153, 157, 168, 171, 197, 204, 205, 208, 216, 217, 218, 220, 221, 224
   Pontus 62, 203
   Tristia 182, 205

Pausanias 38, 55, 61, 62, 68, 104, 146, 152, 163, 168, 172, 173, 178, 180, 181, 195, 204, 205, 217, 221, 222, 223, 224
Periplus of the Erythraean Sea 6, 51, 60, 73, 86
Petosiris 162, 220
Petronius
   Satyricon 128, 114, 205, 211, 215
Pherecydes 101
Philo
   On the Embassy to Caligula 224
Philostratus of Athens
   Apollonius of Tyana 136, 240
   Imagines 148
   Lives of the Sophists 186
Photius
   Biblioteca 221
Phrynicus Monotropos 211
Pindar 5, 205
   Pythian Odes 60, 181, 219, 221
   Nemean Odes 129, 168, 195

Plato 3, 22–4, 26, 30, 31, 32, 41–3, 44, 157, 192, 193
   Critias 68, 71, 102, 104–5
   Euthydemus 181–2
   Ion 225
   Phaedo 1–2, 3, 37, 41, 73, 167, 195, 197, 203, 221
   Republic 42, 133, 200
   Sophist 197
   Timaeus 14, 22, 66, 202, 219
Pliny the Elder
   Natural History 5, 6, 36, 38, 50, 51, 53, 54, 56, 57, 61, 62, 64, 68, 69, 72–5, 79, 80, 81, 82, 89, 91, 92, 101, 109, 119–20, 123, 128, 130, 132, 134–46, 151, 159, 165, 180–1, 182, 192, 193, 196, 199, 200, 201, 202, 203, 204, 205, 206, 207, 208, 209, 211, 212, 213, 214, 215, 216, 217, 219, 220, 225
Pliny the Younger 81, 225
Plutarch 19, 33
   Alexander 183, 209
   Antony 208, 210
   Cimon 195
   Cleverness of Animals 128, 130, 132, 133, 138, 143, 168, 212, 213, 214, 216
   Dinner of the Seven Wise Men 61, 214, 221
   Greek Questions 224
   How to Tell a Flatterer from a Friend 225
   Marius 201, 212
   Natural Phenomena 122, 129
   On Borrowing 209
   On the Tranquility of the Mind 211
   Phocion 140, 215
   Placita 201
   Platonic Questions 201
   Precepts of Stagecraft 211
   Romulus 220
   Table Talk 122
   Theseus 224
   Whether Fire or Water is more Useful 4, 29–32, 194, 220
Polybius 38, 60, 67, 106, 195, 199, 200, 202, 203, 209, 200, 222
Polyclitus 36, 199

Pomponius Mela 59, 81, 82, 161-2, 201, 202, 205, 220
Poseidippos 179, 224
Posidonius of Apamea 4, 9, 28, 45, 49, 51-3, 55, 66, 71, 89, 92, 95, 99, 104, 162, 193, 198, 200, 201, 202, 203, 204, 207
Procopius
  Buildings 72
  History of the Wars 138-9, 202, 215
Propertius 172, 188, 226
Protagoras 20
pseudo-Aristotle
  On Marvelous Things Heard 200
pseudo-Apollodorus
  Bibliotecha 218
pseudo-Erinna 227
pseudo-Hyginus
  Astronomika 181
pseudo-Plutarch
  Placita 201
Ptolemaeus Hephaistos 221
Pythagoras/Pythagoreanism 20, 23, 101
Pytheas of Massilia 4, 36-7, 38, 51, 52

Saga of Erik the Red 199
Sallust
  Jugurtha 112
Sappho 17, 171
Scylax of Caryanda 204
Seleukos of Seleukia 51-2, 54
Semonides
  Women 212
Seneca the Elder
  Susoria 107, 127, 195
Seneca the Younger 5, 28, 36, 52, 54, 70, 74, 100, 104, 105, 192, 204, 206
  Apocolocyntosis 210
  Epistle 201, 202
  NQ 10, 22, 49, 51, 53, 69, 70, 76, 77, 78, 79, 80, 81, 82, 87-8, 89, 98, 99, 101, 123, 133, 162, 180, 181, 193, 196, 197, 200, 202, 203, 204, 205, 207, 208
  On the site and rites of the Egyptians 204
  Phaedra 152, 218
Serenus Sammonicus, Quintus 117, 210
Servius
  ad Aeneid 173, 210, 217, 219, 221

Sextus Empiricus
  Against the Mathematicians 221
Shakespeare, William
  Tempest 180
  Venus and Adonis 206
Silius Italicus
  Punica 202, 210
Simonides 217
Simplicius
  Categories 198
Socrates 9, 41, 42-3, 73, 206
Sophocles 76, 204
  Antigone 158-9
Statius
  Silvae 182, 205, 224, 226
  Thebaid 221
Stobaeus 204
Strabo 5, 6, 10, 36, 37, 38, 45, 47, 49, 50, 51, 52, 53, 55, 59, 60, 61, 63, 64, 65, 66, 67, 68, 71, 72, 73, 78, 80, 101, 103, 104, 108, 114, 120, 122, 123, 128, 137, 165, 179, 180, 195, 196, 199, 200, 201, 202, 203, 204, 208, 212, 214, 215, 220, 221, 224
Strato 29, 73
Suetonius
  Augustus 114, 136, 188
  Julius Caesar 210
  Nero 223
  Tiberius 146
  Titus 208

Tacitus
  Annals 54, 107, 158, 183, 208, 225
  Histories 112
Thales 4, 5, 17, 18-19, 20, 33, 39, 41, 74, 76, 100, 192, 193, 196, 197, 207
Theocritus
  Idyll 226
Theophrastean
  On Weather Signs 88, 205
Theophrastus 9, 70, 87, 203, 205, 221
  Metarsiology 88, 89, 92, 94-5, 97, 98, 99, 207
  On Fire 29
  On Fish 4
  On the Sea 4

On Waters 4
On Winds 206
Theopompus of Chios 78
Thucydides 103, 117–18, 119, 187, 195, 210
Tibullus 224
Timaeus 222
Torlonia Relief 177–8, 186
Trebius Niger 143
Tzetzes, John
  On Lycophron 217, 222

Valerius Flaccus 218
Valerius Maximus 219
Varro 124, 205
  Farming 112, 115, 165, 220
  Latin Language 220, 224, 227
Vegetius 106, 112, 207, 208, 226
Velleius Paterculus 199, 210

Vergil 3, 5, 187
  Aeneid 3, 61, 62, 81, 105, 106, 127, 151, 159, 161, 181, 195, 201, 202, 208, 218, 219, 222
  Georgics 91, 92, 207, 208, 219, 220
Vitruvius 18, 28, 80, 81, 88, 123, 205, 211, 212

Xanthos of Lydia 65
Xenophanes of Colophon 20, 21, 40, 41, 46, 64–5, 93, 94, 95, 180, 192
Xenophon of Athens
  Anabasis 107, 199, 214, 220
Xenophon of Ephesus
  Ephesian Tale 75

Zeno of Elea 27
Zenobius 225

General Index

Abzu 14, 196
Achilles 39, 70, 102, 159, 171, 190, 214
Adrastus 172
Aelius Gallus 61
Aemilius Paullus, Marcus 106
Aeneas 3, 105, 107, 155, 159, 167, 181, 222
Aeolus 159, 171
Aesculapius 123
Agamemnon 64, 220
Agrippa, Marcus Vipsanius 106–7
aither 24, 92, 102, 105, 179, 197
    see also quintessence
Ajax 64, 223
akrostolion/aplustre 178, 186
Alaric I 115
Alexander of Macedon 3, 4, 5–6, 18, 36, 37, 47, 49, 62, 64, 65, 79, 114, 138, 176, 187, 193, 204, 208
Alexander Severus 111
alluvium/alluvial 55, 66, 67, 72, 74, 115, 191, 204
alopekia 217
alum 123
Amazons 8
ambrosia 159, 167
Amenhotep III 75
Amphitrite 59, 61, 153, 171
amulets 117, 120, 136, 139, 171, 188–9, 218
    blue-green diamonds 189
    carbuncle 189
    chalcedony 189
    coral 189
    dryops 189
    emeralds 117
    phallus 189
    sea-blue beryl 189
    shark teeth 139
Anat 150
Anchises 161
anchovies 62
Andromeda 3, 151, 218
anemones 131, 146

Anna Perenna 155
Antigonus Gonatas 175
Antigonus I Monophthalmos 47
Antonine Wall 174, 223
Antoninus Pius 178
Antony, (Marcus Antonius) 106, 107, 115
ants 2
Anubis 163
Aphrodite 33, 128, 132, 152, 176–8, 179, 190, 224
    Epilimenia 176
    Euploia 176, 178, 179, 223
    Galenaia 176
    Pontia 177
    Zephyritis 179
    see also Venus
Apollo 61, 81, 82, 123, 124, 132, 151, 161, 175–6, 190
    Aktios 175
    Apobaterios 176, 180
    Clarion 188
    Delphinios 132
    Ekbasios 175
    Embasios 175
    Grannus 124
    Phoebus 174
    Neosoos 175
    Thyrxuscan 163
aqueducts 77, 203
    Aqua Marcia 158
Ares 158, 188
    see also Mars
Argo 59, 175, 176
Argonauts 72, 175, 181, 182–3
Ariel 180
Arion 133, 168, 172, 221
Arjuna 222
Armilustrium 220
Arminius 54
Arsinoë II Philadelphos 178–9, 183
Artaxerxes 65

Artemis 159, 176, 190
  see also Diana
Ascanius 181
Asklepios 176
Astarte 129
astrometerology 162–3
Athena 3, 123, 154, 173, 176, 184, 185, 187, 215
  see also Minerva
Atlantians 66
Atlas 3, 16, 54, 87, 167, 184
atomism/atomic theory 23, 24, 26–7, 29, 32, 41, 94, 100, 139, 199
Augustus, Gaius Julius Caesar/Gaius Ocatvius (Octavian) 28, 61, 73, 106, 114, 136, 148, 165, 174, 179, 188, 205, 208, 217, 224
Autolycus 49

Baal 149, 150, 217
Babylonians 6, 15, 16, 162–3, 196, 219
Bacchus 158
  see also Dionysus, Liber
baleen 65
balsam 211
barley 209
barnacles 145, 153
Bastet 193
bathysphere 4, 62
bats 208
beans 75
bears 136
beetles 104
Berenike I 178
bitumen (asphalt) 47, 123, 200
blackberries 209
Blessings of the Fleet 185
blowholes 134–5, 137, 214
boars 49, 151, 154, 158, 181, 186, 222
Bubonic Plague 117
bulls 61, 146, 152, 172, 173, 188, 193, 194, 210, 222
buoyancy 49

cabbage 211
Cadmus 182
Caesar, Gaius Julius 108, 115, 162, 165, 180, 182, 196, 199, 219, 221
Caligula (Gaius) 36

canals 15, 55–6, 108, 202–3
capricorns 147, 148, 155, 188, 216–17
Caracalla 117
Carnival 185
Carthaginian Wars 37, 50, 106, 108, 112, 118
Cassander 70
Cassandra 220
Cassiopeia 151
Castor 180, 181
Castori 226
catfish 131, 146
cats 117, 193
celery, wild 211
Centauromachy 175
centaurs 8, 194
centipedes 104
cephalopods 140–3, 151
Cerberus 166
Ceres 178
  see also Demeter
Chaldeans 90
Charybdis 49, 59, 60–1
Cherusci 54
chickenpox 121
Chinese 184
cholera 112, 113, 209
Christ 159, 206
Churchill, Winston 204
Cicero, Quintus 108, 162
Cinna, Lucius Cornelius 201
Circe 150, 195
circumnavigation 36, 45, 52, 121, 200, 204
Claudia Quinta 160
Claudius 36, 108, 138, 221
Cleopatra 186
clouds 19, 40, 59, 74, 86–8, 90–9, 102, 105, 109, 130, 138, 206
Cogidubnus 132
comets 86
congers 131
coral 65, 117, 189, 218
coriander 209
Coriolis force 60
cows 91
crabs 119, 128, 129, 129, 130, 131, 132, 213
Crassus 129
Creon 158
crocodiles 75, 133, 165, 204
crows 91, 186, 207

cumin 140
cuneiform 162–3
currents 3, 30, 35, 38, 39, 43, 50, 54, 59–60,
   61, 62, 64, 75, 102, 138, 145, 191, 202
cuttlefish 131, 141
cyanobacteria 32
Cybele (Magna Mater) 160
Cyclops 173
Cynthia 172
Cyrus 107

Darius 74
deer 158
deforestation 68, 109, 115, 116
Demeter 22, 163, 173, 178, 183, 220, 221
   see also Ceres
Demeter Erinys 173
Demetrios I Poliorcetes 55, 223
Demiurge 23, 157, 197
dendrochronology 86
dew 69, 95, 96, 99
Diana 123, 165
   see also Artemis
diarrhea 113, 115, 118
Dido 155
Dionysius of Syracuse 118
Dionysus 80, 123, 132, 154, 167, 171–2,
   177–8, 185–6, 188, 190
   see also Bacchus, Liber
Dioscuri 87, 162, 171, 172, 180–2, 183, 187,
   189, 190, 225
diphtheria 121
divers/diving 35, 130, 135, 136, 140, 143,
   167, 198, 199, 218
divination 28, 31, 109, 194
   augury 165
   hydromancy 163, 164, 169
   lecanomancy 163–4
   necromancy 164
   scrying 163
   weather signs 161–5
diving cauldron (lebes) 35
dog-rose 119
dogs 119, 120, 143, 151, 208
dolphin 86, 128, 130, 131–5, 136, 137, 138,
   144, 145, 146, 151, 154, 158, 168,
   175, 178, 186, 192, 213, 214
   bottlenose 131
   striped 213

dolphin-riders 168
Domitian, Titus Flavius 208
donkeys 146, 205
Doris 16, 128, 147
dragon 151
dropsy 135
drought 30, 65, 86, 100, 160
Druidism 220
dryads 152
dysentery 113, 118, 209

eagle 178, 188
earthquakes 17, 18, 41, 52, 66, 70, 78, 87,
   100–1, 102, 103, 107, 197, 202, 203,
   208
   Sumatra Andeman earthquake 207
ebola 210
Echidna 150
eclipses 162
eels 190
   moray eels 129
   murenas 131
elephants 159, 178, 193, 208
Eleusinian Mysteries 140, 183
Epicureanism 26–7, 88, 130, 199
epilepsy 136
Encolpius 202
epidemics ("plagues") 112, 117–18, 124,
   138, 151
   Attic 113, 117–18, 210
   Antonine 121, 211
equinox 51–2, 53
estuaries 3, 35, 50
Eudoxus of Cyzicus 45
Eustachian tubes 199
Euthymenes of Massilia 49, 71, 200
ex votos 175, 176, 183

Fabius Maximus Verrucosus Cunctator
   68
fevers 114–16, 118, 119, 135
fire 4, 19, 21–2, 23–4, 25, 27, 29–32, 33, 40,
   71, 79, 88, 92, 93–5, 99, 101, 104,
   105, 150, 157, 159, 161, 168, 194,
   197, 198
fishermen 35, 62, 128, 129, 130, 131, 132,
   135, 139, 146, 153, 161, 166, 173,
   190, 213
fishpond 109

floods/flooding 3, 30, 32, 55, 56, 63, 66, 68, 71, 74, 75, 77, 78, 85, 86, 101, 103, 104, 107–9, 137, 157, 208
  see also Nile
flute (diaulos) music 129, 132, 165, 186
forest-rose 119
fossils 65, 82, 139, 191, 192, 218
four element theory 21–5, 28, 29, 30, 87, 162, 191, 196, 197–8
frankincense 187
frogfish 146
frogs 2, 90–1, 130
frost 95, 96, 97, 99
frostbite 107
Fulvius Paetinus Nobilior, Servius 106
Furius Crassipes 109, 208

Gaia (Earth) 15–16, 21, 150, 153, 187
Gallizenae 161, 190, 220
Ganesha 193
gangrene 115, 120
garlic 211
Gedrosi 138
geese 178
gentian 119
Germanicus Julius Caesar 53, 107, 120, 127, 130, 183
germs/germ theory 112, 125
Gigantomachy 175
Gilgamesh 166, 167
gills 135, 143, 146, 151
Glaucus 150, 153, 161, 167, 171
goats 186
goiters 80
Gordianus III 124
grapes 209
Gulf Stream 60
gum myrrh 211

Hades/Aidoneus 21, 22, 33, 168, 197
Hades (Underworld) 166, 167, 181
Hadrian, Publius Aelius Traianus 4, 86, 107, 123, 180
Hadrian's Wall 174, 223
hail 91, 95, 96, 97–100, 107, 189
hake 190
Hannibal 108
Hanning Speke, John 204
Harmonia 182–3
Hector 159, 219

Hekate Trioditis 189–90
Helen 171, 173, 180, 181, 182, 196
heliocentrism 51, 194, 201
Hephaestus 31, 184
  see also Vulcan
Hera 21–2, 168, 172
Herakles 3, 123, 150, 151, 154, 166, 167, 168, 176, 218
Hercules 123, 164, 202, 219
Hermaphroditus 216
Hermes 154, 188
  see also Mercury
Hermias 179
Hermione 64
Herod the Great 208
Hesione 3, 151, 218
Hesperides Garden 166, 167
Hilaeira 181, 182
hippocampi (seahorses) 145, 147, 148–9, 155, 189, 217
Hippolytus 152
hippopotamus 204
Hoang-ti 184
honey 209
hoopoes 188
horses 3, 61, 80, 146, 147, 172–3, 181, 189, 222
hospitals 124
humors/humoral theory 25, 28, 117, 119, 123
hurricanes 14, 95
hybrid/hybridism 14–15, 128, 146–55, 193–4, 205
hydra 145, 150
hydrophobia (rabies) 119–20
hydrostatics 79
hyena fish 146, 166
Hygieia 123
hygiene 112, 211
Hyperboreans 8, 72, 167
hypocisthis 209

ice 1, 37, 98, 99
Idas 181
Iris 87, 93
Isis 159, 176, 178, 185, 205, 224
  Euploia 178
  Pelagia 178
  Pharia 178, 224
isthmus 56, 63, 64, 103, 223

jackals 193
jellyfish 37, 86, 131
Juba II 6, 75, 81, 196
Julia 205
Jupiter 79, 87, 123, 157, 161, 178,
    179–80
  see also Zeus
Justinian I, Flavius Petrus Sabbatius 138

Kabeiroi 183–4
katabasis 41, 44, 167, 168, 200
ketos/kete 137, 139, 146–7, 151–2, 155,
    214, 216, 218
Kore 168
  see also Nestis, Persephone

Lamassu 193
Laocoön 218
Laomedon 151, 218
latitude 13, 37, 74
latrines 111, 112, 113, 120, 121, 209
laudanum 196
laurel 159, 161
Lavinia 155
laxative 116
Leda 181
lemons 211
leopards 154
leprosy 122–3, 212, 217
Leukippos 199
Leukothea 154–5, 172, 188
Liber 80, 178
  see also Bacchus, Dionysus
lighthouse
    Ostia 177–8
    Pharos 66, 178, 179–80
lightning 40, 87, 93–5, 136, 181
lionfish 146–7
lions 117, 154, 163, 193, 218
lizard fish 146
lizards 143
lobsters (locustae) 131, 145
lodestones 184–5, 225
Ludi Piscatorii 190
Lynkeus 181

mackerel 129
madwort 120
magi 171
magnet/magnetism 184–5, 225

magpies 186
malaria 115–17, 121, 124, 210
maltha 146
Marduk 15, 150, 217
Marius, Gaius 201, 212
marjoram 217
Mars 109, 123, 160
  see also Ares
marshes/marshlands 2, 67, 71, 76, 78, 81,
    82, 115, 167
martichoras 218
measles 121, 211
Medea 72
Medusa 151
Meleager 134
Melicertes 168, 171
Menelaus 3, 62, 64, 136, 153–4, 171, 173,
    174
Mercury 124
  see also Hermes
meridian 52
merman/mermen 153, 154, 161
metamorphosis 3, 128–9, 148, 150, 152,
    153–4, 155, 222
meteorites 86
miasma 157–9
mice 104, 114
microbes 112, 115
microvoids 29
Minerva 123, 174
    Sulis 123–4
  see also Athena
Minoans 131, 141, 150, 159, 213
Minos 167
Minotaur 194
Misenus 159
Mithras 159
Mithridates VI 172, 208
mollusks 65, 130, 143
monism 18–20, 24, 41
monsoons 76
moon 1, 39, 40, 49–52, 54, 71, 86, 91, 92, 93,
    106, 148, 159, 162, 166, 175, 182,
    189, 197, 207, 220
moon-bows 86, 93
moonfish 166
mosquitos 115
murex snail 62, 131, 146
Muses 79, 161–2
mussels 65

mustard seed 211
myrtle 122, 177

naphtha 79
nauplius 130
nausea 122
nautilus 130
Nebhepetra Mentuhotep II 54
Necho II 36, 45
Neleus 205
Neptunalia 226
Neptune 87, 100, 132, 172, 173, 174, 177–8, 182, 218, 223, 226
  see also Poseidon
Nereids 6, 146, 147, 151, 162, 171, 195, 196
Nereus 16, 128, 146, 154, 171
nerites 128
Nero, Claudius Caesar Augustus Germanicus 28, 50, 54, 158, 176, 204
Nestis 21–2, 33, 221
  see also Kore, Persephone
Nestor 172, 173, 187
Newton's Third Law 58
Nike 223
Numa Pompilius 164–5
nymph 16, 78, 155, 158, 165, 168, 216, 217

Octavia 28
October Horse 220
octopods 128–9, 131, 140–3, 145, 149, 150, 215, 216
Odysseus 3, 38, 49, 59, 105, 155, 167, 168, 171–2, 173, 187, 188, 199, 200, 216, 222
Okeanos 14, 16, 21, 39, 153, 154, 168
olive 140, 159, 164, 198
Olympias 183
opium 196
orange 211
Orpheus 184
Osiris 164, 165, 185
Ouranos (Sky) 16, 21, 152, 153, 176
owls 186, 212
oxen 73, 80, 91, 108, 146, 207
oysters 65, 131

Palinurus 3
Pan 148
Pandora 149

panthers 178
parapegmata 89
parasites 113, 115, 116, 121, 210
Parilia 161
Patrocles 78
Patroclus 70, 159
Pegasus 79, 148, 172
Peleus 3, 102, 154
Pelias 205
Peloponnesian War 103, 115, 117–18, 175
Penelope 167
Pericles 118, 174
Peripatetics 26, 29, 47, 48, 57, 58–9, 88, 193
Persephone 22, 167, 168, 183, 197
  see also Kore, Nestis
Perses 87
Perseus 3, 151, 167, 218
Persian War 62, 102–3, 106, 171, 175, 218
Phaeacians 105
Philip of Macedon 77, 183
Phoebe 181, 182
Phorcys 154
Phrontis 174
pigfish 146
pipefish 217
piracy/pirates 62, 132–3, 154, 158, 171, 175, 183, 185–6, 188
Platonism 26, 30, 31, 66, 193
  Middle Platonism 29
  Neoplatonism 157
Pleiades 87
ploiaphesia (Navigium Isidis) 185–6, 187
pneuma 27, 28, 51, 68, 87, 101, 157, 162, 219
Pollux 180, 181
Polyphemus 173
pomegranate 197
Pompey (Gnaeus Pompeius Magnus) 108, 173, 182, 201
Pompey, Sextus 165, 173
pompimos (guide/pilot-fish) 190, 213
poplars 167
poppies 26
porpoise 131, 214
Poseidon 3, 61, 87, 100, 105, 132, 133, 148, 151–2, 153, 154, 163, 168, 171–5, 176, 179, 181, 188, 189, 190, 205, 222, 223
  Asphalios 173, 222
  Hippios 172

General Index

Soter 174
   see also Neptune
prawns 131
Priam 102, 218
Priapus 189
Prometheus 30, 31, 33, 154, 184, 194, 198
propemptikon 187
Proteus 3, 153-4, 179
Prytany 210
Psamanthe 171
Psammetichus 54
psoriasis 123
Ptolemy 165
Ptolemy II Philadelphos 56, 175, 178, 183
Ptolemy IV 177
purification 3, 8, 157, 159-60, 161, 168, 172
pyramids 66

Quartilla 215
quartan fever 116, 210
quintessence 24, 30, 197
   see also aither
quinces 209

radish root 211
radix britannica 120
rain/rainfall 13, 40, 47, 63, 69, 71, 74, 76, 82, 85, 87, 88, 90, 91, 93, 95-9, 100, 105, 106, 108, 111, 113, 163, 164
rainbow wrasses 130
rainbows 86, 87, 92-3
raisins 209
rams 222
ravens 91
rays
   eagle rays 146
   electric rays 131
   manta rays 130
   nubila 130
   ox-rays 131
   stingrays 129, 131, 166
red mullet 189-90
reefs 36, 61-2
Remus 160, 177
Rhea Silvia 160, 161
rivers 2, 3, 5, 6, 7, 8, 13, 14, 35, 36, 38, 40, 42, 43, 44-5, 50, 55, 58, 63, 65, 67, 68, 69, 70-7, 80, 81, 82, 85, 97, 101, 116, 122, 123, 162, 172, 191, 203

tidal 50
underground 77-9
rodents 80
Romulus 160, 177
roosters 210
rough tail/horse mackerel 166

St. Elmo's Fire 87, 180-2
salinity 3, 7, 35, 36, 40, 41, 44, 45, 46-9, 57-9, 60, 62, 67, 73, 78, 82, 104
Salmacis 216
salmonellosis 113
Sanskrit 195, 222
Sataspes 45, 49, 200
satyrs 186
scallops 65
Scaurus, Marcus 151
Scipio Aemilianus 37
scorpions 14, 218
scurvy 120, 121, 211
   grass 211
Scylla 38, 49, 59, 139, 150-1, 152, 155, 217-18
sea bulls 152
sea cows 145
sea cucumbers 137, 146
sea grasses 146, 218
sea lions 146-7
sea robins 145
sea shells 65, 66, 82, 128, 130, 131, 143, 146, 153, 179, 201
sea spiders 145
sea turtles 152
seals 128, 131, 135-6, 137, 138, 147, 153-4, 192, 214
   elephant 146
   monk 135
sealskins 136, 154, 188
seasickness 121-2, 177, 211-12
seaweed 55, 65, 68, 152
seismic activity 66, 70, 86, 102, 103, 104
Serapis 163, 164, 220
Sergestus 62
serpents (snakes, vipers) 14, 36, 81, 104, 128, 150, 154, 189, 218
Servius Tullis 181
sewage/sewers 113, 118, 120, 143
sharks 128, 130, 131, 135, 139-40, 143, 145, 151, 215, 218

blue 139
dogfish (watery dogs/*kuones*) 127, 128, 130, 131, 139, 140, 194, 218
  fox 128
  hammerhead 139
  spiny dogfish (*kentrinas*) 139
  spotted dogfish (*galeotes*) 139
  zebra 146
shearwater 172
sheep 80, 146, 158
shellfish 128, 143
Ship of State 3
Shiva 222
shoals (shallows) 6, 50, 61, 65, 66, 105, 132, 139, 180, 217
Sibyl 220
Sibylline Books 220
Silenus 164
siltation 3, 6, 35, 66–8
Silvanus 123
Skiron 152
skolopendrae 145, 146
Skyllias of Skione 218
snorkels 35
snow 76, 85, 91, 95, 96, 97, 98, 99, 107, 108
Solon 66, 176, 224
solstices 52–3, 145
Sostratos of Knidos 179
Spargi Wreck 187
sparrows 188
spas 123–4, 212
specific density 45, 48, 49, 73, 79, 92
Speke, John Hanning 204
sponge divers 35, 130, 136, 140
springs 7, 14, 15, 22, 35, 44, 64, 70, 71, 73, 79, 80, 82, 101, 102, 164, 166, 168, 200
  hot 37, 111, 122–4, 166, 205
  sulfur 79
squid 130, 131
stars and constellations 18, 42, 54, 87, 89–90, 91, 106, 120, 167, 174, 180–2, 185, 200
  Helen, destroyer of ships 182
  Sirius 87, 106, 119–20
Stoics/Stoicism 4, 26, 27–8, 29, 49, 51, 68, 69, 77, 87, 89, 99, 101, 130, 157, 162, 193, 198, 201
storms on the water 3, 51, 61, 62, 86, 91, 100, 104–7, 109, 121, 127, 155, 171, 172, 174, 177, 178, 179, 181, 182, 183, 188, 189, 190
straits 36, 38, 45, 50–1, 55, 59–60, 61, 62, 63, 65, 106, 226
Sulla Felix, Lucius Cornelius 201
sun halos 86
sundials 208
swallows 186
swamps 42, 44, 115, 202, 210
swans 178, 186
swimmers/swimming 48–9, 105, 119, 123, 130, 134, 147, 153, 158, 214, 218
swordfish 38
Syllaios 61

tamarinds 211
Tarpeia 160
Tartarus 15, 21, 42–4, 69, 150
Telegonos 173
Telemachus 168, 173, 187
tentacles 142–3, 150
tertian fever 116, 210
tetanus 136
Tethys 14, 16, 70
Theoi Megaloi 162, 183, 190
Theophanes of Miletus 108
Theseus 152, 167, 175, 181
Thetis 3, 154, 159, 171, 216
Thoth 162
thunder/thunderbolts 18, 85, 87, 88, 93–5, 99, 100, 105, 106, 163, 206
thunnoskopos 173
Tiamat 14–15, 16, 196, 217
Tiberius Caesar Augustus 53, 146, 208
tides 3, 35, 36, 38, 41, 49–54, 60, 61, 71, 80, 86, 103, 108, 136, 138, 200
Tiresias 167, 222
Titans 3, 16, 54, 154, 194
Titus 208
tortoises/turtles 152
Trajan, Caesar Nerva 107, 108, 180, 188
Trimalchio 128
Triton 153, 155, 159
tritons 146
Trojan War 49, 70, 102, 105, 159, 172, 174
Tryphon 104
tsunami 101–4, 107, 111, 207, 208

Tuccia 150
Tullia 208
tuna 38, 73, 173, 189, 213
Tyndareus 181
Typhoeus/Typhon 15, 60, 148, 150, 152, 155
typhoid fever 112, 113, 118, 124
typhoons 189
Tyro 205

Unmoved Mover 157
Utnapishtim 166

venereal syphilis 121
Veneti 50, 106
Venus 16, 123, 165, 176–8, 185, 187
  see also Aphrodite
Vestal Virgins 160, 165
volcanoes 42, 53, 80, 86, 101, 104, 107, 123
Vulcan 123, 190
  see also Hephaestus

washbasins 159
weasels 104
wetlands 81, 116, 202
whales 6, 127, 130, 131, 134–5, 137–9, 145, 149, 173, 213, 215
  baleen (toothed) 137
  blue (physeter) 137, 152
  orca 137, 138–9, 146, (Porphyrios 138–9, 215)
  ram-fish 146
  sperm 137, 138

wheels 146
whelks 129, 131
whirlpools 36, 59, 60–1, 105, 161, 172, 202
whirlwinds 94, 145
winds 6, 8, 19, 39, 40, 41, 44, 51, 54, 58, 60, 61, 85, 86, 87, 88, 90, 91, 93, 96, 99, 100, 101, 103, 107, 108, 137, 145, 150, 158, 161, 163, 171, 173, 179, 181, 183, 187, 191
  Boreas (north) 47, 58, 89, 105, 107, 167, 171
  Etesians 76, 85
  Eurus (east) 105
  Hellespontian 106
  Iapyx 106
  Notus (south) 47, 58, 105
  Thracian 6
  Zephyr (west) 178, 187
windstorms 18
wolves 151, 158, 163, 177

Xerxes 174

Yam 217
YHWH 150, 196

Zeus 21–2, 64, 87, 104, 148, 150, 154, 165, 168, 172, 176, 179–80, 181, 221
  Apobaterios 176
  Embaterios 180
  Labrandeos 165
  Soter 180, 187, 224
  Stratios 165
  see also Jupiter

www.ingramcontent.com/pod-product-compliance
Lightning Source LLC
Chambersburg PA
CBHW052215300426
44115CB00011B/1692